Iron Metabolism, Redox Balance and Neurological Diseases

Iron Metabolism, Redox Balance and Neurological Diseases

Editor

Yan-Zhong Chang

MDPI

Basel • Beijing • Wuhan • Barcelona • Belgrade • Novi Sad • Cluj • Manchester

Editor
Yan-Zhong Chang
Hebei Normal University
Shijiazhuang
China

Editorial Office
MDPI
St. Alban-Anlage 66
4052 Basel, Switzerland

This is a reprint of articles from the Special Issue published online in the open access journal *Antioxidants* (ISSN 2076-3921) (available at: https://www.mdpi.com/journal/antioxidants/special_issues/iron_metabolism).

For citation purposes, cite each article independently as indicated on the article page online and as indicated below:

Lastname, A.A.; Lastname, B.B. Article Title. *Journal Name* **Year**, *Volume Number*, Page Range.

ISBN 978-3-0365-8844-5 (Hbk)
ISBN 978-3-0365-8845-2 (PDF)
doi.org/10.3390/books978-3-0365-8845-2

Contents

About the Editor

Yan-Zhong Chang

Yan-Zhong Chang, Ph.D., is a Professor and the Director of Laboratory of Molecular Iron Metabolism and Department of Physiology at Hebei Normal University. He obtained his PhD degree in Hong Kong Polytechnic University. From 2008 to 2009, he was a Visiting Professor studying the regulation of brain iron metabolism in Dr. Tracey Rouault's Lab, NICHD. In 2003, Professor Chang founded the Lab of Molecular Iron Metabolism at Hebei Normal University. The major research field focuses on the mechanisms of iron metabolism; the mechanisms and treatment of iron misregulation and redox imbalance in Parkinson's disease, Alzheimer's disease, stroke, mental and emotional disorders as well as heart and motor system diseases; and the preparation and safety evaluation of targeted nanomedicines. Yan-Zhong has led and completed 23 national and provincial fund projects. Currently, he is leading one national fund and two provincial fund projects. To date, as the first author or corresponding author, Prof. Chang has published more than 100 papers in peer-reviewed international journals, and these papers have had more than 5000 citations with an H-index of 40. He was the Editor of 'Brain Iron Metabolism and CNS Diseases', published by Springer (2019). He won the First (2019), Second (2016) and Third (2013) Grade Awards for Natural Science of Hebei Provence.

antioxidants

Editorial

Iron Metabolism, Redox Balance and Neurological Diseases

Guofen Gao and Yan-Zhong Chang *

Ministry of Education Key Laboratory of Molecular and Cellular Biology, The Key Laboratory of Animal Physiology, Biochemistry and Molecular Biology of Hebei Province, Hebei Research Center of the Basic Discipline of Cell Biology, College of Life Sciences, Hebei Normal University, No. 20 Nan'er Huan Eastern Road, Shijiazhuang 050024, China; gaoguofen@hebtu.edu.cn
* Correspondence: yzchang@hebtu.edu.cn

Iron is essential for life, and the dysregulation of iron homeostasis can lead to severe pathological changes in the neurological system. Iron deficiency slows the development of the neural system and causes mental and emotional disorders [1–3], while iron overload is closely related to neurodegenerative diseases such as Alzheimer's disease (AD), Parkinson's disease (PD), and cerebral ischemia [4–8]. Free iron can elicit the generation of reactive oxygen species (ROS), due to its ability to catalyze the Fenton reaction [9]. Some free radicals play an important role as signaling molecules in maintaining the normal function of cells, while excessive ROS cause devastating effects on cells, leading to oxidative stress, inflammation, and ferroptosis, contributing significantly to the pathophysiological mechanisms of neurological diseases [10–13]. In order to protect against harmful effects, the cellular iron content must be precisely controlled, which, at the physiological condition, is tightly regulated by intracellular iron regulatory mechanisms, including hepcidin–ferroportin, transferrin–transferrin receptors, divalent metal transporter 1, ferritin, and iron regulatory proteins (IRP1 and IRP2) [14,15]. Studies have demonstrated that iron dysregulation and redox imbalance are both commonly involved in the occurrence and development of many neurological diseases [11,12,16]. The regulation of the iron metabolism and redox balance has appeared as a potential solution for the treatment of several neurological diseases.

In this Special Issue, "Iron Metabolism, Redox Balance and Neurological Diseases", five original research articles and five scientific review papers are published. These papers highlight the most recent advances in different aspects of iron regulation and redox imbalance in various diseases, including the molecular mechanisms of iron-induced oxidative damage in disease pathogenesis, potential therapeutic targets and approaches for the regulation of iron metabolism and related damages, and challenges to current studies attempting to understand an aberrant iron metabolism in the pathology of different diseases and its potential clinical applications.

In the original research articles, Han et al. revealed a novel role of iron in the maintenance of cell stemness via the Wnt/GSK-3β/β-catenin signaling pathway [17]. The intermediate molecules that mediated the upregulation of ferritin expression and contributed to stem cell viability and differentiation were also identified. The findings of this study provide a theoretical basis for the development of new strategies based on iron regulation in stem cell treatments for neurological diseases.

Bao et al. designed and characterized a mitochondrial-targeted pseudo-mitochondrial membrane potential (PMMP) constructed by antioxidant MitoQ [18], which selectively protected normal cells from radiation-induced damage in glioma radiotherapy, without affecting the efficacy of radiation in inducing autophagy by regulating the cellular energy supply. This study provides insights into the practical applications of PMMP and antioxidant MitoQ in the selective protection of normal cells and tissues in glioma therapy. This treatment strategy may also be applicable to other neurological diseases.

Liu et al. demonstrated that IRP2 not only regulated cellular iron homeostasis, but also medicated tissue iron distribution by managing the involvement of hypoxia-inducible

Citation: Gao, G.; Chang, Y.-Z. Iron Metabolism, Redox Balance and Neurological Diseases. *Antioxidants* **2023**, *12*, 1721. https://doi.org/10.3390/antiox12091721

Received: 30 August 2023
Accepted: 1 September 2023
Published: 5 September 2023

factor 2 (HIF2) and nuclear receptor coactivator 4 (Ncoa4) [19]. This study highlights that HIF2-NCOA4 is a complex axis that contributes to iron metabolic disorders, including anemia, iron-overload disorder, and neurodegeneration, and provides new target molecules for the treatment of diseases with iron dysregulation.

Han et al. investigated the underlying mechanisms of CY-09, a specific inhibitor of the NOD-like receptor protein 3 (NLRP3) inflammasome, on ameliorating AD classical pathology and cognitive impairment in AD mice [20]. Their findings showed that CY-09 effectively reduced fatty acid synthesis and lipid peroxidation and decreased ROS levels in 3 × Tg-AD mice. However, it had no significant effect on restoring the dysregulation of iron metabolism and ferroptosis. The underlying protective mechanism of CY-09 in AD mice may involve the maintenance of the glucose metabolism and ATP production in the brain. This study provides new evidence for targeting NLRP3 inflammasome as a therapeutic strategy for AD, while the contradiction between decreased ferritinophagy and increased ferroptosis in AD remains to be clarified in future research.

Chen et al. revealed that the dopaminergic neuronal death and Parkinsonian symptoms in OTU domain-containing protein 3 (OTUD3) knockout mice might be caused by activating inositol-requiring enzyme 1α (IRE1α) signaling, which mediated endoplasmic reticulum (ER) stress [21]. The OTUD3 was found to regulate the expression of IRP2. Therefore, knockout of OTUD3 could upregulate the content of iron in dopaminergic neurons in the substantia nigra which, in turn, contributes to increased ER stress, and induces neuronal death and PD pathology. This study provides OTUD3 as a therapeutic target for PD treatment by mediating ER stress to restrain iron-induced apoptosis of dopaminergic neurons.

In the review articles, Luo et al. summarized the molecular mechanisms of ferroptosis in glioma cell growth, invasion, migration, and resistance, and introduced potential applications and challenges of manipulating ferroptosis in the development and treatment of gliomas [22]. They also discussed various nanoparticle-based drug delivery systems, and highlighted the therapeutic opportunities of modulating ferroptosis in glioma treatment to improve clinical outcomes. It was emphasized that, although ferroptosis has great advantages in glioma treatment, further explorations are still needed to assess the advantages and disadvantages of targeting ferroptosis and to evaluate its potential value in clinical applications.

Jiménez-Jiménez et al. systematically reviewed the role of antioxidant coenzyme Q10 (CoQ10) in AD and other dementias [23]. Combined with the use of a meta-analysis, they addressed the concentrations of CoQ10 in different tissues of patients with AD and other dementia syndromes, reviewed the therapeutic response to CoQ10 administration in AD experimental models and patients with AD and other dementias, and discussed the possible therapeutic role of CoQ10. Particularly, the clinical improvement and potential application of mitochondrial activation therapy consisting of CoQ10, iron, and vitamin B6 in AD patients was discussed. Despite the promising neuroprotective effects of CoQ10 detected in different models of AD, further long-term studies with a follow-up period are needed to fill the knowledge gaps regarding both the suitability of CoQ10 as a biomarker of AD and the efficacy of treatments with CoQ10 in patients with AD or other dementias.

Holbein et al. described biological iron requirements, iron regulation, and the nature of iron dysregulation in detail in various disease conditions [24], including viral infections, cancer, ferroptotic cell death, inflammatory diseases, diabetes, cardiovascular diseases, neurological diseases, and so on. They concluded that dysregulated iron homeostasis is a common disease etiology. They also reviewed the current findings pertaining to potential new therapies in these diseases, including iron restriction, iron chelators, hepcidin, and agonists, and proposed the potential application of therapeutics affecting iron dysregulation and lowering excess levels of labile reactive iron in disease therapy. In addition, they identified a number of gaps in the current understanding of iron dysregulation in the pathology of diseases, especially related to cause or effect.

Lee et al. reviewed the interplay between intracellular iron homeostasis and neuroinflammation in neurodegenerative diseases [25]. They introduced the physiological functions of various iron transporters, regulators, and iron-containing enzymes, and summarized the interaction between the aberrant expression and dysfunction of these molecules and inflammation. The interplay shows that maintaining the intracellular iron homeostasis is critical to both the cellular redox balance and steady inflammatory homeostasis. On the contrary, dysregulation of iron metabolism in the central nervous system is commonly associated with neuroinflammation, a crucial hallmark of neurodegenerative diseases. This review article also comprehensively discusses the interactions between iron-related molecules and cell signaling molecules under inflammatory pathological conditions, which may contribute to improving the understanding of neurodegenerative diseases.

Gao et al. summarized and elucidated the interplay between the dysregulation of iron metabolism, redox imbalance, and different neurological diseases [26] such as AD, PD, stroke, abnormal neurodevelopment, and neuropsychiatric disorders. The dysregulation of iron metabolism in each disease, particularly the molecules involved, and the possible mechanisms of iron and oxidative stress in the pathogenesis of these diseases were discussed in detail. This article also reviewed the current progress towards targeting iron metabolism in the treatment of neurological diseases, including the use of iron chelators and iron supplements in different neurological diseases, iron chelators in new administration forms, key molecules in brain iron metabolism as targets, and antioxidants and anti-inflammatory reagents as targets. This review highlights the molecular mechanisms, pathogenesis, and treatment strategies of brain iron metabolism disorders in neurological diseases.

In conclusion, the original research and review articles in this Special Issue provide an updated overview of the advances on the mechanisms or treatments of neurological diseases related to iron dysregulation and redox imbalance. These papers offer fresh perspectives on the expanding knowledge and research possibilities in the field of iron metabolism, redox balance, and neurological diseases, and may stimulate future studies to better target the regulation of brain iron metabolism for the prevention and treatment of neurological diseases.

Author Contributions: Conceptualization, Y.-Z.C.; Original draft preparation, G.G.; Review and editing, Y.-Z.C. All authors have read and agreed to the published version of the manuscript.

Funding: This work was supported by the National Natural Science Foundation of China (grant numbers 32070962, 31520103908, and 31471035), the Natural Science Foundation of Hebei Province (grant number E2021205003), and the Natural Science Foundation of Hebei Normal University (grant number L2021Z04).

Conflicts of Interest: The authors declare no conflict of interest.

References

1. Chang, S.; Wang, P.; Han, Y.; Ma, Q.; Liu, Z.; Zhong, S.; Lu, Y.; Chen, R.; Sun, L.; Wu, Q.; et al. Ferrodifferentiation regulates neurodevelopment via ROS generation. *Sci. China Life Sci.* **2023**, *66*, 1841–1857. [CrossRef] [PubMed]
2. Lozoff, B.; Beard, J.; Connor, J.; Barbara, F.; Georgieff, M.; Schallert, T. Long-lasting neural and behavioral effects of iron deficiency in infancy. *Nutr. Rev.* **2006**, *64*, S34–S43; discussion S72–S91. [CrossRef] [PubMed]
3. Wu, Q.; Hao, Q.; Li, H.; Wang, B.; Wang, P.; Jin, X.; Yu, P.; Gao, G.; Chang, Y.Z. Brain iron deficiency and affected contextual fear memory in mice with conditional Ferroportin1 ablation in the brain. *FASEB J.* **2021**, *35*, e21174. [CrossRef] [PubMed]
4. Andersen, H.H.; Johnsen, K.B.; Moos, T. Iron deposits in the chronically inflamed central nervous system and contributes to neurodegeneration. *Cell Mol. Life Sci.* **2014**, *71*, 1607–1622. [CrossRef]
5. Li, J.; Ding, Y.; Zhang, J.; Zhang, Y.; Cui, Y.; Zhang, Y.; Chang, S.; Chang, Y.Z.; Gao, G. Iron overload suppresses hippocampal neurogenesis in adult mice: Implication for iron dysregulation-linked neurological diseases. *CNS Neurosci. Ther.* **2023**. [CrossRef]
6. Peters, D.G.; Connor, J.R.; Meadowcroft, M.D. The relationship between iron dyshomeostasis and amyloidogenesis in Alzheimer's disease: Two sides of the same coin. *Neurobiol. Dis.* **2015**, *81*, 49–65. [CrossRef]
7. Wise, R.M.; Wagener, A.; Fietzek, U.M.; Klopstock, T.; Mosharov, E.V.; Zucca, F.A.; Sulzer, D.; Zecca, L.; Burbulla, L.F. Interactions of dopamine, iron, and alpha-synuclein linked to dopaminergic neuron vulnerability in Parkinson's disease and Neurodegeneration with Brain Iron Accumulation disorders. *Neurobiol. Dis.* **2022**, *175*, 105920. [CrossRef]

8. Tuo, Q.Z.; Lei, P.; Jackman, K.A.; Li, X.L.; Xiong, H.; Li, X.L.; Liuyang, Z.Y.; Roisman, L.; Zhang, S.T.; Ayton, S.; et al. Tau-mediate iron export prevents ferroptotic damage after ischemic stroke. *Mol. Psychiatry* **2017**, *22*, 1520–1530. [CrossRef]
9. Merkofer, M.; Kissner, R.; Hider, R.C.; Brunk, U.T.; Koppenol, W.H. Fenton chemistry and iron chelation under physiologicall relevant conditions: Electrochemistry and kinetics. *Chem. Res. Toxicol.* **2006**, *19*, 1263–1269. [CrossRef]
10. Bush, A.I. Metals and neuroscience. *Curr. Opin. Chem. Biol.* **2000**, *4*, 184–191. [CrossRef]
11. Galaris, D.; Barbouti, A.; Pantopoulos, K. Iron homeostasis and oxidative stress: An intimate relationship. *Biochim. Biophys. Act Mol. Cell Res.* **2019**, *1866*, 118535. [CrossRef] [PubMed]
12. Masaldan, S.; Bush, A.I.; Devos, D.; Rolland, A.S.; Moreau, C. Striking while the iron is hot: Iron metabolism and ferroptosis i neurodegeneration. *Free Radic. Biol. Med.* **2019**, *133*, 221–233. [CrossRef] [PubMed]
13. Sies, H. Hydrogen peroxide as a central redox signaling molecule in physiological oxidative stress: Oxidative eustress. *Redox Bio* **2017**, *11*, 613–619. [CrossRef] [PubMed]
14. Gao, G.; Li, J.; Zhang, Y.; Chang, Y.Z. Cellular Iron Metabolism and Regulation. *Adv. Exp. Med. Biol.* **2019**, *1173*, 21–32. [CrossRef
15. Pantopoulos, K. Iron metabolism and the IRE/IRP regulatory system: An update. *Ann. N. Y. Acad. Sci.* **2004**, *1012*, 1–1 [CrossRef]
16. Wu, J.R.; Tuo, Q.Z.; Lei, P. Ferroptosis, a Recent Defined Form of Critical Cell Death in Neurological Disorders. *J. Mol. Neurosc* **2018**, *66*, 197–206. [CrossRef]
17. Han, Y.; He, Y.; Jin, X.; Xie, J.; Yu, P.; Gao, G.; Chang, S.; Zhang, J.; Chang, Y.Z. CHIR99021 Maintenance of the Cell Stemness b Regulating Cellular Iron Metabolism. *Antioxidants* **2023**, *12*, 377. [CrossRef]
18. Bao, X.; Liu, X.; Wu, Q.; Ye, F.; Shi, Z.; Xu, D.; Zhang, J.; Dou, Z.; Huang, G.; Zhang, H.; et al. Mitochondrial-Targeted Antioxidan MitoQ-Mediated Autophagy: A Novel Strategy for Precise Radiation Protection. *Antioxidants* **2023**, *12*, 453. [CrossRef]
19. Liu, Y.; Li, Y.; Yang, L.; Shen, J.; Zhao, H.; Dong, W.; Chang, Y.; Qiao, T.; Li, K. Stimulation of Hepatic Ferritinophagy Mitigate Irp2 Depletion-Induced Anemia. *Antioxidants* **2023**, *12*, 566. [CrossRef]
20. Han, S.; He, Z.; Hu, X.; Li, X.; Zheng, K.; Huang, Y.; Xiao, P.; Xie, Q.; Ni, J.; Liu, Q. Inhibiting NLRP3 Inflammasome Activation b CY-09 Helps to Restore Cerebral Glucose Metabolism in 3xTg-AD Mice. *Antioxidants* **2023**, *12*, 722. [CrossRef]
21. Chen, L.; Huan, X.; Jia, F.; Zhang, Z.; Bi, M.; Fu, L.; Du, X.; Chen, X.; Yan, C.; Jiao, Q.; et al. Deubiquitylase OTUD3 Mediate Endoplasmic Reticulum Stress through Regulating Fortilin Stability to Restrain Dopaminergic Neurons Apoptosis. *Antioxidant* **2023**, *12*, 809. [CrossRef]
22. Luo, Y.; Tian, G.; Fang, X.; Bai, S.; Yuan, G.; Pan, Y. Ferroptosis and Its Potential Role in Glioma: From Molecular Mechanisms t Therapeutic Opportunities. *Antioxidants* **2022**, *11*, 2123. [CrossRef] [PubMed]
23. Jimenez-Jimenez, F.J.; Alonso-Navarro, H.; Garcia-Martin, E.; Agundez, J.A.G. Coenzyme Q10 and Dementia: A Systemati Review. *Antioxidants* **2023**, *12*, 533. [CrossRef] [PubMed]
24. Holbein, B.E.; Lehmann, C. Dysregulated Iron Homeostasis as Common Disease Etiology and Promising Therapeutic Targe *Antioxidants* **2023**, *12*, 671. [CrossRef] [PubMed]
25. Lee, J.; Hyun, D.H. The Interplay between Intracellular Iron Homeostasis and Neuroinflammation in Neurodegenerative Diseases *Antioxidants* **2023**, *12*, 918. [CrossRef] [PubMed]
26. Gao, G.; You, L.; Zhang, J.; Chang, Y.Z.; Yu, P. Brain Iron Metabolism, Redox Balance and Neurological Diseases. *Antioxidant* **2023**, *12*, 1289. [CrossRef] [PubMed]

antioxidants

Review

Brain Iron Metabolism, Redox Balance and Neurological Diseases

Guofen Gao †, Linhao You †, Jianhua Zhang †, Yan-Zhong Chang * and Peng Yu *

Ministry of Education Key Laboratory of Molecular and Cellular Biology, The Key Laboratory of Animal Physiology, Biochemistry and Molecular Biology of Hebei Province, College of Life Sciences, Hebei Normal University, No. 20 Nan'erhuan Eastern Road, Shijiazhuang 050024, China
* Correspondence: yzchang@hebtu.edu.cn (Y.-Z.C.); yupeng0311@hebtu.edu.cn (P.Y.)
† These authors contributed equally to this work.

Abstract: The incidence of neurological diseases, such as Parkinson's disease, Alzheimer's disease and stroke, is increasing. An increasing number of studies have correlated these diseases with brain iron overload and the resulting oxidative damage. Brain iron deficiency has also been closely linked to neurodevelopment. These neurological disorders seriously affect the physical and mental health of patients and bring heavy economic burdens to families and society. Therefore, it is important to maintain brain iron homeostasis and to understand the mechanism of brain iron disorders affecting reactive oxygen species (ROS) balance, resulting in neural damage, cell death and, ultimately, leading to the development of disease. Evidence has shown that many therapies targeting brain iron and ROS imbalances have good preventive and therapeutic effects on neurological diseases. This review highlights the molecular mechanisms, pathogenesis and treatment strategies of brain iron metabolism disorders in neurological diseases.

Keywords: oxidative stress; Parkinson's disease; Alzheimer's disease; stroke; neurodevelopment; iron chelator

1. Introduction

Iron is important for the physiology of the brain, participating in oxygen transport, energy production and the synthesis of DNA, myelin and neurotransmitters [1]. Brain iron deficiency (ID) impairs the development of neurons and glial cells, leading to abnormal neurodevelopment, which causes mental incapacity [2,3] and is correlated with neuropsychiatric disorders, such as depression and anxiety [4–7], while sufficient iron is crucial for ferrodifferentiation [8]. On the other hand, iron overload in the brain exacerbates the development of neurological diseases, such as Alzheimer's disease (AD), Parkinson's disease (PD) and stroke, because too much iron generates reactive oxygen species (ROS) that can destroy the cell membrane and induce cell death [9,10]. The incidence of neurological diseases has become increasingly higher with the advent of an aging society. Therefore, it is critical to maintain iron homeostasis in the brain and to investigate the underlying regulation mechanisms, which might provide better strategies to prevent and cure neurological diseases.

Recently, it has become much clearer how circulatory iron is absorbed into the brain parenchymal tissue across the blood–brain barrier (BBB), where hepcidin and ceruloplasmin (CP) regulate the iron transport process coordinately with ferroportin 1 (FPN1) [11–14]. In particular, neurodegeneration with brain iron accumulation (NBIA) is a set of rare monogenetic neurodegenerative diseases, which is characterized by iron accumulation in basal ganglia and other related brain regions [15,16]. Targeting iron transport and the regulation pathway to restore iron homeostasis has been applied in the prevention and treatment of NIBA, stroke and other neurological diseases. Iron chelation therapy with deferoxamine (DFO), deferasirox (DFX) and deferiprone (DFP) has been used in the clinic

Citation: Gao, G.; You, L.; Zhang, J.; Chang, Y.-Z.; Yu, P. Brain Iron Metabolism, Redox Balance and Neurological Diseases. *Antioxidants* **2023**, *12*, 1289. https://doi.org/10.3390/antiox12061289

Academic Editor: Stanley Omaye

Received: 1 May 2023
Revised: 10 June 2023
Accepted: 13 June 2023
Published: 16 June 2023

and for animal research [17–22]. As the therapeutic effects of iron chelators are not ideal in the treatment of iron-overload diseases, it is crucial to elaborate the detailed regulatory mechanisms of brain iron metabolism in these diseases to discover new targets and new therapeutic strategies for these neurological diseases. Therefore, in this review, we discuss the molecular mechanisms of brain iron metabolism, pathogenesis of iron-dysregulation-related neurological disorders and the treatment strategies in these neurological diseases.

2. Brain Iron Metabolism

How iron enters the brain has long been a mystery. Given the lack of evidence of iron release from the brain, iron homeostasis is thought to be primarily maintained via the regulation of iron uptake. The brain acquires iron primarily from the blood and cerebrospinal fluid; uptake across the BBB is thought to be the primary pathway. Recently, the molecular mechanisms of brain iron uptake and transport among neurons and different types of glia have slowly been becoming clear (Figure 1).

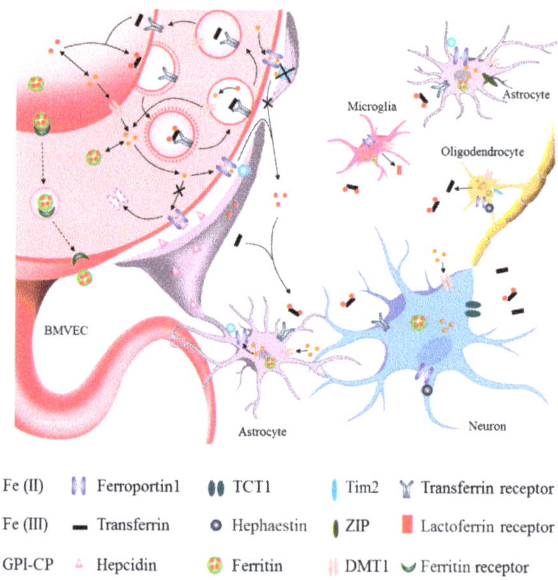

Figure 1. Roles of different cells in brain iron metabolism. The main route of brain iron uptake is where the iron in the blood crosses the blood–brain barrier (BBB) via Tf-TfR1 in the apical surface of brain microvascular epithelial cells (BMVECs) and FPN1 in the basal surface of BMVECs. Iron can also enter the brain through the transcytosis of ferritin by its receptors at BBB. After iron influxes into the brain parenchymal tissue, it can enter astrocytes through their end feet surrounding BBB and then be transferred to neurons. Iron across the BBB can also directly enter the interstitial fluid of the brain and be transferred to neurons and other cells without passing through astrocytes (see black lines). Astrocytes hepcidin secreted through its end feet to directly decrease FPN1 level of BMVECs, which decreased the iron influx into brain tissues. GPI-CP expressed by astrocytes assists FPN1 in releasing iron into the brain. Astrocyte-specific *Cp* knockout blocks iron influx FPN1-CP pathway into the brain (see black lines and crosses). Neurons acquire both trivalent and divalent iron through TfR1, TCT1 and DMT1, while those astrocytes that are not part of the BBB acquire iron via DMT1 and ZIP molecules. Oligodendrocytes mainly uptake iron via DMT1 and Tim2. Oligodendrocytes can secrete Tf, while the activated microglia can secrete Lf. Neurons and glia store iron in ferritin and release iron through FPN1 with the coordination of CP/hephaestin or hepcidin, thereby further promoting cross-talk and interaction with other types of cells.

2.1. Iron Uptake into Brain Parenchyma from Circulatory Iron across the BBB

During infancy, when the BBB has not yet formed, iron is thought to enter the brain directly. In adults with an intact BBB, brain iron uptake mainly occurs in brain microvascular epithelial cells (BMVECs) and depends on transferrin receptor 1 (TfR1) and FPN1 [23,24]. Data have shown that iron in the blood circulation can enter rain parenchymal tissue across the BBB. Transferrin-bound iron in the blood is first taken up by BMVECs by binding TfR1 in the luminal membrane; this is referred to as the Tf-TfR1-dependent iron uptake pathway [23]. However, there is no direct in vivo evidence indicating how iron in BMVECs is released into the brain via the basal surface of these cells. Endothelial cell-specific *Fpn1* knockout mice and conditional knockdown of astrocytic hepcidin mice revealed that endothelial FPN1 acts as a gatekeeper for iron, mediating iron entry into brain tissue from BMVECs [11,25].

The gatekeeper FPN1 is regulated by hepcidin, CP and iron regulatory protein (IRP). Hepcidin is secreted by astrocytes and can decrease the expression of FPN1 in the striatum, cerebral cortex, and hippocampus [12]. Hepcidin can control the entry of iron through FPN1 on BMVECs via a FPN1–hepcidin posttranslational degradation axis [11]. Glycosylphosphatidylinositol-anchored CP (GPI-CP) is mainly expressed in the end feet of astrocytes surrounding the BBB [13,26,27]. Conditional knockout of astrocytic CP caused ID in the brain, providing direct evidence that CP in astrocytes also regulates iron influx into the brain through the BBB [14]. While the level of IRP is negatively regulated by iron levels in the brain, it also regulates *TfR1* and *Fpn1* mRNA via post-transcriptional regulation of the iron-responsive element (IRE)-IRP system [28,29]. These results demonstrate that coordination of Tf-TfR1 and FPN1 plays a critical role in iron efflux into brain parenchyma from BMVECs, and that hepcidin and CP may help maintain iron homeostasis in the brain.

There are other routes of iron entry into the brain [30]. In the non-transferrin-bound iron pathway, which mediates iron entry via BMVECs of the BBB [31,32], ferritin can bind the ferritin receptor or be transcytosed across the BBB [33]. Iron in the cerebrospinal fluid can also enter the brain across the choroid plexus [34,35].

Different brain regions contain different levels of iron, with high iron concentrations in the substantia nigra (SN), red nucleus and globus pallidus [36,37]. It is unclear why such iron heterogenicity exists in the brain. It may be due to differences in iron demand from specific brain nuclei groups or differences in iron absorption, transport or iron release because of the distribution of iron-metabolism-related proteins in different brain regions [38]. Research has shown that iron can be transported along axons from the ventral hippocampus (vHip) to the SN and from the thalamus to the amygdala [39], so it is possible that axonal iron transport may contribute to heterogeneous iron distribution. It is speculated that choroid-plexus-derived transferrin (Tf) in cerebrospinal fluid (CSF) plays a significant role in the export of iron to the blood instead of iron uptake into the brain interstitium because intracerebroventricular injection of [^{59}Fe^{125}I]Tf did not lead to an observable signal in brain regions distant from the CSF [40]. The glymphatic pathway may also be involved in iron release from the brain [41]. Do the different degrees of iron release result in different levels of iron reduction or overload? Further investigation of iron deficiency or accumulation in different brain regions is warranted.

2.2. Iron Uptake and Metabolism in Neurons and Glia

After iron enters the brain parenchyma, it is absorbed by neurons and glial cells for cellular functions. TfR1 and divalent metal transporter 1 (DMT1) are ferric and ferrous iron uptake proteins, respectively; the lactoferrin receptor (LfR) is also responsible for iron uptake. Most brain cells store iron using ferritin and export iron via FPN1 with the assistance of CP and hephaestin (HP) [42,43]. However, different types of brain cells have different dominant iron uptake pathways. Neurons acquire iron through the classical Tf-TfR1-dependent iron uptake pathway and non-transferrin bind iron (NTBI) uptake pathway, such as DMT1 and trivalent cation-specific transporter 1 (TCT1), from brain interstitial fluid [43,44]. Exported iron from neurons is oxidized into the ferric form by astrocytic CP [45]. Astrocytes acquire iron across the interstitial space using NTBI-dependent mechanisms that include citrate, ATP, ascorbic acid, DMT1 and zinc-regulated and iron-regulated transporter-like proteins (ZIP) [46,47]. Astrocytes have also been reported to directly uptake ferrous iron from BMVECs through the end feet [48]. Oligodendrocytes are the only cells that synthesize and release Tf; they acquire iron through DMT1 and the ferritin receptor, T-cell immunoglobulin and mucin domain protein-2 (Tim-2) [49]. Microglia express iron-metabolism-related proteins and are efficient in accumulating iron [50]. Under inflammation, environmental and endogenous stimuli, microglia are activated, resulting in the synthesis and secretion of lactoferrin (Lf), which affects LfR-expressing cells [51,52]. Choroidal epithelia capture iron via TfR1-dependent or TfR1-independent pathways, and they transport iron back to the blood circulation [53]. In the brain, excess iron can be exported back to CSF or interstitial fluid [54].

In addition to strict regulation of iron entry into the brain, iron levels in different brain cell types are also regulated by IRP and hepcidin. When cellular iron levels decrease the expression of IRP increases [55]. This allows IRP to bind the IRE motif in the 5′-untranslated region (UTR) of *ferritin* and *Fpn1* mRNA and the 3′-UTR of *TfR1* and *DMT1* mRNA, which inhibits translation of ferritin and FPN1 and increases the translation of TfR1 and DMT1. This leads to decreased iron storage and iron efflux, as well as increased iron uptake, which induces elevated cellular iron levels to maintain iron homeostasis in neurons and glia [29,56,57]. Altered levels of iron in the brain also affect the expression of hepcidin in astrocytes, which affects the level of FPN1 through the hepcidin–FPN1 regulatory axis [11,12]. Data from IRP2$^{-/-}$ mice and cell lines have demonstrated that IRP and hepcidin coordinately regulate FPN1 expression [28]. Therefore, coordination and crosstalk between IRE-IRP and hepcidin maintain the dynamic balance of iron levels in the brain.

Expression of CP and hepcidin in astrocytes regulates iron metabolism in BMVECs, neurons and other glia, highlighting the communication between different cells in the brain [11,14]. Astrocytes also affect microglia, which, in turn, affect iron metabolism in neurons [58]. Cell communication and the underlying network are complicated; further investigation will improve our understanding of brain iron metabolism and provide strategies to prevent and cure iron-metabolism-disorder-related neurological diseases.

Erythroferrone (ERFE), a protein newly identified by the Ganz group, may inhibit hepcidin and, thus, regulate iron metabolism in response to erythropoietin stimulation in conditions of stress [59,60]. ERFE has also been detected in the brain using real-time PCR [60], but its role in brain iron metabolism remains unclear. The possibility of other roles of ERFE in the brain will also require further investigation.

While there has been significant progress in our understanding of the mechanisms and regulatory pathways of brain iron metabolism in recent decades, many issues remain to be addressed. Data from synchrotron-based X-ray fluorescence elemental mapping demonstrated that the element that accumulates in the SN is iron and not zinc or copper [61], which may explain why dopaminergic neurons are vulnerable to toxins. Microglia are prone to accumulate iron under conditions of stress. The cross-talk between glia and neurons remains unclear; in particular, how they coordinate to maintain iron homeostasis in physiological conditions and how iron is redistributed and accumulated in pathological diseases will require further investigation. Uncovering a detailed picture of brain iron metabolism and understanding the key mechanisms will have profound preventive and therapeutic potential for neurological diseases.

3. Iron, Redox Balance and Oxidative Damage

3.1. Iron Dysregulation Induces ROS Generation

Iron dysregulation can be destructive to cells and tissues. Iron overload, especially labile iron in cells and body fluids, can catalyze the conversion of hydrogen peroxide (H_2O_2) via the Fenton reaction to generate highly reactive hydroxyl radicals and superoxide anion, leading to the generation of ROS [62,63]. Iron and iron derivatives, such as heme or iron–sulfur clusters, are the essential active centers of many enzymes (e.g., lipoxygenases (LOX), cytochrome P450, NADPH oxidases) involved in ROS generation [64]. Moreover, dysregulation of iron metabolism also induces the generation of reactive nitrogen species, including nitrogen monoxide, dioxide and peroxynitrite, while these radicals can conversely regulate cellular iron homeostasis by modulating the binding affinity of IRP and IRE [65]. The drastic increase in oxidative and/or nitrosative radicals can disrupt the redox balance in the cell and lead to biological damage, a condition called oxidative stress and/or nitrosative stress, which is involved in many diseases and medical conditions [65]. In addition to iron overload, studies have also reported that extremely low levels of iron in cells can also trigger an increase in ROS [8,66]. Under prolonged ID, increased levels of H_2O_2 initiate signaling events, resulting in a regulatory loop between H_2O_2 and prolonged ID [65]. ID increases superoxide anion levels, resulting in a significant decrease in catalase activity together with rising levels of dehydroascorbic acid, indicating disruption of redox homeostasis, ultimately triggering programmed cell death [67]. Therefore, iron dysregulation is harmful to cells via increasing ROS levels and resulting oxidative damage.

Neural cells are particularly sensitive to ROS assault because of their intense oxidative metabolism, high consumption of oxygen, and propensity to generate high levels of ROS. Iron levels in different regions of the brain increase with aging, making individuals more prone to age-dependent neurodegenerative diseases [68]. For example, high levels of iron are seen in the cerebral cortex and hippocampus of AD patients, as well as in the dopaminergic neurons of the SN of PD patients [69,70]. Iron induces oxidative damage in proteins and lipids, which is involved in many disease processes, such as synaptic dysfunction, neuroinflammation, and neuronal death, and, thus, is considered an important cause of neurodegenerative diseases [71–73].

3.2. Iron-Induced Neuronal Death

Although ROS are critical for physiological signaling pathways, excess ROS will damage cellular macromolecules, including proteins, lipids and DNA, and the resulting oxidative stress can eventually lead to apoptosis [68,74]. Normally, there are several detoxification systems and antioxidant defense pathways in cells to counter ROS, such as superoxide dismutases, catalases, and glutathione peroxidases (GPx) [75,76]. However, when the generation of ROS drastically exceeds the antioxidant detoxification systems in cells, oxidative stress results in mitochondrial dysfunction, leading to a further increase in ROS formation and Cyt C release [77]. This will trigger the activation of various signaling pathways, such as MAPK, which activates transcription factors, such as the nuclear factor NF-κB, to alter target gene expression, resulting in the upregulation of proapoptotic factors and downregulation of anti-apoptotic factors [64,77]. This, in turn, exaggerates oxidative stress and ultimately leads to programmed cell death.

Ferroptosis, another form of programmed cell death, is primarily caused by iron-dependent lipid peroxidation (Figure 2). In ferroptosis, upon labile iron accumulation, cytosolic lipid oxidation and ROS are increased, while glutathione (GSH) and GPx4 are decreased, and the mitochondria shrink with an increased membrane density, eventually resulting in cell death [78]. Ferroptosis can be activated by blocking xCT antiporter (e.g., by erastin) or GPx4 inhibitor (e.g., RSL3), while it is inhibited by iron chelators (e.g., DFO) and ROS scavengers (e.g., ferrostatin-1). This process is closely regulated by intracellular signaling pathways, including the iron homeostasis regulatory pathway, RAS pathway and cystine transport pathway [79]. Ferroptosis has been linked to the pathological processes of many diseases, including neurodegenerative diseases of the central system [79,80].

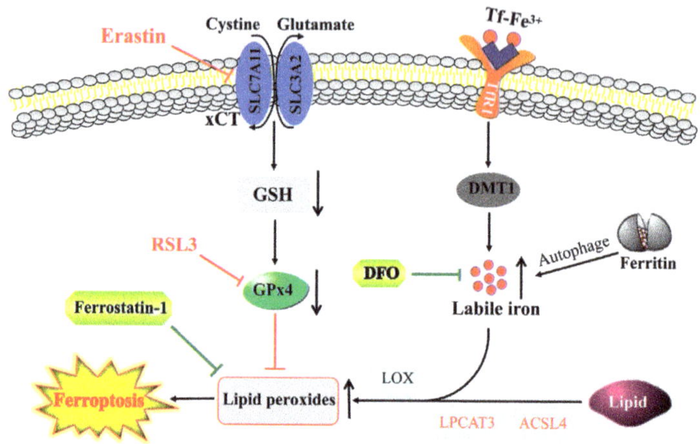

Figure 2. Interplay between ferroptosis and iron homeostasis. Lipid peroxides that induce ferroptosis are produced through auto-oxidation and/or enzymatic activity of LOX on lipid esters generated from lipids via the activity of ACSL4 and lysophosphatidylcholine acyltransferase 3 (LPCAT3). GPx4 blocks ferroptosis by converting lipid peroxides to lipid alcohols, whereas reductions in GSH or GPx4 activity by blocking of xCT antiporter (e.g., by erastin) or inhibiting of GPx4 (e.g., RSL3) can trigger ferroptosis. The increase in labile iron pool in the cytosol via an increased iron uptake through TfR1 and/or autophagic degradation of ferritin can exacerbate ferroptosis via facilitating lipid peroxidation, and, thus, iron chelators, such as DFO and ROS scavengers (e.g., ferrostatin-1), suppress ferroptosis.

4. Role of Iron in Neurological Diseases

Iron overload or deficiency is closely related to ROS levels, both of which participate in and exacerbate the onset and progression of neurological diseases, including PD, AD, stroke, neuropsychiatric disorders and abnormal neurodevelopment (Figure 3).

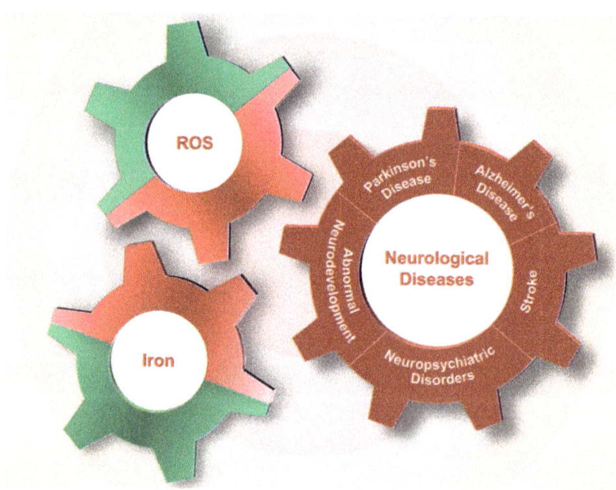

Figure 3. The interplay of iron and ROS in the pathogenesis of neurological diseases. Dysregulation of iron content in neurological system and the associated generation of ROS participate in the pathological processes of Alzheimer's disease (AD), Parkinson's disease (PD), stroke, neuropsychiatric disorders and abnormal neurodevelopment.

4.1. *Iron and Parkinson's Disease*

4.1.1. Iron Dysregulation in PD

PD is one of the most typical neurodegenerative diseases, and its discovery was first described by Parkinson in 1817 [81]. PD is characterized by Lewy bodies (LBs) with α-synuclein protein aggregation, as well as death of dopaminergic neurons in the SN and dopamine (DA) deficiency in the striatum [82,83]. While there is significant evidence demonstrating that excess iron accumulates in the SN pars compacta (SNpc) of PD patients and animal models [37,84–87], it remains unclear whether iron overload is the initial cause or an effect of PD [88]. Although the detailed molecule mechanisms that provoke PD are not fully understood, experimental evidence suggests that iron accumulation occurs earlier [89], and the consequent oxidative damage or mitochondrial dysfunction aggravates the development of PD [90,91]; the evidence also shows that lowering brain iron levels can slow down PD [92–94].

CP is a multi-copper ferroxidase that plays important roles in copper transport and converts toxic ferrous into the nontoxic ferric form. CP facilitates iron release from endothelial cells, neurons and glial cells to maintain iron homeostasis in the brain. Aceruloplasminemia is a recessive neurodegeneration characterized by mutation of the CP gene and marked iron accumulation in the brain [95–97]. Adult *Cp* gene knockout mice ($CP^{-/-}$) show age-dependent iron overload in the central nervous system (CNS) [98,99]; these mice are considered an endogenous iron-overload model. Intracerebroventricular injection of ferric ammonium citrate (FAC), which induces high levels of iron in the brain, is used as an exogenous iron-overload mouse model. Both iron-overload mouse models are more vulnerable to neurotoxin and develop PD following intraperitoneal injection of low-dose 1-methyl-4-phenyl-1,2,3,6-tetrahydropyridine (MPTP). Further investigation demonstrated that iron accumulation is induced by *Cp* gene knockout and exacerbates oxidative stress levels, which promotes apoptosis in the brain. Chelation of iron can decrease brain iron levels and ROS, and apoptosis is reduced in PD mice. Therefore, iron overload in the brain exacerbates dopaminergic neuronal death in the SNpc [90].

Researchers observed a loss of approximately 80% of the ferroxidase activity of CP in the SN of PD cases; $CP^{-/-}$ mice developed parkinsonism and exhibited iron accumulation in the SN [100,101]. Thus, *Cp* deletion induced iron overload and consequent ROS elevation, which induced dopaminergic neuron death and led to the onset of PD.

The expression of IRP2, a key regulator of iron homeostasis, is negatively regulated by high levels of iron; altered IRP2 can impact iron uptake, iron storage and iron release proteins through the IRE-IRP regulatory pathway, which maintains iron homeostasis [57,102]. Dominant expression of IRP2 has been detected in the CNS [103]. *IRP2* gene knockout mice ($IRP2^{-/-}$), reported by the Rouault lab, develop neurodegenerative movement disorders [104] and exhibit excessive iron accumulation in the brain [56,104,105].

Iron overload is observed in the SN of $IRP2^{-/-}$ mice, and low doses of MPTP increase neuronal apoptosis and decrease DA levels by altering iron metabolism, exacerbating parkinsonism symptoms [91]. Levels of the deubiquitylase OTU domain-containing protein 3 (OTUD3) are decreased in PD mice overexpressing A53T α-synuclein. *OTUD3* gene knockout mice showed nigral iron accumulation and dopaminergic neurodegeneration. OTUD3 can stabilize IRP2 to maintain iron homeostasis and prevent PD [106].

Mitochondria ferritin (FtMt) is a protein with ferroxidase activity capable of storing iron in the mitochondria. Although *FtMt* overexpression or deletion does not affect iron levels in the brains of mice, its overexpression induces a slight increase in iron uptake, cytosolic ID and decreases ROS production in SH-SY5Y neuronal cells and significantly blocks iron redistribution in a PD cell model [107]. *FtMt* deletion induced ferroptosis, and its overexpression attenuated ferroptosis during cerebral ischemia/reperfusion [108]. *FtMt* gene knockout promotes ROS generation, and overexpression restricts ROS production in vivo. In vitro, FtMt attenuated oxygen and glucose deprivation and reperfusion-induced iron accumulation in mitochondria [109]. These data revealed that FtMt plays a critical antioxidative role in the progression of PD by regulating ROS.

Nuclear factor erythroid 2-related factor 2 (Nrf2) regulates the expression of antioxidant and detoxification enzymes and participates in many biological functions and disorders including oxeiptosis [110]. *Nrf2* gene knockout decreases the FPN1 level in BMVECs thus decreasing iron entry into the SN and striatum, reducing ROS and decreasing apoptosis of dopaminergic neurons in PD mice [111], which suggests that Nrf2 is neuroprotective against PD via regulation of iron metabolism in the brain.

In addition to the above mechanisms, other molecules and pathways are involved in iron metabolism disorders in PD. Ferroptosis participates in dopaminergic neuronal cell death in PD, and the mutation of other ferroptosis genes has been linked to PD. On the other hand, an autosomal recessive mutation in the PD-related gene, *DJ-1* (*PARK7*), can suppress ferroptosis [112]. Ferroptosis is also characterized by elevated lipid peroxidation and ROS; therefore, iron and the resulting increase in ROS play prominent roles in the pathology of PD. Overall, there are many reports of iron participation or interaction with neurotransmitters and with many PD gene mutations [113–118], which form a vicious cycle that exacerbates the progression of PD.

4.1.2. Iron Overload Exacerbates PD and Related Mechanisms

Iron and α-Synuclein

α-synuclein aggregation is thought to play a key role in the formation of LBs that contribute to PD pathogenesis. Iron and α-synuclein both accumulate in LBs of the remaining dopaminergic neurons of the SN [119]. Evidence has shown that iron promotes α-synuclein aggregation [120]; an IRE motif is predicted in the in 5′-UTR of *α-synuclein* mRNA [121], suggesting IRE-IRP mechanisms might regulate its expression. Further investigation showed that iron regulates the synthesis of α-synuclein through the IRE-IRP pathway at the post-transcriptional level; iron-mediated oxidative stress also regulates α-synuclein at the post-translational level. α-synuclein has also been shown to exhibit ferrireductase activity in the SN and may regulate iron uptake [122,123].

Transgenic PD mice overexpressing a mutant (A53T) human α-synuclein exhibited age-related motor deficits, and their SN was more vulnerable to high dietary iron compared with wild-type mice [124]. Excess iron has been linked to increased oxidative/nitrative stress, which could induce tyrosine nitration. Nitrated α-synuclein has been detected in the LBs of the PD brain [125]. The attachment of nitro molecules to Tyr39, Tyr125, Tyr133 and Tyr136 of α-synuclein causes significant changes in α-synuclein [126]. Nitrated α-synuclein is not readily degraded and is mixed into fibrils, accelerating the formation of fibrils with unmodified α-synuclein [127–129]. In vitro evidence has demonstrated that microglial activation can induce nitric oxide (NO)-dependent oxidative stress in dopaminergic neurons resulting in α-synuclein nitration. Nitrated, aggregated α-synuclein during conditions of oxidative stress induces inflammatory microglial functions [130,131].

Phosphorylation of α-synuclein has been shown in LBs [132]. Phosphorylation of Ser129 is the primary modification of α-synuclein [133]; this mutation is harmful in PD. Mutation of S129 (S129D) increases α-synuclein phosphorylation, the aggregation of which promotes dopaminergic neuronal cell death [134,135]. The S129A mutation prevents α-synuclein phosphorylation and suppresses the loss of dopaminergic neurons [134]. While some reports have indicated that α-synuclein phosphorylation has no toxic effects, it remains clear that iron-induced oxidative stress promotes phosphorylation of α-synuclein. Iron overload has been shown to induce phosphorylation of α-synuclein at S129 and its subsequent aggregation in vitro [114,136]. Furthermore, phosphorylation at Y125 or S129 may increase the binding affinity between ferrous iron and the C-terminal region of α-synuclein [115].

Overall, these results suggest that iron and α-synuclein interact with one another, and their deposition and aggregation may be important factors in the pathology of PD. Blocking iron and α-synuclein interactions may be a useful strategy to prevent and cure PD.

Iron, Dopamine and Neuromelanin

Decreased levels of the neurotransmitter DA are important in PD. DA generates toxic metabolites in the cytoplasm [137]; iron and DA are considered toxic in combination. Physiologically, DA can produce H_2O_2 via monoamine oxidase [113]. H_2O_2 produced in dopaminergic enzymatic processes reacts with the high level of iron in dopaminergic neurons, generating oxidative stress such as hydroxyl free radicals via the Fenton reaction. Hydroxyl radicals can damage membrane phospholipids, proteins and nucleic acids to induce neuronal cell death [138].

Iron is also involved in the oxidation of DA, forming 6-OHDA, which liberates iron from ferritin and produces H_2O_2, therefore, aggravating dopaminergic neuronal cell death and the development of PD. It was reported that high levels of iron inside cells caused ferroptosis, a form of regulated cell death characterized by iron-dependent accumulation of lipid hydroperoxides to lethal levels [78,139].

Neuromelanin, a dark pigment in dopaminergic neurons, binds reactive iron in neurons and plays a neuroprotective role. Loss of neuromelanin is observed in PD patients [89]. When the binding capacity of neuromelanin for iron is decreased, free iron increases in the SN and induces oxidative damage via the Fenton reaction. Increased iron also reacts with α-synuclein to aggravate oxidative stress and protein aggregation, resulting in neurodegeneration and neuronal cell death. Degenerating neurons also release neuromelanin, which activates microglia, further releasing neuromelanin and initiating neuroinflammation and neurodegeneration [116]. Therefore, iron is involved in DA oxidation by interacting with DA metabolites (such as H_2O_2 and neuromelanin) to damage dopaminergic neurons, which accelerates the release of a-synuclein to activate microglia, producing neuroinflammation and participating in the occurrence of PD.

Iron and Parkin

Iron can alter Parkin solubility, resulting in its intracellular aggregation. With the depletion of soluble, functional forms of Parkin, proteasomal activity is impaired with cell damage [140]. In 1998, mutation of *Parkin* was identified in autosomal recessive juvenile parkinsonism; iron staining in the SN of these patients was more intense than that of controls and sporadic PD patients [141]. It was hypothesized that iron accumulation might be related to loss of the *Parkin* gene.

More recently, Parkin was reported to be responsible for ubiquitination of DMT1 (+IRE). Expression of 1B-DMT1 isoforms was decreased in SH-SY5Y cells overexpressing Parkin [117]. When fed an iron-supplemented diet, transgenic mice overexpressing DMT1 showed selective accumulation of iron in the SN; expression of Parkin was also upregulated, likely reflecting a neuroprotective response [142]. Expression of DMT1 (+IRE) was also increased in human lymphocytes containing a homozygous deletion of exon 4 of *Parkin* and in the brains of *Parkin* knockout animals. All these data suggested that there might be a feedback interaction between the abnormal iron level with/without aberrant expressions of iron regulatory molecules and the expression and function of Parkin, thus participating in the progression of PD.

Iron and Leucine-Rich Repeat Kinase 2

Leucine-rich repeat kinase 2 (LRRK2) is involved in inflammation, autophagy, lysosomal processing and vesicular trafficking [143]. Mutations in the *Lrrk2* gene cause PD and inflammatory diseases [144,145]. Increased iron is present in PD patients with the *Lrrk2* mutation [146]. The underlying mechanisms and their contribution to PD pathology have been investigated. LRRK2 has been shown to activate microglia; mutations in *Lrrk2* can induce cytokine release and inflammation in PD [147,148]. Recent investigations have indicated that LRRK2 can phosphorylate Rab GTPases, regulating vesicle traffic. The RabGTPase Rab8a directly interacts with TfR to aid in TfR recycling to the cell membrane in the iron-uptake pathway [149]. Mutation of *Lrrk2* enhances Rab8a phosphorylation, sequestering Rab8a in lysosomes, resulting in the dysregulation of endolysosomal transport, inhibition of Tf-TfR recycling and enhanced cellular iron accumulation [118,150]. Therefore, LRRK2 may modulate iron metabolism, especially iron uptake and storage in microglia in conditions of neuroinflammation [118].

Excess iron has been observed in dopaminergic neurons of the PD brain. FAC can catalyze the phosphorylation of S935 and S1292 of LRRK2, significantly increasing its activity. Active LRRK2 accelerates ferrous iron uptake, indicating a relationship between iron and LRRK2 in dopaminergic neurons [151].

In summary, iron deposition in the SN and LB aggregation is a hallmark of PD. The increased iron induced by *Cp*, *IRP2*, *Nrf2* or *FtMt* gene knockout, and the subsequent increased oxidative stress and their interactions with α-synuclein, DA, neuromelanin, Parkin and LRRK2, all contribute to the development and progression of PD (Figure 4). Thus, targeting iron levels is an important strategy in the prevention and treatment of PD.

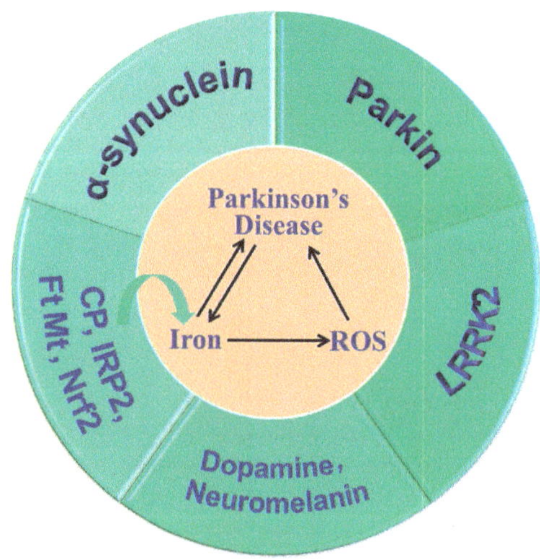

Figure 4. Iron accumulation is important in the pathogenesis of PD. Dysregulation of CP, IRP2, Nrf2 and FtMt alters brain iron levels, which, in turn, affects the expression of iron metabolism proteins. Iron overload and increased ROS aggravate the development and progression of PD, and their interactions with α-synuclein, dopamine, neuromelanin, Parkin and LRRK2 contribute to dopaminergic neuronal cell death and the onset of PD.

4.2. Iron and Alzheimer's Disease

AD is a neurodegenerative disease characterized by progressive memory impairment and cognitive dysfunction. The main pathological features of AD are the intercellular deposition of insoluble β-amyloid (Aβ) plaques and intracellular fiber tangles formed by excessive phosphorylation of the tau protein, as well as neuronal cell loss [152]. Although the cause and exact mechanisms of AD have not been revealed, great progress has been made. It has been suggested that the increase in brain iron and the imbalance of iron metabolism associated with increased ROS generation play an important role in the pathogenesis of AD [153,154]. Therapeutic approaches to decrease brain iron levels or restore iron homeostasis along with the attenuation of oxidative stress show great promise in the treatment of AD [154].

4.2.1. Iron Dysregulation in AD

Iron deposition in Aβ plaques and neuronal tangles in the brains of AD patients have been widely reported. In 1992, a study showed that in brain slices of AD patients, the distribution of iron in senile plaques and the surrounding cells increased significantly, suggesting there was iron deposition and disruption of iron homeostasis in the AD brain [155].

Magnetic resonance imaging showed that the aggregation of Aβ accompanied by the accumulation of iron occurred in the early stages of AD [156]. Compared with healthy individuals, AD patients have increased iron in the cerebral cortex and hippocampus, colocalized with the Aβ plaques [156]. The increased levels of iron in the brain exacerbate the aggregation of Aβ and accelerate neuronal cell death. Symptoms of AD can be attenuated by iron-chelating agents [157].

In addition to the changes in iron distribution in the brains of AD patients, the expression of several key molecules responsible for transporting iron or regulating iron homeostasis is also altered, including Tf, TfR1, DMT1, ferritin, FPN1, CP, IRPs and hepcidin [11,14,69,155,158–160]. Ferritin in senile plaques in the hippocampus and peripheral blood vessels of AD patients is increased [155,161]; ferritin is also increased in the CSF of AD patients [161]. Protein and mRNA levels of FtMt are increased significantly in the cerebral cortex of AD models [162–164]. Expression of TfR1 was shown to be increased in the cerebral cortex and hippocampus of 3-month-old amyloid precursor protein (APP)/PS1 mice [165]. DMT1 in the cerebral cortex and hippocampus of APP/PS1 mice was increased around Aβ plaques [166]. FPN1 in the cortex and hippocampus of AD patients and animal models was significantly decreased [166]. Expression of hepcidin in the brain of AD patients and mouse models was decreased, accompanied by increased neuroinflammation and oxidative damage [58,69,160]. The alterations of these molecules verify the dysregulation of iron metabolism in the pathogenesis of AD.

4.2.2. Iron in Aβ and Tau Pathology

Increased iron directly induces Aβ aggregation, which, in turn, participates in the generation of oxidative stress, contributing to the pathological symptoms of AD [153]. The increased iron can also affect the expression of APP and its subsequent amyloidosis. Under normal conditions, most APP is cleaved by α-secretase and γ-secretase successively, releasing its N-terminal P3 fragment and leaving the APP intracellular domain in the cell membrane, while a small amount of APP undergoes β-secretase (BACE-1) and γ-secretase shearing to produce Aβ [167]. The activation of both α-secretase and BACE1 is regulated by furin, and the transcription of furin is regulated by intracellular iron levels [72,168,169]. When total iron levels are high, furin protein levels decrease [72], which downregulates the activity of α-secretase, resulting in Aβ production. By contrast, in conditions of ID, furin activity increases, increasing α-secretase activity and the stimulation of non-Aβ cleavage of APP [170]. APP translation is also affected by iron levels because of the IRE motif present in the 5′-UTR of APP mRNA, which can be regulated by IRPs upon binding [171]. In conditions of ID, IRPs bind the IRE of *APP* mRNA to inhibit its translation. However, at high iron concentrations, IRPs interact with iron and are released from *APP* mRNA, resulting in increased APP translation, which, in turn, increases Aβ generation [171].

Iron overload can also aggravate tau protein dysfunction and enhance the formation of neuronal fiber tangles. In the brains of AD patients, lipid peroxidation induced by iron overload can promote tau polymerization, which further increases oxidative stress and the formation of tau fibrillary lesions [172]. In vivo experiments have shown that iron overload causes abnormal phosphorylation of tau protein in neurons [173]. Previous studies have shown that a lack of tau can affect the post-translation transport of APP, resulting in its retention in the endoplasmic reticulum [174]. APP exhibits ferrous oxidase activity, which may help iron efflux by stabilizing FPN1 on the cell membrane [175]. Therefore, lack of tau may affect iron release by regulating APP, leading to increased intracellular iron and further aggravating cell damage.

4.2.3. Mechanism of Iron and Oxidative Stress in AD Pathogenesis

Iron overload and the resulting oxidative stress participate in the pathology of AD symptoms. In AD pathogenesis, iron overload has been implicated in mitochondrial dysfunction, neuroinflammation and neuronal cell death (Figure 5).

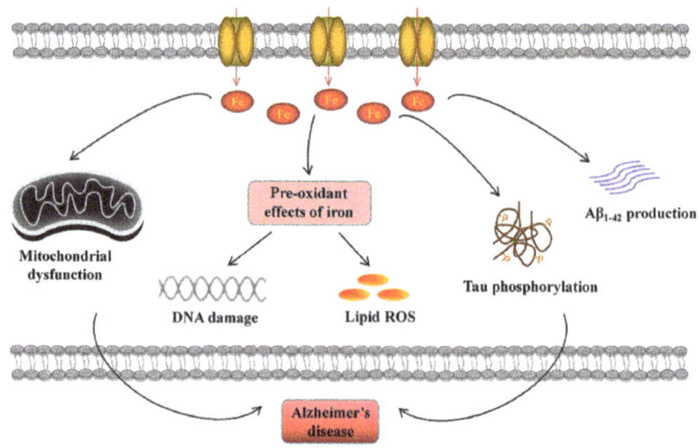

Figure 5. Iron accumulation participates in the pathogenesis of AD. Elevated cellular iron is related to $A\beta_{1\text{-}42}$ production and tau phosphorylation. Excessive iron leads to mitochondrial dysfunction. The pre-oxidant effects of iron induce DNA damage and lipid ROS generation, contributing to cell death.

Mitochondrial dysfunction is a common pathogenic feature of AD [176], evidenced by elevated ROS generation and lipid peroxidation, decreased mitochondrial membrane potential and altered mitochondrial morphology [176]. Sufficient intracellular iron is required for mitochondria-dependent metabolic activities, such as ATP production by the electron transport chain, Fe-S cluster formation and heme biogenesis. Disruptions in iron homeostasis result in mitochondrial dysfunction and energetic failure. ID impairs mitochondrial metabolism and respiratory activity [8], while iron overload promotes the production of damaging ROS during mitochondrial electron transport [177], which triggers oxidative stress in the brain, resulting in neurological damage and disease development [178,179]. Therefore, iron flux in mitochondria must be precisely regulated.

Neuroinflammation is a characteristic of AD and is mainly mediated by the activation of microglia and astrocytes, which release excess inflammatory factors that result in neuronal impairment [176]. Accumulating evidence has shown a relationship between iron levels and neuroinflammation. Elevated neuroinflammation has been reported to contribute to the deleterious impact of iron overload on brain function in aging through astrocytic dysfunction and inflammation [180]. In AD, iron plays a direct role in Aβ-stimulated neuroinflammation. Iron overload promoted activation of NF-κB signaling induced by Aβ and increased secretion of inflammatory factor interleukin (IL)-1β in microglia [181]. Similarly, iron overload in lipopolysaccharide (LPS)-primed peripheral blood mononuclear cells stimulated caspase 1-dependent IL-1β secretion and activated the NOD-like receptor family pyrin domain-containing 3 inflammasome, due to the increased cellular labile iron [182]. Iron overload has also been shown to promote IL-6 secretion through microglia, which, in turn, upregulates the expression of IRP1 and DMT1 and downregulates the expression of FPN1 via C-Jun N-terminal kinase activation [183], aggravating iron overload and inducing oxidative stress and cellular dysfunction.

On the other hand, iron metabolism is also regulated by inflammatory and anti-inflammatory cytokines, such as tumor necrosis factor (TNF)-α and transforming growth factor (TGF)-β1. TNF-α has been shown to promote iron uptake in astrocytes and microglia by promoting DMT1 expression, while TGF-β1 facilitates iron efflux in astrocytes by increasing FPN1 expression, thereby differentially contributing to iron homeostasis [184]. In addition, hepcidin in astrocytes is important in LPS-induced neuroinflammation and neuronal apoptosis [58]. High hepcidin levels are associated with intracellular iron accumulation, as hepcidin binds FPN1, internalizing the receptor and blocking iron release from cells [28,58,185]. These results suggested that neuroinflammation stimulates iron overload by regulating the expression of iron transporters, followed by a positive feedback loop that aggravates neuroinflammation and oxidative stress in the brains of AD patients. As iron overload and its related chronic neuroinflammation contribute to the progression of AD, iron chelators have been investigated as potential agents to alleviate neuroinflammation and ROS generation [186].

Iron-induced cell death is an important cause of neuronal death in AD pathology. Studies have found that there are ~30- to 50-times more DNA fragments in neurons and glial cells in AD patient brains than in normal brains of humans of the same age [187], indicating that apoptosis is one of the main forms of cell death in the AD brain. Iron overload in AD is accompanied by an increase in apoptotic cells; reducing brain iron levels reduced apoptosis in the cortex and hippocampus of AD mice [69,160]. These results suggest that the iron-induced apoptosis pathway plays an important role in neuronal cell death in AD. In addition to apoptosis, ferroptosis is mainly caused by iron-dependent oxidative damage and is thought to be closely regulated by intracellular iron homeostasis [78]. Iron overload in AD reduces the expression of GPx4 and increases expression of acyl-CoA synthetase long chain family member 4 (ACSL4); restoring iron homeostasis ameliorated AD symptoms by inhibiting ferroptosis [69,79], indicating that iron-accumulation-induced ferroptosis is an important characteristic of AD.

4.3. Iron and Stroke

Stroke is the second leading cause of death after cancer, and adults who survive have varying degrees of physical disability. A quarter of people around the world are affected by stroke [188], resulting in a significant burden to society and patients. Ischemic stroke is the most common form of stroke. Due to vascular embolism, continuous occlusion of blood flow leads to irreversible necrosis of nerve cells in ischemic brain tissue, forming the infarction core. Adjacent tissue cells retain some level of metabolic activity, which is referred to as penumbra. At present, the primary treatment is thrombolysis, although reperfusion causes excitoneurotoxicity, Ca^{2+} overload, ROS generation and inflammation, activates innate and adaptive immunity and causes secondary damage to tissues. The mechanism of ischemia/reperfusion (I/R) is complex, and many modes of cell death are involved, including necrosis, apoptosis, autophagy and ferroptosis. Ferroptosis is an important cause of tissue damage and cell death during reperfusion [139]; the use of ferroptosis inhibitors alleviates I/R damage. Our previous study found that brain iron metabolism was disturbed after I/R and iron levels were increased in the cortex and hippocampus. Hepcidin has been shown to regulate iron levels [189] and cells were characterized by brain iron overload. Excess iron can cause lipid peroxidation via the Fenton reaction, which is important in apoptosis [190] and ferroptosis [108]. Therefore, iron plays an important role in neural damage caused by thrombolysis or thrombectomy in ischemic stroke.

4.3.1. Iron Regulation in Ischemic Stroke

Although there have been many reports of iron overload after stroke, the underlying mechanism remains unclear. Previous studies have shown that hepcidin is involved in the regulation of cellular iron overload after stroke [189]. Stroke upregulates hepcidin expression, and hepcidin was shown to bind and internalize to degrade the iron exporter FPN1 [191], which, in turn, leads to iron accumulation in cells. The hypoxic environment created by ischemic stroke also affects intracellular iron and regulates iron-related protein expression, primarily through hypoxia-inducible factor (HIF). Knockdown of *HIF-1α* reduces hypoxia-induced iron accumulation in cells. HIF-1α binds as a transcription factor to the hypoxia-responsive element (HRE) of *Tfr1* and *DMT1*, activating their transcription, which, in turn, increases iron uptake [192]. Glutamate receptors (N-methyl-D-aspartate receptors (NMDARs)) also mediate iron uptake following a stroke. NMDARs induce the Ras protein via NO/dexamethasone, increasing the TfR-dependent iron uptake pathway [193]. In addition to direct the regulation of iron by hypoxia, the BBB is also involved in cerebral blood deposition in I/R. Changes in iron levels in cerebral microvascular endothelial cells can directly affect iron levels in the brain. After oxygen and glucose deprivation and reperfusion (OGD/R) treatment, iron levels in cerebral microvascular endothelial cells were significantly increased [190]. Endothelial cells released intracellular iron into the brain through FPN1, and knockdown of *Fpn1* in endothelial cells decreased iron accumulation in the brain and alleviated oxidative stress, inflammation and cell death after ischemic stroke [25]. The above results suggest that the mechanism of iron metabolism dysregulation in ischemic stroke may not only include the disruption of the BBB and abnormal iron transport across the BBB but also be caused by disorders in iron uptake and release by neural cells in the brain.

4.3.2. Mechanism Underlying Cell Death Induced by Iron Dysregulation

Intracellular iron accumulation after ischemic stroke can aggravate cell death and tissue damage in various ways. Oxidative stress, in particular, is an important factor in excess iron-mediated cell death, including apoptosis and ferroptosis. ROS induces the opening of mitochondrial membrane permeability pores via oxidative damage of lipids and other macromolecules in mitochondria, releasing cytochrome c into the cytoplasm, which activates caspase-3 and triggers apoptosis [194,195]. In the cytoplasm, ROS primarily activates the downstream signaling molecule c-Jun NH2-terminal kinase (JNK) through apoptosis signal-regulating kinase 1 (ASK1) to further phosphorylate the proapoptotic molecule Bak/Bax, resulting in the mitochondrial release of cytochrome c [196–198]. In addition to participating in ROS generation and cell death, iron also plays an important role in energy metabolism. Studies have demonstrated that the overexpression of FtMt, on the one hand, decreases the level of free iron in mitochondria and slows the production of ROS after OGD/R. However, FtMt also enhanced glucose metabolism and the pentose phosphate pathway after OGD/R to promote the synthesis of NADPH and glutathione, thus increasing cellular resistance to oxidative damage [109]. Iron also functions as a coenzyme for many oxidases, such as LOX, NADPH oxidase and xanthine oxidase, which are involved in catalytic lipid oxidation and ROS generation [199] and are another important source of oxidative damage.

Based on the role of iron and ferroptosis in I/R, many compounds and therapeutic strategies targeting iron have been explored and validated in animal models. The iron chelator DFO is widely used in a variety of stroke models and has been shown to alleviate stroke injury in mice [200]. Other iron regulatory proteins have been shown to be protective in animal models. Endothelial cells release intracellular iron into the brain through FPN1; while knockdown of *Fpn1* reduces oxidative stress, inflammation and cell death after stroke, ID is not beneficial during neurological recovery after ischemic stroke [25]. Previous studies have shown an increase in iron in the cerebral cortex during post-stroke recovery, along with a corresponding increase in synaptic plasticity and myelin nerve regeneration, indicating that the recovery process, unlike ischemia, seems to require more iron involvement [201]. Knockdown of hepcidin or peripheral injection of CP during ischemia decreased brain iron levels and improved post-stroke motor capacity in mice [189,202]. Intravenous injection of iron-loaded Tf increased cell mortality, increased ROS production and aggravated damage after OGD [203]. Our previous study found that HIF-1α acts as a transcription factor to activate transcription of FtMt, which preferentially sequesters intracellular iron in mitochondria, diminishing free iron in the cytoplasm. Overexpression of FtMt is protective in both hypoxia and I/R models [108,163]. An increasing number of new therapeutic agents targeting iron to treat ischemic stroke are being explored. Nanoliposomes carrying lycopene have been shown to have therapeutic efficacy following ischemic stroke injury in rats. Lycopene nanoliposomes regulate iron levels after stroke and reduce oxidative stress and apoptosis [204]. Moreover, some ferroptosis inhibitors such as Ferrostain-1 were shown to alleviate I/R injury. Further research on brain iron metabolism imbalance following I/R will improve our understanding of the role of iron and provide new directions for prevention and targeted therapies.

4.4. Iron and Neuropsychiatric Disorders

Neuropsychiatric disorders, such as depression and anxiety, are increasing worldwide, affecting approximately 30% of the general population during their lifetime [205]. Abnormal synthesis and secretion of neurotransmitters, reduced neuroplasticity and impaired neurodevelopment have been linked to the pathogenesis of neuropsychiatric disorders [206], but the specific mechanisms remain unclear. ID has been correlated with behavioral and developmental changes that occur with neuropsychiatric disorders; these changes affect the hippocampus, the striatum and neurotransmitters, such as serotonin, noradrenaline and DA [206]. Analysis of survey studies found a link between iron intake and depression; total iron intake may be inversely associated with depression [207]. A health survey of individuals over 65 found that a higher number of depressive symptoms was associated with lower hemoglobin levels and higher serum TfR levels but not with ferritin levels [208]. Accumulating evidence demonstrates that ID results in increased anxiety and/or depression with social and attentional problems in children [209]. By contrast, iron overload alters anxiety-like behavior and mood [210]. A recent study confirmed that an imbalance of iron metabolism is a cause of anxiety; researchers found that the neural circuit from the vHip to the medial prefrontal cortex (mPFC) to the SN was responsible for brain iron transport and that dysfunction of vHip-mPFC iron transport could induce anxiety-like behaviors [39]. Monoamine metabolism is the most widely studied metabolic pathway, and iron is required for the synthesis of monoamine neurotransmitters. In particular, serotonin plays an important role in depression, anxiety and other neuropsychiatric disorders [211]. ID results in poor brain myelination and impaired monoamine metabolism, and accumulating data have shown that neurotransmitter homeostasis influences emotional behavior [212].

Fear memories are common in humans and can be used to avoid or minimize harm. Few studies have focused on the role of iron metabolism in the development of fear memory. ID in early brain development can lead to long-term neurological damage, including hippocampus-mediated learning and memory deficits [213]. The hippocampus is an important structure for many forms of memory. The dentate gyrus of the hippocampus plays a key role in the acquisition of situational fear memories [214,215]. In addition, hippocampal-dependent learning was shown to be permanently impaired during fear-conditioning experiments in rats with perinatal ID [216]. An investigation into the response of the mice to contextual fear revealed that the formation of fear memory was impeded after neuronal *Fpn1* depletion by reducing brain iron [7]. Hippocampal-dependent memory processes, such as cognitive memory and fear conditioning, are strongly affected by perinatal ID [217]. These studies suggest that the normal development of the nervous system requires a balance of iron levels in the brain; this balance is important for normal nervous system function and can affect fear memories.

4.5. Iron and Abnormal Neurodevelopment

Normal development of the brain is an important process in the establishment of the mammalian nervous system; development involves proliferation, differentiation, migration, synaptogenesis and myelination [218]. During the early stages of mouse embryonic development, neural progenitors divide to give rise to neurons. This is followed by gliogenesis, myelination and synapse construction [219]. Numerous studies have suggested that abnormal brain development is closely related to nervous system disorders, such as microcephaly, schizophrenia, autism spectrum disorder and malformation of cortical development [220–225].

Numerous animal studies have shown that iron is particularly important for many neurodevelopmental processes, especially during pregnancy and infancy due to rapid growth [226]. However, ID is one of the most common nutritional deficiencies, especially in pregnant women and infants [209,227]. Numerous studies have shown that ID in these periods causes neurodevelopment deficits, including impairment of learning and memory, motor skills and emotional regulation, and these deficits are not fully recovered even when iron is restored [4–6,216,228–239]. The mechanisms that may contribute to these impairments include changes in brain energy metabolism, neurotransmitter chemistry, organization and morphology of neuronal networks and the neurobiology of myelination [234,240]. Here, we summarize the major findings from recent decades on the effects of ID on the development and functions of neurons and glia cells (mainly neurons and oligodendrocytes) and highlight new information on the possible mechanisms through which ID affects brain development.

4.5.1. Iron Deficiency Affects Neurodevelopment and Function

Numerous studies have shown that iron is an essential element for cell proliferation by serving as a substrate for enzymes that participate in DNA synthesis, the cell cycle and energy production [241]. Recent studies have revealed the role of iron in maintaining the stemness of embryonic stem cells. Intracellular ID significantly inhibits the proliferation and differentiation of embryonic stem cells/neuronal precursor cells [8,242]. Overall, these data suggest that iron is essential for cell proliferation and differentiation.

Studies in rodents have reported deleterious effects of ID on the structural and morphological development of dendrites and synapses during brain development [243–247]. Furthermore, studies in neurochemistry have shown that ID has significant effects on neuronal DA metabolism [209,233] and the synthesis of monoaminergic neurotransmitters [246,248,249] and growth factors [243]. ID also leads to changes in the synaptic transmission and synaptic function [250,251]. A study using two non-anemic genetic ID mouse models (knockout of *DMT1* or overexpression of dominant negative *TfR1*) showed that neuronal-specific ID dysregulates mammalian target of rapamycin (mTOR) signaling during hippocampal development [252]. Together, this research indicates that ID alters the neuronal structure and morphology, metabolism, synaptic plasticity and structural gene expression and mTOR signaling pathway.

4.5.2. Iron Deficiency Affects Development of Oligodendrocytes and Myelination

Oligodendrocytes are characterized by high intracellular iron and a high rate of oxidative metabolism, which are required for the synthesis and maintenance of myelin [6,253,254]. Studies in mice have also shown that ID negatively affects the development of oligodendrocytes and their myelination [244,255–257]. Impaired myelination has also been reported in studies of gestational and postnatal ID performed in monkeys and piglets, suggesting that iron is essential for oligodendrocyte activity and integrity [258,259].

4.5.3. Mechanisms Underlying ID and Hippocampal Development

Based on the vulnerability of the developing hippocampus to early ID and earlier work showing lasting spatial memory deficits related to the role of the hippocampus in iron deficient rodent brains [216,260–262], more attention has been paid to ID and hippocampal function. ID during late fetal and early postnatal life alters the expression of critical genes involved in hippocampal development and function, including iron metabolism, cell growth, energy metabolism, dendrite morphogenesis and synaptic connectivity in the hippocampus [213,243,252,263–266]. DNA methylation and O-linked-beta-D-N-acetylglucosamine (O-GlcNAc) modifications play important roles in these processes [267,268]. Recently, ID response networks and signatures have been revealed through quantitative proteome and transcriptome dynamics analysis in neuronal cells [269]. Taken together, these studies suggest that ID induces changes at the proteome and transcriptome levels, as well as alterations in post-translational modifications, including phosphorylation signaling and DNA methylation.

Long-term studies show that many of the ID-induced neurodevelopmental deficits during the fetal and early postnatal period cannot be recovered by iron repletion at later stages and eventually lead to sustained impairments [259,270–273], suggesting that ID in the developmental stage results in long-lasting abnormalities, even after iron supplementation. At present, studies support the concept that early ID in critical periods may disrupt the timing of key steps in development, and repletion of iron after these important time points will not rectify anatomic and neurochemical abnormalities. While awaiting prospective trials, it is recommended to screen all gravidas for ID and administer iron supplementation if ID is present with or without anemia.

In conclusion, numerous studies in human and animal models suggest that ID affects brain development and significantly affects the development and function of neurons and oligodendrocytes. At present, the possible mechanisms of ID affecting brain development are derived from studies on animal models of deficits in brain energy metabolism, neurotransmission, neuronal morphology and the myelination of oligodendrocytes. Since ID in pregnant women and pre-school-age children causes poor long-term neurodevelopment outcomes in later life, it is of important scientific, medical and social significance to further clarify the molecular mechanisms of ID affecting brain development, as well as prevention and treatment strategies.

5. Targeting Iron Metabolism in the Treatment of Neurological Diseases

In recent years, research on brain iron metabolism disorders and neurological diseases has shown that increases in brain iron and imbalances in iron metabolism may play essential roles in the pathogenesis of neurological diseases. Therefore, targeting brain iron homeostasis and regulation of iron-metabolism-related molecules for drug development are expected to provide novel ways to treat neurological diseases.

5.1. Iron Chelators

Iron chelators bind iron ions with high affinity, effectively enhancing iron excretion and reducing free iron in the body [274]. Iron chelators have been shown to inhibit lipid peroxidation and reduce ROS levels in neurons, thereby preventing neuronal ferroptosis and apoptosis. Clinically, commonly used iron chelators include DFO, DFX and DFP, which have shown therapeutic promise in preclinical and clinical models of neurological disorders. Decreasing iron levels in the brain with iron chelators has been reported to alleviate the symptoms of AD, PD and stroke [17–22].

5.1.1. Iron Chelators in AD

Clinical application of DFO to treat AD was reported as early as 1991; continuous administration of DFO was found to alleviate cognitive impairment in AD patients [17]. Administration of the iron chelator M30 decreased brain iron accumulation, Aβ accumulation and tau phosphorylation, improving memory deficits in APP/PS1 mice [18]. This improvement may have been achieved by downregulating phosphorylation of cyclin dependent kinase 5 (CDK-5) and increasing phosphorylation of protein kinase-B (PKB/AKT) and glycogen synthase kinase (GSK)-3β [18]. DFO improved the cognitive ability of APP/PS1 mice due to the activation of M2-type microglia and inhibition of activation of M1-type microglia in the hippocampus [275].

5.1.2. Iron Chelators in PD

Overloaded intracellular iron contributes to neuronal cell death in PD via apoptosis and ferroptosis, while DFO can inhibit ferroptosis to protect neurons [19]. In experimental studies, iron chelators have been shown to exhibit neuroprotective effects in vivo against 6-OHDA-induced neurotoxicity in mouse models of PD [20]. Loss of CP iron oxidase activity in the SN of PD patients leads to the accumulation of iron peroxide [100]. Administration of an iron chelator in $CP^{-/-}$ PD mice reversed the accumulation of iron ions caused by the loss of CP, significantly improved the motor ability of mice and reduced the nerve damage caused by MPTP [90,100]. After 8 weeks of pre-administration of the iron chelator clioquinol in a mouse PD model, iron levels in the SN decreased by 30%, and oxidative stress and GSH loss were significantly reduced [94]. Treatment with the iron chelator DFO has been shown to block MPP^+-mediated damage of dopaminergic neurons and prevent iron accumulation and mitochondria dysfunction [276]. Brain iron accumulation exacerbates the pathogenesis of MPTP-induced PD; DFO alleviates PD symptoms by reducing oxidative stress damage caused by elevated brain iron levels [90]. In two phase 2 trials, the high-affinity iron chelator DFO was shown to reduce iron accumulation and improve motor symptoms in PD patients compared with placebo, despite the side effects, such as leukopenia, gastrointestinal discomfort and joint pain [277,278]. A clinical trial evaluating the effects of four different doses of DFP on 140 patients with early-stage PD has yet to be published [278]. A more extensive European multicenter test on the protective effect of DFP on PD patients showed that the iron content in the nigrostriatal pathway was significantly reduced in DFP-treated groups. The mean change in the total movement disorder society-sponsored revision of the unified Parkinson's disease rating scale (MDS-UPDRS) score was 16.7 points in the DFP group and 6.3 points in the placebo group [279]. Surprisingly, DFP without DA treatment increased the patient's disability. This may be related to the fact that iron is a cofactor in DA synthesis. Iron is required to assist in the synthesis of DA, but excess iron can cause the death of dopaminergic neurons [278]. Another long-term clinical trial used an iron chelator combined with DA to avoid the drawbacks of iron chelators alone, but further research is needed.

5.1.3. Iron Chelators in Stroke

Iron-overloaded animals are more affected by middle cerebral artery occlusion [280], whereas iron chelation or depletion reduces I/R-induced brain injury [21,22]. DFO has been shown to inhibit lipid peroxidation and hydroxyl radical production via the Fenton reaction and to decrease cerebral I/R-associated brain injury [281,282]. DFO decreases excitatory amino acid levels and improves the histological outcome in the hippocampus of neonatal rats after hypoxia–ischemia [283]. Gerbils fed a low-iron diet for 8 weeks had decreased brain iron levels, neurologic deficits and brain edema after cerebral I/R [22]. Treatment with DFO resulted in decreased brain edema following I/R [22]. DFO treatment attenuated oxidative damage and cell loss induced by oxygen–glucose deprivation followed by reoxygenation in a cell model of cerebral I/R [108,190]. Most stroke-related clinical trials have focused on the treatment of intracerebral hemorrhage. Thus, 294 participants with intracranial hemorrhage were recruited to participate in the safety and efficacy evaluation of DFO and placebo, which showed that deferoxamine mesylate was safe, and DFO treatment significantly improved clinical outcomes [284]. In two clinical trials, DFO treatment appeared to accelerate recovery [285] and reduce hematoma volume [286] in patients with cerebral hemorrhage.

In conclusion, iron chelators are commonly used to reduce the level of iron in the brain, which attenuate oxidative damage, inhibit neuronal ferroptosis and apoptosis and effectively relieve the symptoms of AD, PD and stroke. However, the clinical application of iron chelators still needs to better explore drug administration to improve the therapeutic effect.

5.2. *Iron Chelators in New Administration Forms*

DFO, a hydrophilic drug that binds extracellular iron in a ratio of 1:1, has low oral availability, poor BBB permeability and a short half-life [287]. By contrast, DFX and DFP have higher oral bioavailability and intracellular iron affinity [287,288]. The main advantage of DFP is that it can cross the BBB and chelate iron in cells in the brain [288]. However, the binding ratio of DFP to iron is 3:1, which is lower than that of DFX to iron (2:1), so DFP is less likely to consume stored iron in the body [287,289]. DFP tends to mechanically cross the cell membrane, form complexes with iron, leave the cell and redistribute iron to Tf for recycling [290,291]. Therefore, the use of iron chelators alone in the treatment of neurological diseases is limited.

New forms of administration are being developed to reduce the side effects of DFO and improve its ability to penetrate the BBB. A substantial body of preclinical evidence and early clinical data has demonstrated that intranasal delivery of DFO and other iron chelators has strong disease-modifying impacts in AD, PD and ischemic stroke [292]. Administration of DFO to APP/PS1 mice through the nasal cavity significantly diminished iron-induced tau phosphorylation, APP expression and Aβ accumulation, improving the cognitive decline in mice [293,294]. This result may be due to the inhibitory effect of DFO on iron-induced tau phosphorylation through CDK5 and GSK-3β pathways [293], as well as upregulation of HIF-1α expression via activation of the MAPK/P38 pathway and HIF-1α-mediated regulation of iron metabolism [294]. This, then, ultimately decreased iron levels in the CA3 region of the hippocampus [294]. Intranasal drug delivery allows for direct targeting of drugs to the brain, bypassing the BBB and minimizing systemic adverse effects [295,296]. Improvement in motor deficits and dopaminergic neuronal survival with non-invasive intranasal delivery of DFO in 6-OHDA-induced PD has been reported [296]. Intranasal administration of DFO decreased pathological α-synuclein formation at the terminal level and slowed PD progression [292]. Intranasal administration targets DFO to the brain and reduces systemic exposure; intranasal DFO has also been shown to prevent and treat stroke damage after middle cerebral artery occlusion in rats [297].

Nanodrug delivery systems are also being used to increase the efficiencies of drugs such as iron chelators in the brain to treat neurological diseases [298]. Targeted brain delivery of rabies virus glycoprotein 29-modified DFO-loaded nanoparticles developed by our team, which can cross the BBB through receptor-mediated endocytosis, significantly increased entry of DFO into the brain and prolonged the half-life of DFO [92]. Administration of these nanoparticles significantly decreased iron content and oxidative stress levels in the SN and striatum of PD mice and effectively reduced dopaminergic neuron damage and reversed neurobehavioral deficits, without causing any overt adverse effects in the brain or other organs [92]. DFO-loaded nanoparticles are also being investigated to target decreased brain iron levels in AD and ischemic stroke.

5.3. Key Molecules of Brain Iron Metabolism as Targets

Studies have found that long-term use of iron chelators can cause side effects [279]. An increasing number of studies have been conducted on the regulation of key molecules of iron metabolism in the brain as therapeutic targets. Experiments in SH-SY5Y cells stably overexpressing the human APP Swedish mutation revealed that decreasing expression of the iron intake protein DMT1 can decrease iron flow into cells and, thus, reduce Aβ secretion [299]. FtMt overexpression can restore Aβ-induced changes in iron and iron-metabolism-related proteins and has a neuroprotective effect on Aβ-induced neurotoxicity [164]. Specifically, increasing the level of FtMt in the brain may be a novel strategy to prevent or treat AD. CP overexpression in the brain of mice via injection of a *Cp* gene plasmid into the lateral ventricle diminished brain iron and hippocampal cell apoptosis, reducing Aβ-induced memory dysfunction in mice [159], providing a theoretical basis for the development of CP as an effective treatment for AD. Conditional knockout of astrocyte *Cp* significantly decreased brain iron levels; iron was deposited in BMVECs, resulting in diminished iron levels in neurons and glial cells [14]. In terms of alleviating iron deposition in the brains of elderly mice, astrocyte *Cp* knockout reduced tau phosphorylation and Aβ deposition and alleviated ROS-MAPK-pathway-mediated apoptosis, thus improving cognitive function [14]. Overexpression of hepcidin in astrocytes downregulated FPN1 in BMVECs, inhibited iron entry into the brain [11], decreased iron levels in the brain and neurons of APP/PS1 mice and reduced oxidative stress and neuroinflammation, ultimately reducing neuronal cell death of APP/PS1 mice and alleviating the symptoms of AD [69,160]. Finally, overexpression of hepcidin in astrocytes delayed the pathological process of AD and effectively improved the spatial cognitive ability of aged mice [69,160].

Overexpression of ferritin in dopaminergic neurons significantly decreased iron levels in the SN and alleviated oxidative stress damage in dopaminergic neurons in MPTP-induced PD models. Overexpression of ferritin heavy chain (FTH1) inhibits ferroptosis and mitochondrial dysfunction in the 6-OHDA model of PD through decreased iron accumulation and ferritinophagy [300]. *Nrf2* knockout prevented entry of iron into the brain, reduced ROS levels and apoptosis of dopaminergic neurons in the SN and improved the exercise ability of elderly mice [111]. FtMt has been shown to inhibit erastin-induced ferroptosis by regulating iron homeostasis and reducing lipid peroxidation levels [301]. Overexpression of FtMt inhibits mitochondrial damage, decreases ROS generation and lipid peroxidation and alleviates 6-OHDA-induced neuronal damage [107]. Overexpression of FtMt suppresses MPTP-induced cell damage in PD by regulating iron metabolism and attenuating oxidative stress [302].

Overexpression of FtMt attenuates cerebral I/R injury by inhibiting iron-mediated ferroptosis [108]. Overexpression of FtMt enhances BBB integrity following ischemic stroke in mice by maintaining iron homeostasis in endothelial cells [190]. FtMt alleviates apoptosis by enhancing mitochondrial bioenergetics and stimulating glucose metabolism in cerebral I/R [109]. Knockout of *Fpn1* in BMVECs can significantly reduce the injury caused by acute cerebral ischemia; the underlying mechanism has been linked to a reduction in iron oxidative stress and the inflammatory response and a reduction in iron-mediated cell death and apoptosis [25]. Regulating the expressions of critical molecules in iron metabolism, such as FtMt, CP and hepcidin, can effectively restore the brain iron homeostasis, reduce ROS and, thus, alleviate the symptoms of AD, PD and stroke. Regulation of the expressions of these critical molecules in brain iron metabolism is expected to be a potential new therapeutic strategy for these diseases.

5.4. Iron Supplements

ID affects neurotransmitter homeostasis and neurodevelopment and has been linked to the pathogenesis of neuropsychiatric diseases. The use of iron supplements is expected to play a positive role in these diseases. Intranasal administration of nanoliposome-encapsulated FAC successfully increased brain iron content [303]. ID reduces cortical plasticity and delays neurological recovery after ischemic stroke [201]. The use of iron supplements can promote endogenous repair in ischemic stroke [201]. *IRP* knockout decreased iron levels in embryonic stem cells and inhibited stem cell proliferation and differentiation by increasing ROS production and decreasing iron–sulfur cluster proteins [8]. With iron supplements, stem cells differentiated normally [8].

5.5. Antioxidants and Anti-Inflammatory Reagents Regulate Iron Metabolism as Targets

In addition to chelation of excess iron with iron chelators and maintenance of brain iron homeostasis by regulating key molecules in brain iron metabolism, antioxidants and anti-inflammatory reagents can also regulate iron metabolism and influence lipid peroxidation and neuroinflammation, thus showing great potential in the treatment of different neurological diseases.

Vitamin E treatment can reduce oxidative stress and lipid peroxidation, improve mitochondrial function, attenuate intracellular iron accumulation and recover cell morphology of fibroblasts in PLA2G6-associated neurodegeneration [304]. Melatonin is a free radical scavenger and has the property of iron chelating, which can effectively inhibit iron overload-mediated oxidative stress and ameliorate oxidation/nitrosation injuries [305]. The iron levels, oxidative stress markers and inflammatory markers were determined and compared in 40 PD patients and 46 controls. It was found that while the iron level was disturbed in PD patients, the content of their antioxidants, such as plasma vitamin C, was lower, and the oxidative stress and the inflammation levels were increased [306]. This indicates that the low level of antioxidants is corrected with the production of free radicals leading to the neurodegeneration in PD [306]. On the contrary, increasing levels of the antioxidant vitamin C may help improve neurological conditions. Coenzyme Q10 (CoQ10) is a lipophilic antioxidant that can reduce lipid peroxidation levels [307]. Ferroptosis suppressor protein 1 (FSP1) can catalyze the CoQ10 reduction to ubiquinol by NADPH restoring the antioxidative effects of CoQ10 [308]. The FSP1/CoQ10 pathway prevents irreversible ferroptosis by reducing lipid peroxides [278]. A multicenter RCT was reported to increase CoQ10 activity and slow the functional decline in PD [278,309]. Subsequent larger clinical studies have shown that the treatment effect of CoQ10 in PD patients is not obvious [278]. Therefore, optimizing the dosages and combinations of antioxidants and considering the potential interactions with other treatments are needed in developing antioxidants as therapeutic strategies.

Tea flavonoids (catechins) have been reported to possess the activities of divalent metal chelating, antioxidant and anti-inflammatory, with the advantage of penetration of BBB [310], showing protective effects in different neurological diseases [311]. The bioactive components of green tea, red wine, arctic root and dwarf periwinkle have been shown to have neuroprotective, antioxidant, anti-inflammatory and iron-chelating potential. They may treat neurological diseases at the cellular level by decreasing microglia activation, attenuating damage from ROS, chelating iron and promoting cell growth [312].

To date, most of the research on different neurological diseases focuses on the manifestations and pathogenesis of a single disease to study the treatment strategies. Drugs on the market and in development also tend to target a single neurological disorder or symptom, lacking the ability to explore the common causes of different neurological diseases. The studies have shown that oxidative stress injury and ferroptosis of neurons caused by dysregulation of brain iron metabolism are common issues in the occurrence and development of different neurological diseases. Therefore, targeting brain iron metabolism and designing drugs or therapeutic strategies for the common etiology of different neurological diseases may reduce or inhibit the occurrence and development of these neurological diseases at the source. However, targeting brain iron metabolism to treat these diseases may also have shortcomings and face certain challenges. The insufficient targeting of iron chelators to the brain may affect peripheral iron metabolism, leading to disorder in the systemic iron metabolism and damage to peripheral organs. Moreover, the currently identified targets that can regulate iron metabolism do not exhibit brain-specific expression patterns, and most of them are still in the laboratory stage, lacking clinical data. Thus, further explorations are needed to accurately target specific brain regions and improve delivery efficiency. Therefore, further exploration is needed to accurately target specific brain regions and improve delivery efficiency in the development of drugs that regulate brain iron metabolism.

6. Conclusions and Prospects

In this review, we have summarized and elucidated the interplay between dysregulation of iron metabolism, redox imbalance and different neurological diseases. We focused on the mechanisms of iron-induced oxidative damage in disease pathogenesis and proposed the broad application of targeting the regulation of brain iron metabolism to treat neurological diseases. However, the current research faces certain challenges. The mechanism of iron release from brain tissue is unknown, the specific iron metabolism pathways in different nerve cells remain unclear and the role of oxidative stress in the induction of neural damage is not fully understood. Furthermore, translational studies and clinical trials on the optimal use of iron chelators and regulators in targeting iron metabolism in neurological disease are relatively few. Untangling these issues in the future will aid in our ability to better target the regulation of brain iron metabolism for the prevention and treatment of neurological diseases.

Author Contributions: G.G. wrote Section 3 (Iron, redox balance and oxidative damage), Section 4.2 (Iron and Alzheimer's disease) and Section 6 (Conclusions and prospects). L.Y. wrote Section 4.4 (Iron and neuropsychiatric disorders) and Section 5 (Targeting iron metabolism in the treatment of neurological diseases). J.Z. wrote Section 4.5 (Iron and abnormal neurodevelopment). Y.-Z.C. wrote Section 4.3 (Iron and stroke). P.Y. wrote Section 1 (Introduction), Section 2 (Brain iron metabolism) and Section 4.1 (Iron and Parkinson's disease). Y.-Z.C. and P.Y. conceived the review and edited the manuscript. All authors have read and agreed to the published version of the manuscript.

Funding: This research was funded by the National Natural Science Foundation of China (grant numbers 31970905, 31520103908, 32070962, 31471035, 32100617 and 31900739) and the Science and Technology Project of Hebei Education Department (grant numbers ZD2022004 and ZD2021327).

Institutional Review Board Statement: Not applicable.

Informed Consent Statement: Not applicable.

Data Availability Statement: Data is contained within the article.

Acknowledgments: The authors would like to thank Chengsuo Liu (Hebei Normal University, China) for drawing the schematic image (Figure 3).

Conflicts of Interest: The authors declare no conflict of interest.

References

1. Hentze, M.W.; Muckenthaler, M.U.; Galy, B.; Camaschella, C. Two to tango: Regulation of Mammalian iron metabolism. *Cell* **2010**, *142*, 24–38. [CrossRef]
2. Faucheux, B.A.; Nillesse, N.; Damier, P.; Spik, G.; Mouatt-Prigent, A.; Pierce, A.; Leveugle, B.; Kubis, N.; Hauw, J.J.; Agid, Y.; et al. Expression of lactoferrin receptors is increased in the mesencephalon of patients with Parkinson disease. *Proc. Natl. Acad. Sci. USA* **1995**, *92*, 9603–9607. [CrossRef]
3. Peters, D.G.; Connor, J.R.; Meadowcroft, M.D. The relationship between iron dyshomeostasis and amyloidogenesis in Alzheimer's disease: Two sides of the same coin. *Neurobiol. Dis.* **2015**, *81*, 49–65. [CrossRef]
4. Riggins, T.; Miller, N.C.; Bauer, P.J.; Georgieff, M.K.; Nelson, C.A. Consequences of low neonatal iron status due to maternal diabetes mellitus on explicit memory performance in childhood. *Dev. Neuropsychol.* **2009**, *34*, 762–779. [CrossRef] [PubMed]
5. Lozoff, B.; Beard, J.; Connor, J.; Barbara, F.; Georgieff, M.; Schallert, T. Long-lasting neural and behavioral effects of iron deficiency in infancy. *Nutr. Rev.* **2006**, *64*, S34–S43. discussion S72–S91. [CrossRef] [PubMed]
6. Connor, J.R.; Menzies, S.L. Relationship of iron to oligodendrocytes and myelination. *Glia* **1996**, *17*, 83–93. [CrossRef]
7. Wu, Q.; Hao, Q.; Li, H.; Wang, B.; Wang, P.; Jin, X.; Yu, P.; Gao, G.; Chang, Y.Z. Brain iron deficiency and affected contextual fear memory in mice with conditional Ferroportin1 ablation in the brain. *FASEB J.* **2021**, *35*, e21174. [CrossRef] [PubMed]
8. Chang, S.; Wang, P.; Han, Y.; Ma, Q.; Liu, Z.; Zhong, S.; Lu, Y.; Chen, R.; Sun, L.; Wu, Q.; et al. Ferrodifferentiation regulates neurodevelopment via ROS generation. *Sci. China Life Sci.* **2023**; *online ahead of print*. [CrossRef]
9. Ke, Y.; Ming Qian, Z. Iron misregulation in the brain: A primary cause of neurodegenerative disorders. *Lancet Neurol.* **2003**, *2*, 246–253. [CrossRef]
10. Oshiro, S.; Morioka, M.S.; Kikuchi, M. Dysregulation of iron metabolism in Alzheimer's disease, Parkinson's disease, and amyotrophic lateral sclerosis. *Adv. Pharmacol. Sci.* **2011**, *2011*, 378278. [CrossRef]
11. You, L.; Yu, P.P.; Dong, T.; Guo, W.; Chang, S.; Zheng, B.; Ci, Y.; Wang, F.; Yu, P.; Gao, G.; et al. Astrocyte-derived hepcidin controls iron traffic at the blood–brain-barrier via regulating ferroportin 1 of microvascular endothelial cells. *Cell Death Dis.* **2022**, *13*, 667. [CrossRef] [PubMed]
12. Wang, S.M.; Fu, L.J.; Duan, X.L.; Crooks, D.R.; Yu, P.; Qian, Z.M.; Di, X.J.; Li, J.; Rouault, T.A.; Chang, Y.Z. Role of hepcidin in murine brain iron metabolism. *Cell. Mol. Life Sci.* **2010**, *67*, 123–133. [CrossRef]
13. Jeong, S.Y.; David, S. Glycosylphosphatidylinositol-anchored ceruloplasmin is required for iron efflux from cells in the central nervous system. *J. Biol. Chem.* **2003**, *278*, 27144–27148. [CrossRef]
14. Li, Z.D.; Li, H.; Kang, S.; Cui, Y.G.; Zheng, H.; Wang, P.; Han, K.; Yu, P.; Chang, Y.Z. The divergent effects of astrocyte ceruloplasmin on learning and memory function in young and old mice. *Cell Death Dis.* **2022**, *13*, 1006. [CrossRef]
15. Levi, S.; Tiranti, V. Neurodegeneration with Brain Iron Accumulation Disorders: Valuable Models Aimed at Understanding the Pathogenesis of Iron Deposition. *Pharmaceuticals* **2019**, *12*, 27. [CrossRef]
16. Wise, R.M.; Wagener, A.; Fietzek, U.M.; Klopstock, T.; Mosharov, E.V.; Zucca, F.A.; Sulzer, D.; Zecca, L.; Burbulla, L.F. Interactions of dopamine, iron, and alpha-synuclein linked to dopaminergic neuron vulnerability in Parkinson's disease and Neurodegeneration with Brain Iron Accumulation disorders. *Neurobiol. Dis.* **2022**, *175*, 105920. [CrossRef]
17. Crapper McLachlan, D.R.; Dalton, A.J.; Kruck, T.P.; Bell, M.Y.; Smith, W.L.; Kalow, W.; Andrews, D.F. Intramuscular desferrioxamine in patients with Alzheimer's disease. *Lancet* **1991**, *337*, 1304–1308. [CrossRef]
18. Kupershmidt, L.; Amit, T.; Bar-Am, O.; Youdim, M.B.; Weinreb, O. The novel multi-target iron chelating-radical scavenging compound M30 possesses beneficial effects on major hallmarks of Alzheimer's disease. *Antioxid. Redox Signal.* **2012**, *17*, 860–877. [CrossRef]
19. Zeng, X.; An, H.; Yu, F.; Wang, K.; Zheng, L.; Zhou, W.; Bao, Y.; Yang, J.; Shen, N.; Huang, D. Benefits of Iron Chelators in the Treatment of Parkinson's Disease. *Neurochem. Res.* **2021**, *46*, 1239–1251. [CrossRef] [PubMed]
20. Youdim, M.B.; Stephenson, G.; Ben Shachar, D. Ironing iron out in Parkinson's disease and other neurodegenerative diseases with iron chelators: A lesson from 6-hydroxydopamine and iron chelators, desferal and VK-28. *Ann. N. Y. Acad. Sci.* **2004**, *1012*, 306–325. [CrossRef] [PubMed]
21. Palmer, C.; Roberts, R.L.; Bero, C. Deferoxamine posttreatment reduces ischemic brain injury in neonatal rats. *Stroke* **1994**, *25*, 1039–1045. [CrossRef]
22. Patt, A.; Horesh, I.R.; Berger, E.M.; Harken, A.H.; Repine, J.E. Iron depletion or chelation reduces ischemia/reperfusion-induced edema in gerbil brains. *J. Pediatr. Surg.* **1990**, *25*, 224–227; discussion 227–228. [CrossRef]
23. Moos, T.; Morgan, E.H. Transferrin and transferrin receptor function in brain barrier systems. *Cell. Mol. Neurobiol.* **2000**, *20*, 77–95. [CrossRef]

4. Yang, W.M.; Jung, K.J.; Lee, M.O.; Lee, Y.S.; Lee, Y.H.; Nakagawa, S.; Niwa, M.; Cho, S.S.; Kim, D.W. Transient expression of iron transport proteins in the capillary of the developing rat brain. *Cell. Mol. Neurobiol.* **2011**, *31*, 93–99. [CrossRef]

5. Zheng, H.; Guo, X.; Kang, S.; Li, Z.; Tian, T.; Li, J.; Wang, F.; Yu, P.; Chang, S.; Chang, Y.Z. Cdh5-mediated Fpn1 deletion exerts neuroprotective effects during the acute phase and inhibitory effects during the recovery phase of ischemic stroke. *Cell Death Dis.* **2023**, *14*, 161. [CrossRef]

6. Klomp, L.W.; Farhangrazi, Z.S.; Dugan, L.L.; Gitlin, J.D. Ceruloplasmin gene expression in the murine central nervous system. *J. Clin. Investig.* **1996**, *98*, 207–215. [CrossRef]

7. Patel, B.N.; Dunn, R.J.; David, S. Alternative RNA splicing generates a glycosylphosphatidylinositol-anchored form of ceruloplasmin in mammalian brain. *J. Biol. Chem.* **2000**, *275*, 4305–4310. [CrossRef]

8. Wang, L.; Liu, X.; You, L.H.; Ci, Y.Z.; Chang, S.; Yu, P.; Gao, G.; Chang, Y.Z. Hepcidin and iron regulatory proteins coordinately regulate ferroportin 1 expression in the brain of mice. *J. Cell. Physiol.* **2019**, *234*, 7600–7607. [CrossRef] [PubMed]

9. Yu, P.; Qian, Z.M.; Duan, X.L.; Chang, Y.Z. The structure of DMT1 and its regulation. *Chin. J. Neuroanat.* **2004**, *20*, 205–209.

0. Yu, P.; Chang, Y.Z. Brain Iron Metabolism and Regulation. *Adv. Exp. Med. Biol.* **2019**, *1173*, 33–44. [CrossRef] [PubMed]

1. Deane, R.; Zheng, W.; Zlokovic, B.V. Brain capillary endothelium and choroid plexus epithelium regulate transport of transferrin-bound and free iron into the rat brain. *J. Neurochem.* **2004**, *88*, 813–820. [CrossRef]

2. Thirupathi, A.; Chang, Y.Z. Brain Iron Metabolism and CNS Diseases. *Adv. Exp. Med. Biol.* **2019**, *1173*, 1–19. [CrossRef] [PubMed]

3. Fisher, J.; Devraj, K.; Ingram, J.; Slagle-Webb, B.; Madhankumar, A.B.; Liu, X.; Klinger, M.; Simpson, I.A.; Connor, J.R. Ferritin: A novel mechanism for delivery of iron to the brain and other organs. *Am. J. Physiol. Cell Physiol.* **2007**, *293*, C641–C649. [CrossRef]

4. Qian, Z.M.; Chang, Y.Z.; Zhu, L.; Yang, L.; Du, J.R.; Ho, K.P.; Wang, Q.; Li, L.Z.; Wang, C.Y.; Ge, X.; et al. Development and iron-dependent expression of hephaestin in different brain regions of rats. *J. Cell. Biochem.* **2007**, *102*, 1225–1233. [CrossRef] [PubMed]

5. Rouault, T.A.; Zhang, D.L.; Jeong, S.Y. Brain iron homeostasis, the choroid plexus, and localization of iron transport proteins. *Metab. Brain Dis.* **2009**, *24*, 673–684. [CrossRef] [PubMed]

6. Gutteridge, J.M. Iron and oxygen radicals in brain. *Ann. Neurol.* **1992**, *32*, S16–S21. [CrossRef]

7. Gałazka-Friedman, J.; Friedman, A. Controversies about iron in parkinsonian and control substantia nigra. *Acta Neurobiol. Exp. (Wars)* **1997**, *57*, 217–225. [PubMed]

8. Lu, L.N.; Qian, Z.M.; Wu, K.C.; Yung, W.H.; Ke, Y. Expression of Iron Transporters and Pathological Hallmarks of Parkinson's and Alzheimer's Diseases in the Brain of Young, Adult, and Aged Rats. *Mol. Neurobiol.* **2017**, *54*, 5213–5224. [CrossRef]

9. Wang, Z.; Zeng, Y.N.; Yang, P.; Jin, L.Q.; Xiong, W.C.; Zhu, M.Z.; Zhang, J.Z.; He, X.; Zhu, X.H. Axonal iron transport in the brain modulates anxiety-related behaviors. *Nat. Chem. Biol.* **2019**, *15*, 1214–1222. [CrossRef]

0. Moos, T. Brain iron homeostasis. *Dan. Med. Bull.* **2002**, *49*, 279–301.

1. Xu, J.Q.; Liu, Q.Q.; Huang, S.Y.; Duan, C.Y.; Lu, H.B.; Cao, Y.; Hu, J.Z. The lymphatic system: A therapeutic target for central nervous system disorders. *Neural Regen. Res.* **2023**, *18*, 1249–1256. [CrossRef]

2. Hahn, P.; Qian, Y.; Dentchev, T.; Chen, L.; Beard, J.; Harris, Z.L.; Dunaief, J.L. Disruption of ceruloplasmin and hephaestin in mice causes retinal iron overload and retinal degeneration with features of age-related macular degeneration. *Proc. Natl. Acad. Sci. USA* **2004**, *101*, 13850–13855. [CrossRef]

3. Ke, Y.; Qian, Z.M. Brain iron metabolism: Neurobiology and neurochemistry. *Prog. Neurobiol.* **2007**, *83*, 149–173. [CrossRef]

4. Attieh, Z.K.; Mukhopadhyay, C.K.; Seshadri, V.; Tripoulas, N.A.; Fox, P.L. Ceruloplasmin ferroxidase activity stimulates cellular iron uptake by a trivalent cation-specific transport mechanism. *J. Biol. Chem.* **1999**, *274*, 1116–1123. [CrossRef]

5. Andersen, H.H.; Johnsen, K.B.; Moos, T. Iron deposits in the chronically inflamed central nervous system and contributes to neurodegeneration. *Cell. Mol. Life Sci.* **2014**, *71*, 1607–1622. [CrossRef]

6. Lane, D.J.; Robinson, S.R.; Czerwinska, H.; Bishop, G.M.; Lawen, A. Two routes of iron accumulation in astrocytes: Ascorbate-dependent ferrous iron uptake via the divalent metal transporter (DMT1) plus an independent route for ferric iron. *Biochem. J.* **2010**, *432*, 123–132. [CrossRef]

7. Xu, H.; Wang, Y.; Song, N.; Wang, J.; Jiang, H.; Xie, J. New Progress on the Role of Glia in Iron Metabolism and Iron-Induced Degeneration of Dopamine Neurons in Parkinson's Disease. *Front. Mol. Neurosci.* **2017**, *10*, 455. [CrossRef] [PubMed]

8. Oshiro, S.; Kawahara, M.; Kuroda, Y.; Zhang, C.; Cai, Y.; Kitajima, S.; Shirao, M. Glial cells contribute more to iron and aluminum accumulation but are more resistant to oxidative stress than neuronal cells. *Biochim. Biophys. Acta* **2000**, *1502*, 405–414. [CrossRef] [PubMed]

9. Todorich, B.; Zhang, X.; Slagle-Webb, B.; Seaman, W.E.; Connor, J.R. Tim-2 is the receptor for H-ferritin on oligodendrocytes. *J. Neurochem.* **2008**, *107*, 1495–1505. [CrossRef] [PubMed]

0. Bishop, G.M.; Dang, T.N.; Dringen, R.; Robinson, S.R. Accumulation of non-transferrin-bound iron by neurons, astrocytes, and microglia. *Neurotox. Res.* **2011**, *19*, 443–451. [CrossRef]

1. Fillebeen, C.; Ruchoux, M.M.; Mitchell, V.; Vincent, S.; Benaïssa, M.; Pierce, A. Lactoferrin is synthesized by activated microglia in the human substantia nigra and its synthesis by the human microglial CHME cell line is upregulated by tumor necrosis factor alpha or 1-methyl-4-phenylpyridinium treatment. *Brain Res. Mol. Brain Res.* **2001**, *96*, 103–113. [CrossRef] [PubMed]

2. Urrutia, P.; Aguirre, P.; Esparza, A.; Tapia, V.; Mena, N.P.; Arredondo, M.; González-Billault, C.; Núñez, M.T. Inflammation alters the expression of DMT1, FPN1 and hepcidin, and it causes iron accumulation in central nervous system cells. *J. Neurochem.* **2013**, *126*, 541–549. [CrossRef] [PubMed]

53. Gao, G.; You, L.; Chang, Y. Iron metabolism in Parkinson's disease. In *Oxidative Stress and Redox Signalling in Parkinson's Disease*; The Royal Society of Chemistry: London, UK, 2017; pp. 255–276.
54. Bradbury, M.W. Transport of iron in the blood–brain-cerebrospinal fluid system. *J. Neurochem.* **1997**, *69*, 443–454. [CrossRef] [PubMed]
55. Siddappa, A.J.; Rao, R.B.; Wobken, J.D.; Casperson, K.; Leibold, E.A.; Connor, J.R.; Georgieff, M.K. Iron deficiency alters iron regulatory protein and iron transport protein expression in the perinatal rat brain. *Pediatr. Res.* **2003**, *53*, 800–807. [CrossRef]
56. Pantopoulos, K. Iron metabolism and the IRE/IRP regulatory system: An update. *Ann. N. Y. Acad. Sci.* **2004**, *1012*, 1–13. [CrossRef]
57. Muckenthaler, M.U.; Galy, B.; Hentze, M.W. Systemic iron homeostasis and the iron-responsive element/iron-regulatory protein (IRE/IRP) regulatory network. *Annu. Rev. Nutr.* **2008**, *28*, 197–213. [CrossRef]
58. You, L.H.; Yan, C.Z.; Zheng, B.J.; Ci, Y.Z.; Chang, S.Y.; Yu, P.; Gao, G.F.; Li, H.Y.; Dong, T.Y.; Chang, Y.Z. Astrocyte hepcidin is a key factor in LPS-induced neuronal apoptosis. *Cell Death Dis.* **2017**, *8*, e2676. [CrossRef]
59. Kautz, L.; Jung, G.; Du, X.; Gabayan, V.; Chapman, J.; Nasoff, M.; Nemeth, E.; Ganz, T. Erythroferrone contributes to hepcidin suppression and iron overload in a mouse model of β-thalassemia. *Blood* **2015**, *126*, 2031–2037. [CrossRef]
60. Kautz, L.; Jung, G.; Valore, E.V.; Rivella, S.; Nemeth, E.; Ganz, T. Identification of erythroferrone as an erythroid regulator of iron metabolism. *Nat. Genet.* **2014**, *46*, 678–684. [CrossRef]
61. Ashraf, A.; Michaelides, C.; Walker, T.A.; Ekonomou, A.; Suessmilch, M.; Sriskanthanathan, A.; Abraha, S.; Parkes, A.; Parkes, H.G.; Geraki, K.; et al. Regional Distributions of Iron, Copper and Zinc and Their Relationships with Glia in a Normal Aging Mouse Model. *Front. Aging Neurosci.* **2019**, *11*, 351. [CrossRef]
62. Merkofer, M.; Kissner, R.; Hider, R.C.; Brunk, U.T.; Koppenol, W.H. Fenton chemistry and iron chelation under physiologically relevant conditions: Electrochemistry and kinetics. *Chem. Res. Toxicol.* **2006**, *19*, 1263–1269. [CrossRef] [PubMed]
63. Cabantchik, Z.I. Labile iron in cells and body fluids: Physiology, pathology, and pharmacology. *Front. Pharmacol.* **2014**, *5*, 45. [CrossRef]
64. Galaris, D.; Barbouti, A.; Pantopoulos, K. Iron homeostasis and oxidative stress: An intimate relationship. *Biochim. Biophys. Acta Mol. Cell Res.* **2019**, *1866*, 118535. [CrossRef] [PubMed]
65. von der Mark, C.; Ivanov, R.; Eutebach, M.; Maurino, V.G.; Bauer, P.; Brumbarova, T. Reactive oxygen species coordinate the transcriptional responses to iron availability in Arabidopsis. *J. Exp. Bot.* **2021**, *72*, 2181–2195. [CrossRef]
66. Koskenkorva-Frank, T.S.; Weiss, G.; Koppenol, W.H.; Burckhardt, S. The complex interplay of iron metabolism, reactive oxygen species, and reactive nitrogen species: Insights into the potential of various iron therapies to induce oxidative and nitrosative stress. *Free Radic. Biol. Med.* **2013**, *65*, 1174–1194. [CrossRef]
67. Tewari, R.K.; Hadacek, F.; Sassmann, S.; Lang, I. Iron deprivation-induced reactive oxygen species generation leads to non-autolytic PCD in Brassica napus leaves. *Environ. Exp. Bot.* **2013**, *91*, 74–83. [CrossRef]
68. Bush, A.I. Metals and neuroscience. *Curr. Opin. Chem. Biol.* **2000**, *4*, 184–191. [CrossRef]
69. Xu, Y.; Zhang, Y.; Zhang, J.H.; Han, K.; Zhang, X.; Bai, X.; You, L.H.; Yu, P.; Shi, Z.; Chang, Y.Z.; et al. Astrocyte hepcidin ameliorates neuronal loss through attenuating brain iron deposition and oxidative stress in APP/PS1 mice. *Free Radic. Biol. Med.* **2020**, *158*, 84–95. [CrossRef] [PubMed]
70. Jiang, H.; Wang, J.; Rogers, J.; Xie, J.X. Brain Iron Metabolism Dysfunction in Parkinson's Disease. *Mol. Neurobiol.* **2017**, *54*, 3078–3101. [CrossRef]
71. Masaldan, S.; Bush, A.I.; Devos, D.; Rolland, A.S.; Moreau, C. Striking while the iron is hot: Iron metabolism and ferroptosis in neurodegeneration. *Free Radic. Biol. Med.* **2019**, *133*, 221–233. [CrossRef]
72. Zhang, Y.; Bai, X.; Zhang, Y.; Yao, S.; Cui, Y.; You, L.H.; Yu, P.; Chang, Y.Z.; Gao, G. Hippocampal Iron Accumulation Impairs Synapses and Memory via Suppressing Furin Expression and Downregulating BDNF Maturation. *Mol. Neurobiol.* **2022**, *59*, 5574–5590. [CrossRef] [PubMed]
73. Choi, D.H.; Kwon, K.C.; Hwang, D.J.; Koo, J.H.; Um, H.S.; Song, H.S.; Kim, J.S.; Jang, Y.; Cho, J.Y. Treadmill Exercise Alleviates Brain Iron Dyshomeostasis Accelerating Neuronal Amyloid-β Production, Neuronal Cell Death, and Cognitive Impairment in Transgenic Mice Model of Alzheimer's Disease. *Mol. Neurobiol.* **2021**, *58*, 3208–3223. [CrossRef] [PubMed]
74. Sies, H.; Berndt, C.; Jones, D.P. Oxidative Stress. *Annu. Rev. Biochem.* **2017**, *86*, 715–748. [CrossRef] [PubMed]
75. Rhee, S.G.; Yang, K.S.; Kang, S.W.; Woo, H.A.; Chang, T.S. Controlled elimination of intracellular $H_{(2)}O_{(2)}$: Regulation of peroxiredoxin, catalase, and glutathione peroxidase via post-translational modification. *Antioxid. Redox Signal.* **2005**, *7*, 619–626. [CrossRef]
76. Lillig, C.H.; Berndt, C.; Holmgren, A. Glutaredoxin systems. *Biochim. Biophys. Acta* **2008**, *1780*, 1304–1317. [CrossRef]
77. Wilson, D.F. Oxidative phosphorylation: Regulation and role in cellular and tissue metabolism. *J. Physiol.* **2017**, *595*, 7023–7038. [CrossRef]
78. Dixon, S.J.; Lemberg, K.M.; Lamprecht, M.R.; Skouta, R.; Zaitsev, E.M.; Gleason, C.E.; Patel, D.N.; Bauer, A.J.; Cantley, A.M.; Yang, W.S.; et al. Ferroptosis: An iron-dependent form of nonapoptotic cell death. *Cell* **2012**, *149*, 1060–1072. [CrossRef]
79. Wu, J.R.; Tuo, Q.Z.; Lei, P. Ferroptosis, a Recent Defined Form of Critical Cell Death in Neurological Disorders. *J. Mol. Neurosci.* **2018**, *66*, 197–206. [CrossRef]

80. Zuo, Y.; Xie, J.; Li, X.; Li, Y.; Thirupathi, A.; Zhang, J.; Yu, P.; Gao, G.; Chang, Y.; Shi, Z. Ferritinophagy-Mediated Ferroptosis Involved in Paraquat-Induced Neurotoxicity of Dopaminergic Neurons: Implication for Neurotoxicity in PD. *Oxid. Med. Cell. Longev.* **2021**, *2021*, 9961628. [CrossRef]
81. Parkinson, J. *An Essay on the Shaking Palsy*; Whittingham & Rowland, Neely & Jones: London, UK, 1817.
82. Kalia, L.V.; Lang, A.E. Parkinson's disease. *Lancet* **2015**, *386*, 896–912. [CrossRef]
83. Przedborski, S. The two-century journey of Parkinson disease research. *Nat. Rev. Neurosci.* **2017**, *18*, 251–259. [CrossRef] [PubMed]
84. Sofic, E.; Riederer, P.; Heinsen, H.; Beckmann, H.; Reynolds, G.P.; Hebenstreit, G.; Youdim, M.B. Increased iron (III) and total iron content in post mortem substantia nigra of parkinsonian brain. *J. Neural Transm.* **1988**, *74*, 199–205. [CrossRef] [PubMed]
85. Dexter, D.T.; Wells, F.R.; Agid, F.; Agid, Y.; Lees, A.J.; Jenner, P.; Marsden, C.D. Increased nigral iron content in postmortem parkinsonian brain. *Lancet* **1987**, *2*, 1219–1220. [CrossRef]
86. He, Y.; Thong, P.S.; Lee, T.; Leong, S.K.; Mao, B.Y.; Dong, F.; Watt, F. Dopaminergic cell death precedes iron elevation in MPTP-injected monkeys. *Free. Radic. Biol. Med.* **2003**, *35*, 540–547. [CrossRef] [PubMed]
87. He, Y.; Lee, T.; Leong, S.K. Time course of dopaminergic cell death and changes in iron, ferritin and transferrin levels in the rat substantia nigra after 6-hydroxydopamine (6-OHDA) lesioning. *Free Radic. Res.* **1999**, *31*, 103–112. [CrossRef] [PubMed]
88. Ndayisaba, A.; Kaindlstorfer, C.; Wenning, G.K. Iron in Neurodegeneration—Cause or Consequence? *Front. Neurosci.* **2019**, *13*, 180. [CrossRef]
89. Biondetti, E.; Santin, M.D.; Valabrègue, R.; Mangone, G.; Gaurav, R.; Pyatigorskaya, N.; Hutchison, M.; Yahia-Cherif, L.; Villain, N.; Habert, M.O.; et al. The spatiotemporal changes in dopamine, neuromelanin and iron characterizing Parkinson's disease. *Brain* **2021**, *144*, 3114–3125. [CrossRef]
90. You, L.H.; Li, F.; Wang, L.; Zhao, S.E.; Wang, S.M.; Zhang, L.L.; Zhang, L.H.; Duan, X.L.; Yu, P.; Chang, Y.Z. Brain iron accumulation exacerbates the pathogenesis of MPTP-induced Parkinson's disease. *Neuroscience* **2015**, *284*, 234–246. [CrossRef]
91. Ci, Y.Z.; Li, H.; You, L.H.; Jin, Y.; Zhou, R.; Gao, G.; Hoi, M.P.M.; Wang, C.; Chang, Y.Z.; Yu, P. Iron overload induced by IRP2 gene knockout aggravates symptoms of Parkinson's disease. *Neurochem. Int.* **2020**, *134*, 104657. [CrossRef]
92. You, L.; Wang, J.; Liu, T.; Zhang, Y.; Han, X.; Wang, T.; Guo, S.; Dong, T.; Xu, J.; Anderson, G.J.; et al. Targeted Brain Delivery of Rabies Virus Glycoprotein 29-Modified Deferoxamine-Loaded Nanoparticles Reverses Functional Deficits in Parkinsonian Mice. *ACS Nano* **2018**, *12*, 4123–4139. [CrossRef]
93. Weinreb, O.; Mandel, S.; Youdim, M.B.H.; Amit, T. Targeting dysregulation of brain iron homeostasis in Parkinson's disease by iron chelators. *Free. Radic. Biol. Med.* **2013**, *62*, 52–64. [CrossRef] [PubMed]
94. Kaur, D.; Yantiri, F.; Rajagopalan, S.; Kumar, J.; Mo, J.Q.; Boonplueang, R.; Viswanath, V.; Jacobs, R.; Yang, L.; Beal, M.F.; et al. Genetic or pharmacological iron chelation prevents MPTP-induced neurotoxicity in vivo: A novel therapy for Parkinson's disease. *Neuron* **2003**, *37*, 899–909. [CrossRef] [PubMed]
95. Loréal, O.; Turlin, B.; Pigeon, C.; Moisan, A.; Ropert, M.; Morice, P.; Gandon, Y.; Jouanolle, A.M.; Vérin, M.; Hider, R.C.; et al. Aceruloplasminemia: New clinical, pathophysiological and therapeutic insights. *J. Hepatol.* **2002**, *36*, 851–856. [CrossRef]
96. Harris, Z.L.; Klomp, L.W.; Gitlin, J.D. Aceruloplasminemia: An inherited neurodegenerative disease with impairment of iron homeostasis. *Am. J. Clin. Nutr.* **1998**, *67*, 972s–977s. [CrossRef]
97. Hellman, N.E.; Gitlin, J.D. Ceruloplasmin metabolism and function. *Annu. Rev. Nutr.* **2002**, *22*, 439–458. [CrossRef]
98. Jeong, S.Y.; David, S. Age-related changes in iron homeostasis and cell death in the cerebellum of ceruloplasmin-deficient mice. *J. Neurosci.* **2006**, *26*, 9810–9819. [CrossRef]
99. Patel, B.N.; Dunn, R.J.; Jeong, S.Y.; Zhu, Q.; Julien, J.P.; David, S. Ceruloplasmin regulates iron levels in the CNS and prevents free radical injury. *J. Neurosci.* **2002**, *22*, 6578–6586. [CrossRef]
100. Ayton, S.; Lei, P.; Duce, J.A.; Wong, B.X.; Sedjahtera, A.; Adlard, P.A.; Bush, A.I.; Finkelstein, D.I. Ceruloplasmin dysfunction and therapeutic potential for Parkinson disease. *Ann. Neurol.* **2013**, *73*, 554–559. [CrossRef]
101. Ayton, S.; Lei, P.; Adlard, P.A.; Volitakis, I.; Cherny, R.A.; Bush, A.I.; Finkelstein, D.I. Iron accumulation confers neurotoxicity to a vulnerable population of nigral neurons: Implications for Parkinson's disease. *Mol. Neurodegener.* **2014**, *9*, 27. [CrossRef] [PubMed]
102. De Domenico, I.; McVey Ward, D.; Kaplan, J. Regulation of iron acquisition and storage: Consequences for iron-linked disorders. *Nat. Rev. Mol. Cell Biol.* **2008**, *9*, 72–81. [CrossRef] [PubMed]
103. Meyron-Holtz, E.G.; Ghosh, M.C.; Iwai, K.; LaVaute, T.; Brazzolotto, X.; Berger, U.V.; Land, W.; Ollivierre-Wilson, H.; Grinberg, A.; Love, P.; et al. Genetic ablations of iron regulatory proteins 1 and 2 reveal why iron regulatory protein 2 dominates iron homeostasis. *EMBO J.* **2004**, *23*, 386–395. [CrossRef]
104. LaVaute, T.; Smith, S.; Cooperman, S.; Iwai, K.; Land, W.; Meyron-Holtz, E.; Drake, S.K.; Miller, G.; Abu-Asab, M.; Tsokos, M.; et al. Targeted deletion of the gene encoding iron regulatory protein-2 causes misregulation of iron metabolism and neurodegenerative disease in mice. *Nat. Genet.* **2001**, *27*, 209–214. [CrossRef]
105. Ghosh, M.C.; Zhang, D.L.; Rouault, T.A. Iron misregulation and neurodegenerative disease in mouse models that lack iron regulatory proteins. *Neurobiol. Dis.* **2015**, *81*, 66–75. [CrossRef]
106. Jia, F.; Li, H.; Jiao, Q.; Li, C.; Fu, L.; Cui, C.; Jiang, H.; Zhang, L. Deubiquitylase OTUD3 prevents Parkinson's disease through stabilizing iron regulatory protein 2. *Cell Death Dis.* **2022**, *13*, 418. [CrossRef] [PubMed]

107. Shi, Z.H.; Nie, G.; Duan, X.L.; Rouault, T.; Wu, W.S.; Ning, B.; Zhang, N.; Chang, Y.Z.; Zhao, B.L. Neuroprotective mechanism of mitochondrial ferritin on 6-hydroxydopamine-induced dopaminergic cell damage: Implication for neuroprotection in Parkinson's disease. *Antioxid. Redox Signal.* **2010**, *13*, 783–796. [CrossRef] [PubMed]
108. Wang, P.; Cui, Y.; Ren, Q.; Yan, B.; Zhao, Y.; Yu, P.; Gao, G.; Shi, H.; Chang, S.; Chang, Y.Z. Mitochondrial ferritin attenuates cerebral ischaemia/reperfusion injury by inhibiting ferroptosis. *Cell Death Dis.* **2021**, *12*, 447. [CrossRef] [PubMed]
109. Wang, P.; Cui, Y.; Liu, Y.; Li, Z.; Bai, H.; Zhao, Y.; Chang, Y.Z. Mitochondrial ferritin alleviates apoptosis by enhancing mitochondrial bioenergetics and stimulating glucose metabolism in cerebral ischemia reperfusion. *Redox Biol.* **2022**, *57*, 102475 [CrossRef] [PubMed]
110. Holze, C.; Michaudel, C.; Mackowiak, C.; Haas, D.A.; Benda, C.; Hubel, P.; Pennemann, F.L.; Schnepf, D.; Wettmarshausen, J.; Braun, M.; et al. Oxeiptosis, a ROS-induced caspase-independent apoptosis-like cell-death pathway. *Nat. Immunol.* **2018**, *19*, 130–140. [CrossRef]
111. Han, K.; Jin, X.; Guo, X.; Cao, G.; Tian, S.; Song, Y.; Zuo, Y.; Yu, P.; Gao, G.; Chang, Y.Z. Nrf2 knockout altered brain iron deposition and mitigated age-related motor dysfunction in aging mice. *Free Radic. Biol. Med.* **2021**, *162*, 592–602. [CrossRef]
112. Wang, Z.L.; Yuan, L.; Li, W.; Li, J.Y. Ferroptosis in Parkinson's disease: Glia-neuron crosstalk. *Trends Mol. Med.* **2022**, *28*, 258–269 [CrossRef] [PubMed]
113. Hare, D.J.; Double, K.L. Iron and dopamine: A toxic couple. *Brain* **2016**, *139*, 1026–1035. [CrossRef]
114. Perfeito, R.; Lázaro, D.F.; Outeiro, T.F.; Rego, A.C. Linking alpha-synuclein phosphorylation to reactive oxygen species formation and mitochondrial dysfunction in SH-SY5Y cells. *Mol. Cell. Neurosci.* **2014**, *62*, 51–59. [CrossRef]
115. Lu, Y.; Prudent, M.; Fauvet, B.; Lashuel, H.A.; Girault, H.H. Phosphorylation of α-Synuclein at Y125 and S129 alters its metal binding properties: Implications for understanding the role of α-Synuclein in the pathogenesis of Parkinson's Disease and related disorders. *ACS Chem. Neurosci.* **2011**, *2*, 667–675. [CrossRef]
116. Zucca, F.A.; Segura-Aguilar, J.; Ferrari, E.; Muñoz, P.; Paris, I.; Sulzer, D.; Sarna, T.; Casella, L.; Zecca, L. Interactions of iron, dopamine and neuromelanin pathways in brain aging and Parkinson's disease. *Prog. Neurobiol.* **2017**, *155*, 96–119. [CrossRef]
117. Roth, J.A.; Singleton, S.; Feng, J.; Garrick, M.; Paradkar, P.N. Parkin regulates metal transport via proteasomal degradation of the 1B isoforms of divalent metal transporter 1. *J. Neurochem.* **2010**, *113*, 454–464. [CrossRef]
118. Mamais, A.; Kluss, J.H.; Bonet-Ponce, L.; Landeck, N.; Langston, R.G.; Smith, N.; Beilina, A.; Kaganovich, A.; Ghosh, M.C.; Pellegrini, L.; et al. Mutations in LRRK2 linked to Parkinson disease sequester Rab8a to damaged lysosomes and regulate transferrin-mediated iron uptake in microglia. *PLoS Biol.* **2021**, *19*, e3001480. [CrossRef]
119. Castellani, R.; Siedlak, S.; Perry, G.; Smith, M. Sequestration of iron by Lewy bodies in Parkinson's disease. *Acta Neuropathol.* **2000**, *100*, 111–114. [CrossRef]
120. Golts, N.; Snyder, H.; Frasier, M.; Theisler, C.; Choi, P.; Wolozin, B. Magnesium inhibits spontaneous and iron-induced aggregation of alpha-synuclein. *J. Biol. Chem.* **2002**, *277*, 16116–16123. [CrossRef] [PubMed]
121. Friedlich, A.L.; Tanzi, R.E.; Rogers, J.T. The 5′-untranslated region of Parkinson's disease alpha-synuclein messengerRNA contains a predicted iron responsive element. *Mol. Psychiatry* **2007**, *12*, 222–223. [CrossRef] [PubMed]
122. McDowall, J.S.; Ntai, I.; Honeychurch, K.C.; Hart, J.P.; Colin, P.; Schneider, B.L.; Brown, D.R. Alpha-synuclein ferrireductase activity is detectible in vivo, is altered in Parkinson's disease and increases the neurotoxicity of DOPAL. *Mol. Cell. Neurosci.* **2017**, *85*, 1–11. [CrossRef] [PubMed]
123. Chen, B.; Wen, X.; Jiang, H.; Wang, J.; Song, N.; Xie, J. Interactions between iron and α-synuclein pathology in Parkinson's disease. *Free. Radic. Biol. Med.* **2019**, *141*, 253–260. [CrossRef] [PubMed]
124. Jia, F.; Song, N.; Wang, W.; Du, X.; Chi, Y.; Jiang, H. High Dietary Iron Supplement Induces the Nigrostriatal Dopaminergic Neurons Lesion in Transgenic Mice Expressing Mutant A53T Human Alpha-Synuclein. *Front. Aging Neurosci.* **2018**, *10*, 97. [CrossRef] [PubMed]
125. Giasson, B.I.; Duda, J.E.; Murray, I.V.; Chen, Q.; Souza, J.M.; Hurtig, H.I.; Ischiropoulos, H.; Trojanowski, J.Q.; Lee, V.M. Oxidative damage linked to neurodegeneration by selective alpha-synuclein nitration in synucleinopathy lesions. *Science* **2000**, *290*, 985–989. [CrossRef]
126. Reynolds, M.R.; Berry, R.W.; Binder, L.I. Nitration in neurodegeneration: Deciphering the "Hows" "nYs". *Biochemistry* **2007**, *46*, 7325–7336. [CrossRef] [PubMed]
127. Hodara, R.; Norris, E.H.; Giasson, B.I.; Mishizen-Eberz, A.J.; Lynch, D.R.; Lee, V.M.; Ischiropoulos, H. Functional consequences of alpha-synuclein tyrosine nitration: Diminished binding to lipid vesicles and increased fibril formation. *J. Biol. Chem.* **2004**, *279*, 47746–47753. [CrossRef]
128. Burai, R.; Ait-Bouziad, N.; Chiki, A.; Lashuel, H.A. Elucidating the Role of Site-Specific Nitration of α-Synuclein in the Pathogenesis of Parkinson's Disease via Protein Semisynthesis and Mutagenesis. *J. Am. Chem. Soc.* **2015**, *137*, 5041–5052. [CrossRef]
129. Kleinknecht, A.; Popova, B.; Lázaro, D.F.; Pinho, R.; Valerius, O.; Outeiro, T.F.; Braus, G.H. C-Terminal Tyrosine Residue Modifications Modulate the Protective Phosphorylation of Serine 129 of α-Synuclein in a Yeast Model of Parkinson's Disease. *PLoS Genet.* **2016**, *12*, e1006098. [CrossRef]
130. Shavali, S.; Combs, C.K.; Ebadi, M. Reactive macrophages increase oxidative stress and alpha-synuclein nitration during death of dopaminergic neuronal cells in co-culture: Relevance to Parkinson's disease. *Neurochem. Res.* **2006**, *31*, 85–94. [CrossRef]

31. Reynolds, A.D.; Glanzer, J.G.; Kadiu, I.; Ricardo-Dukelow, M.; Chaudhuri, A.; Ciborowski, P.; Cerny, R.; Gelman, B.; Thomas, M.P.; Mosley, R.L.; et al. Nitrated alpha-synuclein-activated microglial profiling for Parkinson's disease. *J. Neurochem.* **2008**, *104*, 1504–1525. [CrossRef]
32. Fujiwara, H.; Hasegawa, M.; Dohmae, N.; Kawashima, A.; Masliah, E.; Goldberg, M.S.; Shen, J.; Takio, K.; Iwatsubo, T. alpha-Synuclein is phosphorylated in synucleinopathy lesions. *Nat. Cell Biol.* **2002**, *4*, 160–164. [CrossRef]
33. Anderson, J.P.; Walker, D.E.; Goldstein, J.M.; de Laat, R.; Banducci, K.; Caccavello, R.J.; Barbour, R.; Huang, J.; Kling, K.; Lee, M.; et al. Phosphorylation of Ser-129 is the dominant pathological modification of alpha-synuclein in familial and sporadic Lewy body disease. *J. Biol. Chem.* **2006**, *281*, 29739–29752. [CrossRef] [PubMed]
34. Chen, L.; Feany, M.B. Alpha-synuclein phosphorylation controls neurotoxicity and inclusion formation in a Drosophila model of Parkinson disease. *Nat. Neurosci.* **2005**, *8*, 657–663. [CrossRef] [PubMed]
35. Huang, B.; Wu, S.; Wang, Z.; Ge, L.; Rizak, J.D.; Wu, J.; Li, J.; Xu, L.; Lv, L.; Yin, Y.; et al. Phosphorylated α-Synuclein Accumulations and Lewy Body-like Pathology Distributed in Parkinson's Disease-Related Brain Areas of Aged Rhesus Monkeys Treated with MPTP. *Neuroscience* **2018**, *379*, 302–315. [CrossRef] [PubMed]
36. Takahashi, M.; Ko, L.W.; Kulathingal, J.; Jiang, P.; Sevlever, D.; Yen, S.H. Oxidative stress-induced phosphorylation, degradation and aggregation of alpha-synuclein are linked to upregulated CK2 and cathepsin D. *Eur. J. Neurosci.* **2007**, *26*, 863–874. [CrossRef]
37. Xu, J.; Kao, S.Y.; Lee, F.J.; Song, W.; Jin, L.W.; Yankner, B.A. Dopamine-dependent neurotoxicity of alpha-synuclein: A mechanism for selective neurodegeneration in Parkinson disease. *Nat. Med.* **2002**, *8*, 600–606. [CrossRef]
38. Dias, V.; Junn, E.; Mouradian, M.M. The role of oxidative stress in Parkinson's disease. *J. Park. Dis.* **2013**, *3*, 461–491. [CrossRef]
39. Stockwell, B.R.; Friedmann Angeli, J.P.; Bayir, H.; Bush, A.I.; Conrad, M.; Dixon, S.J.; Fulda, S.; Gascón, S.; Hatzios, S.K.; Kagan, V.E.; et al. Ferroptosis: A Regulated Cell Death Nexus Linking Metabolism, Redox Biology, and Disease. *Cell* **2017**, *171*, 273–285. [CrossRef]
40. Holemans, S.; Javoy-Agid, F.; Agid, Y.; De Paermentier, F.; Laterre, E.C.; Maloteaux, J.M. Sulfonylurea binding sites in normal human brain and in Parkinson's disease, progressive supranuclear palsy and Huntington's disease. *Brain Res.* **1994**, *642*, 327–333. [CrossRef]
41. Takanashi, M.; Mochizuki, H.; Yokomizo, K.; Hattori, N.; Mori, H.; Yamamura, Y.; Mizuno, Y. Iron accumulation in the substantia nigra of autosomal recessive juvenile parkinsonism (ARJP). *Park. Relat. Disord.* **2001**, *7*, 311–314. [CrossRef]
42. Zhang, C.W.; Tai, Y.K.; Chai, B.H.; Chew, K.C.M.; Ang, E.T.; Tsang, F.; Tan, B.W.Q.; Hong, E.T.E.; Asad, A.B.A.; Chuang, K.H.; et al. Transgenic Mice Overexpressing the Divalent Metal Transporter 1 Exhibit Iron Accumulation and Enhanced Parkin Expression in the Brain. *Neuromolecular Med.* **2017**, *19*, 375–386. [CrossRef]
43. Kluss, J.H.; Mamais, A.; Cookson, M.R. LRRK2 links genetic and sporadic Parkinson's disease. *Biochem. Soc. Trans.* **2019**, *47*, 651–661. [CrossRef]
44. Barrett, J.C.; Hansoul, S.; Nicolae, D.L.; Cho, J.H.; Duerr, R.H.; Rioux, J.D.; Brant, S.R.; Silverberg, M.S.; Taylor, K.D.; Barmada, M.M.; et al. Genome-wide association defines more than 30 distinct susceptibility loci for Crohn's disease. *Nat. Genet.* **2008**, *40*, 955–962. [CrossRef]
45. Zimprich, A.; Biskup, S.; Leitner, P.; Lichtner, P.; Farrer, M.; Lincoln, S.; Kachergus, J.; Hulihan, M.; Uitti, R.J.; Calne, D.B.; et al. Mutations in LRRK2 cause autosomal-dominant parkinsonism with pleomorphic pathology. *Neuron* **2004**, *44*, 601–607. [CrossRef]
46. Pyatigorskaya, N.; Sharman, M.; Corvol, J.C.; Valabregue, R.; Yahia-Cherif, L.; Poupon, F.; Cormier-Dequaire, F.; Siebner, H.; Klebe, S.; Vidailhet, M.; et al. High nigral iron deposition in LRRK2 and Parkin mutation carriers using R2* relaxometry. *Mov. Disord.* **2015**, *30*, 1077–1084. [CrossRef]
47. Gillardon, F.; Schmid, R.; Draheim, H. Parkinson's disease-linked leucine-rich repeat kinase 2(R1441G) mutation increases proinflammatory cytokine release from activated primary microglial cells and resultant neurotoxicity. *Neuroscience* **2012**, *208*, 41–48. [CrossRef] [PubMed]
48. Moehle, M.S.; Webber, P.J.; Tse, T.; Sukar, N.; Standaert, D.G.; DeSilva, T.M.; Cowell, R.M.; West, A.B. LRRK2 inhibition attenuates microglial inflammatory responses. *J. Neurosci.* **2012**, *32*, 1602–1611. [CrossRef] [PubMed]
49. Hattula, K.; Furuhjelm, J.; Tikkanen, J.; Tanhuanpää, K.; Laakkonen, P.; Peränen, J. Characterization of the Rab8-specific membrane traffic route linked to protrusion formation. *J. Cell Sci.* **2006**, *119*, 4866–4877. [CrossRef]
50. Rivero-Ríos, P.; Romo-Lozano, M.; Madero-Pérez, J.; Thomas, A.P.; Biosa, A.; Greggio, E.; Hilfiker, S. The G2019S variant of leucine-rich repeat kinase 2 (LRRK2) alters endolysosomal trafficking by impairing the function of the GTPase RAB8A. *J. Biol. Chem.* **2019**, *294*, 4738–4758. [CrossRef]
51. Jia, R.; Liu, Y.; Shuai, K.; Zhou, C.; Chen, L.; Zhu, L.; Wu, X.M. The Relationship between Iron and LRRK2 in a 6-OHDA-Induced Parkinson's Disease Model. *Int. J. Mol. Sci.* **2023**, *24*, 3709. [CrossRef] [PubMed]
52. Selkoe, D.J. Amyloid beta-protein and the genetics of Alzheimer's disease. *J. Biol. Chem.* **1996**, *271*, 18295–18298. [CrossRef]
53. Lane, D.J.R.; Ayton, S.; Bush, A.I. Iron and Alzheimer's Disease: An Update on Emerging Mechanisms. *J. Alzheimers Dis.* **2018**, *64*, S379–S395. [CrossRef]
54. Jan, A.T.; Azam, M.; Rahman, S.; Almigeiti, A.M.S.; Choi, D.H.; Lee, E.J.; Haq, Q.M.R.; Choi, I. Perspective Insights into Disease Progression, Diagnostics, and Therapeutic Approaches in Alzheimer's Disease: A Judicious Update. *Front. Aging Neurosci.* **2017**, *9*, 356. [CrossRef] [PubMed]

155. Connor, J.R.; Menzies, S.L.; St Martin, S.M.; Mufson, E.J. A histochemical study of iron, transferrin, and ferritin in Alzheimer's diseased brains. *J. Neurosci. Res.* **1992**, *31*, 75–83. [CrossRef] [PubMed]
156. Meadowcroft, M.D.; Connor, J.R.; Smith, M.B.; Yang, Q.X. MRI and Histological Analysis of Beta-Amyloid Plaques in Both Human Alzheimer's Disease and APP/PS1 Transgenic Mice. *J. Magn. Reson. Imaging* **2009**, *29*, 997–1007. [CrossRef]
157. Chen, L.L.; Fan, Y.G.; Zhao, L.X.; Zhang, Q.; Wang, Z.Y. The metal ion hypothesis of Alzheimer's disease and the anti neuroinflammatory effect of metal chelators. *Bioorg. Chem.* **2023**, *131*, 106301. [CrossRef]
158. Ge, X.; Zhang, Y.; Zuo, Y.; Israr, M.; Li, B.; Yu, P.; Gao, G.; Chang, Y.Z.; Shi, Z. Transcriptomic analysis reveals the molecular mechanism of Alzheimer-related neuropathology induced by sevoflurane in mice. *J. Cell. Biochem.* **2019**, *120*, 17555–17565 [CrossRef]
159. Zhao, Y.S.; Zhang, L.H.; Yu, P.P.; Gou, Y.J.; Zhao, J.; You, L.H.; Wang, Z.Y.; Zheng, X.; Yan, L.J.; Yu, P.; et al. Ceruloplasmin, a Potential Therapeutic Agent for Alzheimer's Disease. *Antioxid. Redox Signal.* **2018**, *28*, 1323–1337. [CrossRef] [PubMed]
160. Zhang, X.; Gou, Y.J.; Zhang, Y.; Li, J.; Han, K.; Xu, Y.; Li, H.; You, L.H.; Yu, P.; Chang, Y.Z.; et al. Hepcidin overexpression in astrocytes alters brain iron metabolism and protects against amyloid-β induced brain damage in mice. *Cell Death Discov.* **2020**, *6*, 113. [CrossRef]
161. Kuiper, M.A.; Mulder, C.; van Kamp, G.J.; Scheltens, P.; Wolters, E.C. Cerebrospinal fluid ferritin levels of patients with Parkinson's disease, Alzheimer's disease, and multiple system atrophy. *J. Neural Transm. Park. Dis. Dement. Sect.* **1994**, *7*, 109–114. [CrossRef
162. Gao, G.; Chang, Y.Z. Mitochondrial ferritin in the regulation of brain iron homeostasis and neurodegenerative diseases. *Front. Pharmacol.* **2014**, *5*, 19. [CrossRef]
163. Wu, Q.; Wu, W.S.; Su, L.; Zheng, X.; Wu, W.Y.; Santambrogio, P.; Gou, Y.J.; Hao, Q.; Wang, P.N.; Li, Y.R.; et al. Mitochondrial Ferritin Is a Hypoxia-Inducible Factor 1alpha-Inducible Gene That Protects from Hypoxia-Induced Cell Death in Brain. *Antioxid. Redox Signal.* **2019**, *30*, 198–212. [CrossRef] [PubMed]
164. Wu, W.S.; Zhao, Y.S.; Shi, Z.H.; Chang, S.Y.; Nie, G.J.; Duan, X.L.; Zhao, S.M.; Wu, Q.; Yang, Z.L.; Zhao, B.L.; et al. Mitochondrial ferritin attenuates β-amyloid-induced neurotoxicity: Reduction in oxidative damage through the Erk/P38 mitogen-activated protein kinase pathways. *Antioxid. Redox Signal.* **2013**, *18*, 158–169. [CrossRef] [PubMed]
165. Lu, C.D.; Ma, J.K.; Luo, Z.Y.; Tai, Q.X.; Wang, P.; Guan, P.P. Transferrin is responsible for mediating the effects of iron ions on the regulation of anterior pharynx-defective-1alpha/beta and Presenilin 1 expression via PGE(2) and PGD(2) at the early stage of Alzheimer's Disease. *Aging* **2018**, *10*, 3117–3135. [CrossRef]
166. Dong, X.; Gao, W.; Shao, T.; Xie, H.; Bai, J.; Zhao, J.; Chai, X. Age-related changes of brain iron load changes in the frontal cortex in APPswe/PS1DeltaE9 transgenic mouse model of Alzheimer's disease. *J. Trace Elem. Med. Biol.* **2015**, *30*, 118–123. [CrossRef]
167. Walsh, D.M.; Klyubin, I.; Fadeeva, J.V.; Rowan, M.J.; Selkoe, D.J. Amyloid-beta oligomers: Their production, toxicity and therapeutic inhibition. *Biochem. Soc. Trans.* **2002**, *30*, 552–557. [CrossRef]
168. Guillemot, J.; Canuel, M.; Essalmani, R.; Prat, A.; Seidah, N.G. Implication of the proprotein convertases in iron homeostasis: Proprotein convertase 7 sheds human transferrin receptor 1 and furin activates hepcidin. *Hepatology* **2013**, *57*, 2514–2524. [CrossRef]
169. Zhang, Y.; Gao, X.; Bai, X.; Yao, S.; Chang, Y.Z.; Gao, G. The emerging role of furin in neurodegenerative and neuropsychiatric diseases. *Transl. Neurodegener.* **2022**, *11*, 39. [CrossRef]
170. Silvestri, L.; Camaschella, C. A potential pathogenetic role of iron in Alzheimer's disease. *J. Cell. Mol. Med.* **2008**, *12*, 1548–1550 [CrossRef]
171. Rogers, J.T.; Randall, J.D.; Cahill, C.M.; Eder, P.S.; Huang, X.; Gunshin, H.; Leiter, L.; McPhee, J.; Sarang, S.S.; Utsuki, T.; et al. An iron-responsive element type II in the 5′-untranslated region of the Alzheimer's amyloid precursor protein transcript. *J. Biol. Chem.* **2002**, *277*, 45518–45528. [CrossRef]
172. Gamblin, T.C.; King, M.E.; Kuret, J.; Berry, R.W.; Binder, L.I. Oxidative regulation of fatty acid-induced tau polymerization. *Biochemistry* **2000**, *39*, 14203–14210. [CrossRef] [PubMed]
173. Chen, M.; Zheng, J.; Liu, G.; Zeng, C.; Xu, E.; Zhu, W.; Anderson, G.J.; Chen, H. High Dietary Iron Disrupts Iron Homeostasis and Induces Amyloid-β and Phospho-τ Expression in the Hippocampus of Adult Wild-Type and APP/PS1 Transgenic Mice. *J. Nutr.* **2019**, *149*, 2247–2254. [CrossRef]
174. Multhaup, G.; Huber, O.; Buee, L.; Galas, M.C. Amyloid Precursor Protein (APP) Metabolites APP Intracellular Fragment (AICD), Abeta42, and Tau in Nuclear Roles. *J. Biol. Chem.* **2015**, *290*, 23515–23522. [CrossRef]
175. Duce, J.A.; Tsatsanis, A.; Cater, M.A.; James, S.A.; Robb, E.; Wikhe, K.; Leong, S.L.; Perez, K.; Johanssen, T.; Greenough, M.A.; et al. Iron-export ferroxidase activity of beta-amyloid precursor protein is inhibited by zinc in Alzheimer's disease. *Cell* **2010**, *142*, 857–867. [CrossRef]
176. Onukwufor, J.O.; Dirksen, R.T.; Wojtovich, A.P. Iron Dysregulation in Mitochondrial Dysfunction and Alzheimer's Disease. *Antioxidants* **2022**, *11*, 692. [CrossRef] [PubMed]
177. Duan, G.; Li, J.; Duan, Y.; Zheng, C.; Guo, Q.; Li, F.; Zheng, J.; Yu, J.; Zhang, P.; Wan, M.; et al. Mitochondrial Iron Metabolism: The Crucial Actors in Diseases. *Molecules* **2022**, *28*, 29. [CrossRef] [PubMed]
178. Ward, R.J.; Dexter, D.T.; Crichton, R.R. Ageing, neuroinflammation and neurodegeneration. *Front. Biosci. (Schol. Ed.)* **2015**, *7*, 189–204. [CrossRef]

179. Wang, P.; Wu, Q.; Wu, W.; Li, H.; Guo, Y.; Yu, P.; Gao, G.; Shi, Z.; Zhao, B.; Chang, Y.Z. Mitochondrial Ferritin Deletion Exacerbates β-Amyloid-Induced Neurotoxicity in Mice. *Oxid. Med. Cell. Longev.* **2017**, *2017*, 1020357. [CrossRef] [PubMed]
180. Salami, A.; Papenberg, G.; Sitnikov, R.; Laukka, E.J.; Persson, J.; Kalpouzos, G. Elevated neuroinflammation contributes to the deleterious impact of iron overload on brain function in aging. *Neuroimage* **2021**, *230*, 117792. [CrossRef] [PubMed]
181. Nnah, I.C.; Lee, C.H.; Wessling-Resnick, M. Iron potentiates microglial interleukin-1beta secretion induced by amyloid-beta. *J. Neurochem.* **2020**, *154*, 177–189. [CrossRef]
182. Nakamura, K.; Kawakami, T.; Yamamoto, N.; Tomizawa, M.; Fujiwara, T.; Ishii, T.; Harigae, H.; Ogasawara, K. Activation of the NLRP3 inflammasome by cellular labile iron. *Exp. Hematol.* **2016**, *44*, 116–124. [CrossRef]
183. Zhou, S.; Du, X.; Xie, J.; Wang, J. Interleukin-6 regulates iron-related proteins through c-Jun N-terminal kinase activation in BV2 microglial cell lines. *PLoS ONE* **2017**, *12*, e0180464. [CrossRef] [PubMed]
184. Rathore, K.I.; Redensek, A.; David, S. Iron homeostasis in astrocytes and microglia is differentially regulated by TNF-alpha and TGF-beta1. *Glia* **2012**, *60*, 738–750. [CrossRef] [PubMed]
185. Nemeth, E.; Tuttle, M.S.; Powelson, J.; Vaughn, M.B.; Donovan, A.; Ward, D.M.; Ganz, T.; Kaplan, J. Hepcidin regulates cellular iron efflux by binding to ferroportin and inducing its internalization. *Science* **2004**, *306*, 2090–2093. [CrossRef] [PubMed]
186. Bolognin, S.; Drago, D.; Messori, L.; Zatta, P. Chelation therapy for neurodegenerative diseases. *Med. Res. Rev.* **2009**, *29*, 547–570. [CrossRef]
187. Radi, E.; Formichi, P.; Battisti, C.; Federico, A. Apoptosis and oxidative stress in neurodegenerative diseases. *J. Alzheimers Dis.* **2014**, *42* (Suppl. S3), S125–S152. [CrossRef] [PubMed]
188. Campbell, B.C.V.; Khatri, P. Stroke. *Lancet* **2020**, *396*, 129–142. [CrossRef]
189. Ding, H.; Yan, C.Z.; Shi, H.; Zhao, Y.S.; Chang, S.Y.; Yu, P.; Wu, W.S.; Zhao, C.Y.; Chang, Y.Z.; Duan, X.L. Hepcidin is involved in iron regulation in the ischemic brain. *PLoS ONE* **2011**, *6*, e25324. [CrossRef]
190. Wang, P.; Ren, Q.; Shi, M.; Liu, Y.; Bai, H.; Chang, Y.Z. Overexpression of Mitochondrial Ferritin Enhances Blood–brain Barrier Integrity Following Ischemic Stroke in Mice by Maintaining Iron Homeostasis in Endothelial Cells. *Antioxidants* **2022**, *11*, 1257. [CrossRef]
191. Zhang, D.-L.; Rouault, T.A. How does hepcidin hinder ferroportin activity? *Blood* **2018**, *131*, 839–840. [CrossRef]
192. Yang, L.; Wang, D.; Wang, X.T.; Lu, Y.P.; Zhu, L. The roles of hypoxia-inducible Factor-1 and iron regulatory protein 1 in iron uptake induced by acute hypoxia. *Biochem. Biophys. Res. Commun.* **2018**, *507*, 128–135. [CrossRef]
193. Cheah, J.H.; Kim, S.F.; Hester, L.D.; Clancy, K.W.; Patterson, S.E., 3rd; Papadopoulos, V.; Snyder, S.H. NMDA receptor-nitric oxide transmission mediates neuronal iron homeostasis via the GTPase Dexras1. *Neuron* **2006**, *51*, 431–440. [CrossRef]
194. Riedl, S.J.; Salvesen, G.S. The apoptosome: Signalling platform of cell death. *Nat. Rev. Mol. Cell Biol.* **2007**, *8*, 405–413. [CrossRef]
195. Ott, M.; Robertson, J.D.; Gogvadze, V.; Zhivotovsky, B.; Orrenius, S. Cytochrome c release from mitochondria proceeds by a two-step process. *Proc. Natl. Acad. Sci. USA* **2002**, *99*, 1259–1263. [CrossRef]
196. Mantzaris, M.D.; Bellou, S.; Skiada, V.; Kitsati, N.; Fotsis, T.; Galaris, D. Intracellular labile iron determines H_2O_2-induced apoptotic signaling via sustained activation of ASK1/JNK-p38 axis. *Free Radic. Biol. Med.* **2016**, *97*, 454–465. [CrossRef]
197. Ichijo, H.; Nishida, E. Induction of apoptosis by ASK1, a mammalian MAPKKK that activates SAPK/JNK and p38 signaling pathways. *Science* **1997**, *275*, 90–94. [CrossRef]
198. Saitoh, M.; Nishitoh, H. Mammalian thioredoxin is a direct inhibitor of apoptosis signal-regulating kinase (ASK) 1. *EMBO J.* **1998**, *17*, 2596–2606. [CrossRef] [PubMed]
199. Kuhn, H.; Humeniuk, L.; Kozlov, N.; Roigas, S.; Adel, S.; Heydeck, D. The evolutionary hypothesis of reaction specificity of mammalian ALOX15 orthologs. *Prog. Lipid Res.* **2018**, *72*, 55–74. [CrossRef] [PubMed]
200. Li, Y.X.; Ding, S.J.; Xiao, L.; Guo, W.; Zhan, Q. Desferoxamine preconditioning protects against cerebral ischemia in rats by inducing expressions of hypoxia inducible factor 1 alpha and erythropoietin. *Neurosci. Bull.* **2008**, *24*, 89–95. [CrossRef]
201. Guo, X.; Jin, X.; Han, K.; Kang, S.; Tian, S.; Lv, X.; Feng, M.; Zheng, H.; Zuo, Y.; Xu, G.; et al. Iron promotes neurological function recovery in mice with ischemic stroke through endogenous repair mechanisms. *Free Radic. Biol. Med.* **2022**, *182*, 59–72. [CrossRef] [PubMed]
202. Tuo, Q.Z.; Lei, P.; Jackman, K.A.; Li, X.L.; Xiong, H.; Li, X.L.; Liuyang, Z.Y.; Roisman, L.; Zhang, S.T.; Ayton, S.; et al. Tau-mediated iron export prevents ferroptotic damage after ischemic stroke. *Mol. Psychiatry* **2017**, *22*, 1520–1530. [CrossRef] [PubMed]
203. DeGregorio-Rocasolano, N.; Marti-Sistac, O.; Ponce, J.; Castello-Ruiz, M.; Millan, M.; Guirao, V.; Garcia-Yebenes, I.; Salom, J.B.; Ramos-Cabrer, P.; Alborch, E.; et al. Iron-loaded transferrin (Tf) is detrimental whereas iron-free Tf confers protection against brain ischemia by modifying blood Tf saturation and subsequent neuronal damage. *Redox Biol.* **2018**, *15*, 143–158. [CrossRef]
204. Zhao, Y.; Xin, Z.; Li, N.; Chang, S.; Chen, Y.; Geng, L.; Chang, H.; Shi, H.; Chang, Y.Z. Nano-liposomes of lycopene reduces ischemic brain damage in rodents by regulating iron metabolism. *Free Radic. Biol. Med.* **2018**, *124*, 1–11. [CrossRef]
205. Bandelow, B.; Michaelis, S. Epidemiology of anxiety disorders in the 21st century. *Dialogues Clin. Neurosci.* **2015**, *17*, 327–335. [CrossRef]
206. Shah, H.E.; Bhawnani, N.; Ethirajulu, A.; Alkasabera, A.; Onyali, C.B.; Anim-Koranteng, C.; Mostafa, J.A. Iron Deficiency-Induced Changes in the Hippocampus, Corpus Striatum, and Monoamines Levels That Lead to Anxiety, Depression, Sleep Disorders, and Psychotic Disorders. *Cureus* **2021**, *13*, e18138. [CrossRef]

207. Li, Z.; Wang, W.; Xin, X.; Song, X.; Zhang, D. Association of total zinc, iron, copper and selenium intakes with depression in the US adults. *J. Affect. Disord.* **2018**, *228*, 68–74. [CrossRef]
208. Stewart, R.; Hirani, V. Relationship between depressive symptoms, anemia, and iron status in older residents from a national survey population. *Psychosom. Med.* **2012**, *74*, 208–213. [CrossRef] [PubMed]
209. Beard, J.L.; Connor, J.R. Iron status and neural functioning. *Annu. Rev. Nutr.* **2003**, *23*, 41–58. [CrossRef]
210. Sobotka, T.J.; Whittaker, P.; Sobotka, J.M.; Brodie, R.E.; Quander, D.Y.; Robl, M.; Bryant, M.; Barton, C.N. Neurobehavioral dysfunctions associated with dietary iron overload. *Physiol. Behav.* **1996**, *59*, 213–219. [CrossRef] [PubMed]
211. Philibert, R.A.; Sandhu, H.; Hollenbeck, N.; Gunter, T.; Adams, W.; Madan, A. The relationship of 5HTT (SLC6A4) methylation and genotype on mRNA expression and liability to major depression and alcohol dependence in subjects from the Iowa Adoption Studies. *Am. J. Med. Genet. B. Neuropsychiatr. Genet.* **2008**, *147*, 543–549. [CrossRef] [PubMed]
212. Kim, J.; Wessling-Resnick, M. Iron and mechanisms of emotional behavior. *J. Nutr. Biochem.* **2014**, *25*, 1101–1107. [CrossRef]
213. Bastian, T.W.; von Hohenberg, W.C.; Mickelson, D.J.; Lanier, L.M.; Georgieff, M.K. Iron Deficiency Impairs Developing Hippocampal Neuron Gene Expression, Energy Metabolism, and Dendrite Complexity. *Dev. Neurosci.* **2016**, *38*, 264–276. [CrossRef] [PubMed]
214. Denny, C.A.; Kheirbek, M.A.; Alba, E.L.; Tanaka, K.F.; Brachman, R.A.; Laughman, K.B.; Tomm, N.K.; Turi, G.F.; Losonczy, A.; Hen, R. Hippocampal memory traces are differentially modulated by experience, time, and adult neurogenesis. *Neuron* **2014**, *83*, 189–201. [CrossRef] [PubMed]
215. Liu, X.; Ramirez, S.; Pang, P.T.; Puryear, C.B.; Govindarajan, A.; Deisseroth, K.; Tonegawa, S. Optogenetic stimulation of a hippocampal engram activates fear memory recall. *Nature* **2012**, *484*, 381–385. [CrossRef]
216. McEchron, M.D.; Cheng, A.Y.; Liu, H.; Connor, J.R.; Gilmartin, M.R. Perinatal nutritional iron deficiency permanently impairs hippocampus-dependent trace fear conditioning in rats. *Nutr. Neurosci.* **2005**, *8*, 195–206. [CrossRef] [PubMed]
217. Georgieff, M.K. Iron deficiency in pregnancy. *Am. J. Obstet. Gynecol.* **2020**, *223*, 516–524. [CrossRef]
218. Rice, D.; Barone, S., Jr. Critical periods of vulnerability for the developing nervous system: Evidence from humans and animal models. *Environ. Health Perspect.* **2000**, *108* (Suppl. S3), 511–533. [CrossRef]
219. Di Bella, D.J.; Habibi, E.; Stickels, R.R.; Scalia, G.; Brown, J.; Yadollahpour, P.; Yang, S.M.; Abbate, C.; Biancalani, T.; Macosko, E.Z.; et al. Molecular logic of cellular diversification in the mouse cerebral cortex. *Nature* **2021**, *595*, 554–559. [CrossRef]
220. Evsyukova, I.; Plestant, C.; Anton, E. Integrative mechanisms of oriented neuronal migration in the developing brain. *Annu. Rev. Cell Dev. Biol.* **2013**, *29*, 299–353. [CrossRef]
221. Gilmore, E.C.; Walsh, C.A. Genetic causes of microcephaly and lessons for neuronal development. *Wiley Interdiscip. Rev. Dev. Biol.* **2013**, *2*, 461–478. [CrossRef]
222. Gleeson, J.G.; Allen, K.M.; Fox, J.W.; Lamperti, E.D.; Berkovic, S.; Scheffer, I.; Cooper, E.C.; Dobyns, W.B.; Minnerath, S.R.; Ross, M.E. Doublecortin, a brain-specific gene mutated in human X-linked lissencephaly and double cortex syndrome, encodes a putative signaling protein. *Cell* **1998**, *92*, 63–72. [CrossRef]
223. Juric-Sekhar, G.; Hevner, R.F. Malformations of cerebral cortex development: Molecules and mechanisms. *Annu. Rev. Pathol.* **2019**, *14*, 293–318. [CrossRef]
224. Kahn, R.S.; Sommer, I.E.; Murray, R.M.; Meyer-Lindenberg, A.; Weinberger, D.R.; Cannon, T.D.; O'Donovan, M.; Correll, C.U.; Kane, J.M.; van Os, J.; et al. Schizophrenia. *Nat. Rev. Dis. Prim.* **2015**, *1*, 15067. [CrossRef]
225. Hazlett, H.C.; Gu, H.; Munsell, B.C.; Kim, S.H.; Styner, M.; Wolff, J.J.; Elison, J.T.; Swanson, M.R.; Zhu, H.; Botteron, K.N.; et al. Early brain development in infants at high risk for autism spectrum disorder. *Nature* **2017**, *542*, 348–351. [CrossRef]
226. McCann, S.; Perapoch Amado, M.; Moore, S.E. The Role of Iron in Brain Development: A Systematic Review. *Nutrients* **2020**, *12*, 2001. [CrossRef]
227. McLean, E.; Cogswell, M.; Egli, I.; Wojdyla, D.; de Benoist, B. Worldwide prevalence of anaemia, WHO Vitamin and Mineral Nutrition Information System, 1993–2005. *Public. Health Nutr.* **2009**, *12*, 444–454. [CrossRef] [PubMed]
228. Fretham, S.J.; Carlson, E.S.; Georgieff, M.K. The role of iron in learning and memory. *Adv. Nutr.* **2011**, *2*, 112–121. [CrossRef] [PubMed]
229. Carter, R.C.; Jacobson, J.L.; Burden, M.J.; Armony-Sivan, R.; Dodge, N.C.; Angelilli, M.L.; Lozoff, B.; Jacobson, S.W. Iron deficiency anemia and cognitive function in infancy. *Pediatrics* **2010**, *126*, e427–e434. [CrossRef]
230. Lozoff, B. Iron deficiency and child development. *Food Nutr. Bull.* **2007**, *28*, S560–S571. [CrossRef]
231. Basu, S.; Kumar, D.; Anupurba, S.; Verma, A.; Kumar, A. Effect of maternal iron deficiency anemia on fetal neural development. *J. Perinatol.* **2018**, *38*, 233–239. [CrossRef]
232. Berglund, S.K.; Torres-Espínola, F.J.; García-Valdés, L.; Segura, M.T.; Martínez-Zaldívar, C.; Padilla, C.; Rueda, R.; Pérez García, M.; McArdle, H.J.; Campoy, C. The impacts of maternal iron deficiency and being overweight during pregnancy on neurodevelopment of the offspring. *Br. J. Nutr.* **2017**, *118*, 533–540. [CrossRef] [PubMed]
233. Li, Y.; Kim, J.; Buckett, P.D.; Böhlke, M.; Maher, T.J.; Wessling-Resnick, M. Severe postnatal iron deficiency alters emotional behavior and dopamine levels in the prefrontal cortex of young male rats. *J. Nutr.* **2011**, *141*, 2133–2138. [CrossRef]
234. Beard, J. Recent evidence from human and animal studies regarding iron status and infant development. *J. Nutr.* **2007**, *137*, 524s–530s. [CrossRef]
235. de los Monteros, A.E.; Korsak, R.A.; Tran, T.; Vu, D.; de Vellis, J.; Edmond, J. Dietary iron and the integrity of the developing rat brain: A study with the artificially-reared rat pup. *Cell. Mol. Biol.* **2000**, *46*, 501–515.

36. Beard, J.L.; Wiesinger, J.A.; Connor, J.R. Pre- and postweaning iron deficiency alters myelination in Sprague-Dawley rats. *Dev. Neurosci.* **2003**, *25*, 308–315. [CrossRef]
37. Ortiz, E.; Pasquini, J.M.; Thompson, K.; Felt, B.; Butkus, G.; Beard, J.; Connor, J.R. Effect of manipulation of iron storage, transport, or availability on myelin composition and brain iron content in three different animal models. *J. Neurosci. Res.* **2004**, *77*, 681–689. [CrossRef]
38. Kwik-Uribe, C.L.; Golub, M.S.; Keen, C.L. Chronic marginal iron intakes during early development in mice alter brain iron concentrations and behavior despite postnatal iron supplementation. *J. Nutr.* **2000**, *130*, 2040–2048. [CrossRef] [PubMed]
39. Markova, V.; Holm, C.; Pinborg, A.B.; Thomsen, L.L.; Moos, T. Impairment of the Developing Human Brain in Iron Deficiency: Correlations to Findings in Experimental Animals and Prospects for Early Intervention Therapy. *Pharmaceuticals* **2019**, *12*, 120. [CrossRef] [PubMed]
40. Beard, J. Iron deficiency alters brain development and functioning. *J. Nutr.* **2003**, *133*, 1468s–1472s. [CrossRef] [PubMed]
41. Lin, L.; Chen, H.; Zhao, R.; Zhu, M.; Nie, G. Nanomedicine targets iron metabolism for cancer therapy. *Cancer Sci.* **2022**, *113*, 828. [CrossRef] [PubMed]
42. Han, Y.; He, Y.; Jin, X.; Xie, J.; Yu, P.; Gao, G.; Chang, S.; Zhang, J.; Chang, Y.Z. CHIR99021 Maintenance of the Cell Stemness by Regulating Cellular Iron Metabolism. *Antioxidants* **2023**, *12*, 377. [CrossRef] [PubMed]
43. Tran, P.V.; Carlson, E.S.; Fretham, S.J.; Georgieff, M.K. Early-life iron deficiency anemia alters neurotrophic factor expression and hippocampal neuron differentiation in male rats. *J. Nutr.* **2008**, *138*, 2495–2501. [CrossRef]
44. Jorgenson, L.A.; Sun, M.; O'Connor, M.; Georgieff, M.K. Fetal iron deficiency disrupts the maturation of synaptic function and efficacy in area CA1 of the developing rat hippocampus. *Hippocampus* **2005**, *15*, 1094–1102. [CrossRef]
45. Brunette, K.E.; Tran, P.V.; Wobken, J.D. Gestational and neonatal iron deficiency alters apical dendrite structure of CA1 pyramidal neurons in adult rat hippocampus. *Dev. Neurosci.* **2010**, *32*, 238–248. [CrossRef] [PubMed]
46. Callahan, L.S.; Thibert, K.A.; Wobken, J.D.; Georgieff, M.K. Early-life iron deficiency anemia alters the development and long-term expression of parvalbumin and perineuronal nets in the rat hippocampus. *Dev. Neurosci.* **2013**, *35*, 427–436. [CrossRef] [PubMed]
47. Greminger, A.R.; Lee, D.L.; Shrager, P.; Mayer-Pröschel, M. Gestational iron deficiency differentially alters the structure and function of white and gray matter brain regions of developing rats. *J. Nutr.* **2014**, *144*, 1058–1066. [CrossRef] [PubMed]
48. Felt, B.T.; Beard, J.L.; Schallert, T.; Shao, J.; Aldridge, J.W.; Connor, J.R.; Georgieff, M.K.; Lozoff, B. Persistent neurochemical and behavioral abnormalities in adulthood despite early iron supplementation for perinatal iron deficiency anemia in rats. *Behav. Brain Res.* **2006**, *171*, 261–270. [CrossRef]
49. Unger, E.L.; Paul, T.; Murray-Kolb, L.E.; Felt, B.; Jones, B.C.; Beard, J.L. Early iron deficiency alters sensorimotor development and brain monoamines in rats. *J. Nutr.* **2007**, *137*, 118–124. [CrossRef]
50. Mihaila, C.; Schramm, J.; Strathmann, F.G.; Lee, D.L.; Gelein, R.M.; Luebke, A.E.; Mayer-Pröschel, M. Identifying a window of vulnerability during fetal development in a maternal iron restriction model. *PLoS ONE* **2011**, *6*, e17483. [CrossRef]
51. Greminger, A.R.; Mayer-Pröschel, M. Identifying the threshold of iron deficiency in the central nervous system of the rat by the auditory brainstem response. *ASN Neuro* **2015**, *7*, 1759091415569911. [CrossRef]
52. Fretham, S.J.; Carlson, E.S.; Georgieff, M.K. Neuronal-specific iron deficiency dysregulates mammalian target of rapamycin signaling during hippocampal development in nonanemic genetic mouse models. *J. Nutr.* **2013**, *143*, 260–266. [CrossRef]
53. Benkovic, S.A.; Connor, J.R. Ferritin, transferrin, and iron in selected regions of the adult and aged rat brain. *J. Comp. Neurol.* **1993**, *338*, 97–113. [CrossRef]
54. Connor, J.R.; Pavlick, G.; Karli, D.; Menzies, S.L.; Palmer, C. A histochemical study of iron-positive cells in the developing rat brain. *J. Comp. Neurol.* **1995**, *355*, 111–123. [CrossRef]
55. Horiquini-Barbosa, E.; Gibb, R.; Kolb, B.; Bray, D.; Lachat, J.J. Tactile stimulation partially prevents neurodevelopmental changes in visual tract caused by early iron deficiency. *Brain Res.* **2017**, *1657*, 130–139. [CrossRef]
56. Rosato-Siri, M.V.; Marziali, L.; Guitart, M.E.; Badaracco, M.E.; Puntel, M.; Pitossi, F.; Correale, J.; Pasquini, J.M. Iron Availability Compromises Not Only Oligodendrocytes But Also Astrocytes and Microglial Cells. *Mol. Neurobiol.* **2018**, *55*, 1068–1081. [CrossRef]
57. Guitart, M.E.; Vence, M.; Correale, J.; Pasquini, J.M.; Rosato-Siri, M.V. Ontogenetic oligodendrocyte maturation through gestational iron deprivation: The road not taken. *Glia* **2019**, *67*, 1760–1774. [CrossRef]
58. Rao, R.; Ennis, K.; Oz, G.; Lubach, G.R.; Georgieff, M.K.; Coe, C.L. Metabolomic analysis of cerebrospinal fluid indicates iron deficiency compromises cerebral energy metabolism in the infant monkey. *Neurochem. Res.* **2013**, *38*, 573–580. [CrossRef]
59. Mudd, A.T.; Fil, J.E.; Knight, L.C.; Dilger, R.N. Dietary Iron Repletion following Early-Life Dietary Iron Deficiency Does Not Correct Regional Volumetric or Diffusion Tensor Changes in the Developing Pig Brain. *Front. Neurol.* **2017**, *8*, 735. [CrossRef] [PubMed]
60. Felt, B.T.; Lozoff, B. Brain iron and behavior of rats are not normalized by treatment of iron deficiency anemia during early development. *J. Nutr.* **1996**, *126*, 693–701. [CrossRef] [PubMed]
61. Lozoff, B.; Jimenez, E.; Hagen, J.; Mollen, E.; Wolf, A.W. Poorer behavioral and developmental outcome more than 10 years after treatment for iron deficiency in infancy. *Pediatrics* **2000**, *105*, E51. [CrossRef] [PubMed]

262. Uecker, A.; Nadel, L. Spatial locations gone awry: Object and spatial memory deficits in children with fetal alcohol syndrome. *Neuropsychologia* **1996**, *34*, 209–223. [CrossRef]
263. Carlson, E.S.; Stead, J.D.H.; Neal, C.R. Perinatal iron deficiency results in altered developmental expression of genes mediating energy metabolism and neuronal morphogenesis in hippocampus. *Hippocampus* **2007**, *17*, 679–691. [CrossRef] [PubMed]
264. Bastian, T.W.; von Hohenberg, W.C.; Georgieff, M.K.; Lanier, L.M. Chronic Energy Depletion due to Iron Deficiency Impairs Dendritic Mitochondrial Motility during Hippocampal Neuron Development. *J. Neurosci.* **2019**, *39*, 802–813. [CrossRef] [PubMed]
265. Bastian, T.W.; Rao, R.; Tran, P.V.; Georgieff, M.K. The Effects of Early-Life Iron Deficiency on Brain Energy Metabolism. *Neurosci. Insights* **2020**, *15*, 2633105520935104. [CrossRef] [PubMed]
266. Perng, V.; Li, C.; Klocke, C.R.; Navazesh, S.E.; Pinneles, D.K.; Lein, P.J.; Ji, P. Iron Deficiency and Iron Excess Differently Affect Dendritic Architecture of Pyramidal Neurons in the Hippocampus of Piglets. *J. Nutr.* **2021**, *151*, 235–244. [CrossRef] [PubMed]
267. Lien, Y.C.; Condon, D.E.; Georgieff, M.K.; Simmons, R.A.; Tran, P.V. Dysregulation of Neuronal Genes by Fetal-Neonatal Iron Deficiency Anemia Is Associated with Altered DNA Methylation in the Rat Hippocampus. *Nutrients* **2019**, *11*, 1191. [CrossRef] [PubMed]
268. Erber, L.N.; Luo, A.; Gong, Y.; Beeson, M.; Tu, M.; Tran, P.; Chen, Y. Iron Deficiency Reprograms Phosphorylation Signaling and Reduces O-GlcNAc Pathways in Neuronal Cells. *Nutrients* **2021**, *13*, 179. [CrossRef]
269. Erber, L.; Liu, S.; Gong, Y.; Tran, P.; Chen, Y. Quantitative Proteome and Transcriptome Dynamics Analysis Reveals Iron Deficiency Response Networks and Signature in Neuronal Cells. *Molecules* **2022**, *27*, 484. [CrossRef]
270. DeMaman, A.S.; Melo, P.; Homem, J.M.; Tavares, M.A.; Lachat, J.J. Effectiveness of iron repletion in the diet for the optic nerve development of anaemic rats. *Eye* **2010**, *24*, 901–908. [CrossRef]
271. Unger, E.L.; Hurst, A.R.; Georgieff, M.K.; Schallert, T.; Rao, R.; Connor, J.R.; Kaciroti, N.; Lozoff, B.; Felt, B. Behavior and monoamine deficits in prenatal and perinatal iron deficiency are not corrected by early postnatal moderate-iron or high-iron diets in rats. *J. Nutr.* **2012**, *142*, 2040–2049. [CrossRef]
272. Georgieff, M.K. Long-term brain and behavioral consequences of early iron deficiency. *Nutr. Rev.* **2011**, *69* (Suppl. S1), S43–S48. [CrossRef]
273. Mudd, A.T.; Fil, J.E.; Knight, L.C.; Lam, F.; Liang, Z.P.; Dilger, R.N. Early-Life Iron Deficiency Reduces Brain Iron Content and Alters Brain Tissue Composition Despite Iron Repletion: A Neuroimaging Assessment. *Nutrients* **2018**, *10*, 135. [CrossRef] [PubMed]
274. Adlard, P.A.; Bush, A.I. Metals and Alzheimer's Disease: How Far Have We Come in the Clinic? *J. Alzheimers Dis.* **2018**, *62*, 1369–1379. [CrossRef] [PubMed]
275. Zhang, Y.; He, M.L. Deferoxamine enhances alternative activation of microglia and inhibits amyloid beta deposits in APP/PS1 mice. *Brain Res.* **2017**, *1677*, 86–92. [CrossRef]
276. Zhang, S.; Wang, J.; Song, N.; Xie, J.; Jiang, H. Up-regulation of divalent metal transporter 1 is involved in 1-methyl-4-phenylpyridinium (MPP(+))-induced apoptosis in MES23.5 cells. *Neurobiol. Aging* **2009**, *30*, 1466–1476. [CrossRef] [PubMed]
277. Martin-Bastida, A.; Ward, R.J.; Newbould, R.; Piccini, P.; Sharp, D.; Kabba, C.; Patel, M.C.; Spino, M.; Connelly, J.; Tricta, F.; et al. Brain iron chelation by deferiprone in a phase 2 randomised double-blinded placebo controlled clinical trial in Parkinson's disease. *Sci. Rep.* **2017**, *7*, 1398. [CrossRef] [PubMed]
278. Lin, K.J.; Chen, S.D.; Lin, K.L.; Liou, C.W.; Lan, M.Y.; Chuang, Y.C.; Wang, P.W.; Lee, J.J.; Wang, F.S.; Lin, H.Y.; et al. Iron Brain Menace: The Involvement of Ferroptosis in Parkinson Disease. *Cells* **2022**, *11*, 3829. [CrossRef] [PubMed]
279. Devos, D.; Labreuche, J.; Rascol, O.; Corvol, J.C.; Duhamel, A.; Guyon Delannoy, P.; Poewe, W.; Compta, Y.; Pavese, N.; Růžička, E.; et al. Trial of Deferiprone in Parkinson's Disease. *N. Engl. J. Med.* **2022**, *387*, 2045–2055. [CrossRef]
280. Castellanos, M.; Puig, N.; Carbonell, T.; Castillo, J.; Martinez, J.; Rama, R.; Davalos, A. Iron intake increases infarct volume after permanent middle cerebral artery occlusion in rats. *Brain Res.* **2002**, *952*, 1–6. [CrossRef]
281. Shadid, M.; Buonocore, G.; Groenendaal, F.; Moison, R.; Ferrali, M.; Berger, H.M.; van Bel, F. Effect of deferoxamine and allopurinol on non-protein-bound iron concentrations in plasma and cortical brain tissue of newborn lambs following hypoxia-ischemia. *Neurosci. Lett.* **1998**, *248*, 5–8. [CrossRef]
282. Sarco, D.P.; Becker, J.; Palmer, C.; Sheldon, R.A.; Ferriero, D.M. The neuroprotective effect of deferoxamine in the hypoxic-ischemic immature mouse brain. *Neurosci. Lett.* **2000**, *282*, 113–116. [CrossRef]
283. Papazisis, G.; Pourzitaki, C.; Sardeli, C.; Lallas, A.; Amaniti, E.; Kouvelas, D. Deferoxamine decreases the excitatory amino acid levels and improves the histological outcome in the hippocampus of neonatal rats after hypoxia-ischemia. *Pharmacol. Res.* **2008**, *57*, 73–78. [CrossRef]
284. Selim, M.; Foster, L.D.; Moy, C.S.; Xi, G.; Hill, M.D.; Morgenstern, L.B.; Greenberg, S.M.; James, M.L.; Singh, V.; Clark, W.M.; et al. Deferoxamine mesylate in patients with intracerebral haemorrhage (i-DEF): A multicentre, randomised, placebo-controlled, double-blind phase 2 trial. *Lancet Neurol.* **2019**, *18*, 428–438. [CrossRef]
285. Foster, L.; Robinson, L.; Yeatts, S.D.; Conwit, R.A.; Shehadah, A.; Lioutas, V.; Selim, M. Effect of Deferoxamine on Trajectory of Recovery After Intracerebral Hemorrhage: A Post Hoc Analysis of the i-DEF Trial. *Stroke* **2022**, *53*, 2204–2210. [CrossRef] [PubMed]

286. Wei, C.; Wang, J.; Foster, L.D.; Yeatts, S.D.; Moy, C.; Mocco, J.; Selim, M. Effect of Deferoxamine on Outcome According to Baseline Hematoma Volume: A Post Hoc Analysis of the i-DEF Trial. *Stroke* **2022**, *53*, 1149–1156. [CrossRef] [PubMed]
287. Neufeld, E.J. Oral chelators deferasirox and deferiprone for transfusional iron overload in thalassemia major: New data, new questions. *Blood* **2006**, *107*, 3436–3441. [CrossRef] [PubMed]
288. Jakaria, M.; Belaidi, A.A.; Bush, A.I.; Ayton, S. Ferroptosis as a mechanism of neurodegeneration in Alzheimer's disease. *J. Neurochem.* **2021**, *159*, 804–825. [CrossRef]
289. Vlachodimitropoulou, E.; Chen, Y.L.; Garbowski, M.; Koonyosying, P.; Psaila, B.; Sola-Visner, M.; Cooper, N.; Hider, R.; Porter, J. Eltrombopag: A powerful chelator of cellular or extracellular iron(III) alone or combined with a second chelator. *Blood* **2017**, *130*, 1923–1933. [CrossRef]
290. Boddaert, N.; Le Quan Sang, K.H.; Rotig, A.; Leroy-Willig, A.; Gallet, S.; Brunelle, F.; Sidi, D.; Thalabard, J.C.; Munnich, A.; Cabantchik, Z.I. Selective iron chelation in Friedreich ataxia: Biologic and clinical implications. *Blood* **2007**, *110*, 401–408. [CrossRef]
291. Sohn, Y.S.; Breuer, W.; Munnich, A.; Cabantchik, Z.I. Redistribution of accumulated cell iron: A modality of chelation with therapeutic implications. *Blood* **2008**, *111*, 1690–1699. [CrossRef]
292. Kosyakovsky, J.; Fine, J.M.; Frey, W.H., 2nd; Hanson, L.R. Mechanisms of Intranasal Deferoxamine in Neurodegenerative and Neurovascular Disease. *Pharmaceuticals* **2021**, *14*, 95. [CrossRef]
293. Guo, C.; Wang, P.; Zhong, M.L.; Wang, T.; Huang, X.S.; Li, J.Y.; Wang, Z.Y. Deferoxamine inhibits iron induced hippocampal tau phosphorylation in the Alzheimer transgenic mouse brain. *Neurochem. Int.* **2013**, *62*, 165–172. [CrossRef]
294. Guo, C.; Zhang, Y.X.; Wang, T.; Zhong, M.L.; Yang, Z.H.; Hao, L.J.; Chai, R.; Zhang, S. Intranasal deferoxamine attenuates synapse loss via up-regulating the P38/HIF-1alpha pathway on the brain of APP/PS1 transgenic mice. *Front. Aging Neurosci.* **2015**, *7*, 104. [CrossRef]
295. Fine, J.M.; Stroebel, B.M.; Faltesek, K.A.; Terai, K.; Haase, L.; Knutzen, K.E.; Kosyakovsky, J.; Bowe, T.J.; Fuller, A.K.; Frey, W.H.; et al. Intranasal delivery of low-dose insulin ameliorates motor dysfunction and dopaminergic cell death in a 6-OHDA rat model of Parkinson's Disease. *Neurosci. Lett.* **2020**, *714*, 134567. [CrossRef]
296. Fine, J.M.; Forsberg, A.C.; Renner, D.B.; Faltesek, K.A.; Mohan, K.G.; Wong, J.C.; Arneson, L.C.; Crow, J.M.; Frey, W.H., 2nd; Hanson, L.R. Intranasally-administered deferoxamine mitigates toxicity of 6-OHDA in a rat model of Parkinson's disease. *Brain Res.* **2014**, *1574*, 96–104. [CrossRef]
297. Hanson, L.R.; Roeytenberg, A.; Martinez, P.M.; Coppes, V.G.; Sweet, D.C.; Rao, R.J.; Marti, D.L.; Hoekman, J.D.; Matthews, R.B.; Frey, W.H., 2nd; et al. Intranasal deferoxamine provides increased brain exposure and significant protection in rat ischemic stroke. *J. Pharmacol. Exp. Ther.* **2009**, *330*, 679–686. [CrossRef]
298. Zhao, Y.P.; Tian, S.Y.; Zhang, J.; Cheng, X.; Huang, W.P.; Cao, G.L.; Chang, Y.Z.; Wang, H.; Nie, G.J.; Qiu, W. Regulation of neuroinflammation with GLP-1 receptor targeting nanostructures to alleviate Alzheimer's symptoms in the disease models. *Nano Today* **2022**, *44*, 101457. [CrossRef]
299. Zheng, W.; Xin, N.; Chi, Z.H.; Zhao, B.L.; Zhang, J.; Li, J.Y.; Wang, Z.Y. Divalent metal transporter 1 is involved in amyloid precursor protein processing and Abeta generation. *FASEB J.* **2009**, *23*, 4207–4217. [CrossRef] [PubMed]
300. Tian, Y.; Lu, J.; Hao, X.; Li, H.; Zhang, G.; Liu, X.; Li, X.; Zhao, C.; Kuang, W.; Chen, D.; et al. FTH1 Inhibits Ferroptosis Through Ferritinophagy in the 6-OHDA Model of Parkinson's Disease. *Neurotherapeutics* **2020**, *17*, 1796–1812. [CrossRef] [PubMed]
301. Wang, Y.Q.; Chang, S.Y.; Wu, Q.; Gou, Y.J.; Jia, L.; Cui, Y.M.; Yu, P.; Shi, Z.H.; Wu, W.S.; Gao, G.; et al. The Protective Role of Mitochondrial Ferritin on Erastin-Induced Ferroptosis. *Front. Aging Neurosci.* **2016**, *8*, 308. [CrossRef] [PubMed]
302. You, L.H.; Li, Z.; Duan, X.L.; Zhao, B.L.; Chang, Y.Z.; Shi, Z.H. Mitochondrial ferritin suppresses MPTP-induced cell damage by regulating iron metabolism and attenuating oxidative stress. *Brain Res.* **2016**, *1642*, 33–42. [CrossRef]
303. Guo, X.; Zheng, H.; Guo, Y.; Wang, Y.; Anderson, G.J.; Ci, Y.; Yu, P.; Geng, L.; Chang, Y.Z. Nasal delivery of nanoliposome-encapsulated ferric ammonium citrate can increase the iron content of rat brain. *J. Nanobiotechnology* **2017**, *15*, 42. [CrossRef]
304. Villalón-García, I.; Álvarez-Córdoba, M.; Povea-Cabello, S.; Talaverón-Rey, M.; Villanueva-Paz, M.; Luzón-Hidalgo, R.; Suárez-Rivero, J.M.; Suárez-Carrillo, A.; Munuera-Cabeza, M.; Salas, J.J.; et al. Vitamin E prevents lipid peroxidation and iron accumulation in PLA2G6-Associated Neurodegeneration. *Neurobiol. Dis.* **2022**, *165*, 105649. [CrossRef]
305. Romero, A.; Ramos, E.; de Los Ríos, C.; Egea, J.; Del Pino, J.; Reiter, R.J. A review of metal-catalyzed molecular damage: Protection by melatonin. *J. Pineal Res.* **2014**, *56*, 343–370. [CrossRef] [PubMed]
306. Medeiros, M.S.; Schumacher-Schuh, A.; Cardoso, A.M.; Bochi, G.V.; Baldissarelli, J.; Kegler, A.; Santana, D.; Chaves, C.M.; Schetinger, M.R.; Moresco, R.N.; et al. Iron and Oxidative Stress in Parkinson's Disease: An Observational Study of Injury Biomarkers. *PLoS ONE* **2016**, *11*, e0146129. [CrossRef]
307. Santoro, M.M. The Antioxidant Role of Non-mitochondrial CoQ10: Mystery Solved! *Cell Metab.* **2020**, *31*, 13–15. [CrossRef] [PubMed]
308. Bersuker, K.; Hendricks, J.M.; Li, Z.; Magtanong, L.; Ford, B.; Tang, P.H.; Roberts, M.A.; Tong, B.; Maimone, T.J.; Zoncu, R.; et al. The CoQ oxidoreductase FSP1 acts parallel to GPX4 to inhibit ferroptosis. *Nature* **2019**, *575*, 688–692. [CrossRef] [PubMed]
309. Shults, C.W.; Oakes, D.; Kieburtz, K.; Beal, M.F.; Haas, R.; Plumb, S.; Juncos, J.L.; Nutt, J.; Shoulson, I.; Carter, J.; et al. Effects of coenzyme Q10 in early Parkinson disease: Evidence of slowing of the functional decline. *Arch. Neurol.* **2002**, *59*, 1541–1550. [CrossRef]

310. Mandel, S.; Amit, T.; Reznichenko, L.; Weinreb, O.; Youdim, M.B. Green tea catechins as brain-permeable, natural iron chelators-antioxidants for the treatment of neurodegenerative disorders. *Mol. Nutr. Food Res.* **2006**, *50*, 229–234. [CrossRef]
311. Mandel, S.A.; Avramovich-Tirosh, Y.; Reznichenko, L.; Zheng, H.; Weinreb, O.; Amit, T.; Youdim, M.B. Multifunctional activities of green tea catechins in neuroprotection. Modulation of cell survival genes, iron-dependent oxidative stress and PKC signaling pathway. *Neurosignals* **2005**, *14*, 46–60. [CrossRef]
312. Morgan, L.A.; Grundmann, O. Preclinical and Potential Applications of Common Western Herbal Supplements as Complementary Treatment in Parkinson's Disease. *J. Diet. Suppl.* **2017**, *14*, 453–466. [CrossRef]

antioxidants

MDPI

Review

The Interplay between Intracellular Iron Homeostasis and Neuroinflammation in Neurodegenerative Diseases

Jaewang Lee and Dong-Hoon Hyun *

Department of Life Science, Ewha Womans University, Seoul 03760, Republic of Korea
* Correspondence: hyundh@ewha.ac.kr; Tel.: +82-2-3277-6635

Abstract: Iron is essential for life. Many enzymes require iron for appropriate function. However, dysregulation of intracellular iron homeostasis produces excessive reactive oxygen species (ROS) via the Fenton reaction and causes devastating effects on cells, leading to ferroptosis, an iron-dependent cell death. In order to protect against harmful effects, the intracellular system regulates cellular iron levels through iron regulatory mechanisms, including hepcidin–ferroportin, divalent metal transporter 1 (DMT1)–transferrin, and ferritin–nuclear receptor coactivator 4 (NCOA4). During iron deficiency, DMT1–transferrin and ferritin–NCOA4 systems increase intracellular iron levels via endosomes and ferritinophagy, respectively. In contrast, repleting extracellular iron promotes cellular iron absorption through the hepcidin–ferroportin axis. These processes are regulated by the iron-regulatory protein (IRP)/iron-responsive element (IRE) system and nuclear factor erythroid 2-related factor 2 (Nrf2). Meanwhile, excessive ROS also promotes neuroinflammation by activating the nuclear factor kappa-light-chain-enhancer of activated B cells (NF-κB). NF-κB forms inflammasomes, inhibits silent information regulator 2-related enzyme 1 (SIRT1), and induces pro-inflammatory cytokines (IL-6, TNF-α, and IL-1β). Furthermore, 4-hydroxy-2,3-*trans*-nonenal (4-HNE), the end-product of ferroptosis, promotes the inflammatory response by producing amyloid-beta (Aβ) fibrils and neurofibrillary tangles in Alzheimer's disease, and alpha-synuclein aggregation in Parkinson's disease. This interplay shows that intracellular iron homeostasis is vital to maintain inflammatory homeostasis. Here, we review the role of iron homeostasis in inflammation based on recent findings.

Keywords: intracellular iron homeostasis; neuroinflammation; neurodegenerative diseases; Nrf2; NF-κB; ferroptosis; 4-HNE

Citation: Lee, J.; Hyun, D.-H. The Interplay between Intracellular Iron Homeostasis and Neuroinflammation in Neurodegenerative Diseases. *Antioxidants* **2023**, *12*, 918. https://doi.org/10.3390/antiox12040918

Academic Editor: Yan-Zhong Chang

Received: 28 February 2023
Revised: 23 March 2023
Accepted: 7 April 2023
Published: 12 April 2023

1. Introduction

Iron is a mineral nutrient essential for the survival of living organisms. It is a cofactor of many vital enzymes and has a crucial role as a heme component in transferring molecular oxygen to cells. Iron is known as the most abundant transition metal in the brain. However, iron does not exist in the brain at birth [1]. Instead, iron levels are drastically increased during adolescence and then maintained at constant levels [2]. Excessive iron can increase the labile iron pool (LIP), raising the levels of intracellular reactive oxygen species (ROS) [3–6], and iron depletion can promote the dysfunction of iron-dependent enzymes. Disruption of iron regulation is known to be involved in the pathogenesis of various neurodegenerative disorders [7–10]. Most of the total brain iron exists in the glial cells, such as astrocytes, oligodendrocytes, and microglia, rather than in the neurons [11] and is bound to ferritin, an iron storage protein [12]. Consequently, neurons are more vulnerable than glial cells to alterations in the iron balance.

Iron usually exists in two forms in the body: ferrous iron (Fe^{2+}) and ferric iron (Fe^{3+}) [13]. Fe^{3+} binds to transferrin (Tf), a bilobal protein, and forms the diferric Tf (Fe_2Tf) complex, which circulates in the body [14]. In enterocytes, duodenal cytochrome B (DcytB) reduces Fe^{3+} of Fe_2Tf to Fe^{2+}, and divalent metal transporter 1 (DMT1) imports Fe^{2+} into the cells [15]. Transferrin receptor (TfR), one of the iron transporters, can also import Fe^{2+} into cells by forming vesicles, and then iron is stored complexed with ferritin which is composed of ferritin heavy chain 1 (FTH1) and ferritin light chain (FTL) [16–18].

When more cellular iron is required, the cellular iron-regulatory protein (IRP)/iron-responsive element (IRE) system facilitates TfR expression, and stored iron (ferritin) is released through nuclear receptor coactivator 4 (NCOA4) activation [19,20]. By contrast, iron depletion increases iron storage and ferroportin 1 (FPN1) expression to reduce labile iron. This counteraction can exquisitely regulate cellular iron levels. Intracellular iron is trafficked throughout the body and transferred to many enzymes by iron carrier proteins including poly(rC)-RNA-binding protein 1 (PCBP1) or poly(rC)-RNA-binding protein 2 (PCBP2) [21–23]. Iron is an essential cofactor for iron-dependent enzymes that require iron–sulfur clusters for proper function, which transfer an electron to targets [24]. The conjugation of iron with proteins generates cellular energy, promotes DNA synthesis and repair, and transmits oxygen to other cells. However, iron can also impair cellular function due to free radical generation by iron redox cycling [25]. Iron-catalyzed reaction products can induce mutations in the active site of an enzyme, causing carcinogenesis [26]. The Fenton reaction is the main source of free radicals in cells. During the Fenton reaction Fe^{2+} reacts with hydrogen peroxide (H_2O_2), producing Fe^{3+}, hydroxyl radical (•OH), and hydroxyl ion. In turn, •OH is able to initiate lipid peroxidation by abstracting a hydrogen atom from a polyunsaturated fatty acid (PUFA) with bis-allylic hydrogens (–CH=CH-CH_2-CH=CH_2–) in the central methylene group to yield their corresponding hydroperoxides [5]. This process culminates in cell death via ferroptosis, a newly defined iron-dependent cell death [5,27]. Thus, the iron balance must be finely regulated at the cellular level.

Cells have an antioxidant system to protect against nucleophiles. Antioxidants eliminate detrimental ROS by functioning as electrophiles. Interestingly, unlike other organs, neurons do not have enough antioxidant proteins, despite their functional importance in life [28]. For example, nuclear factor erythroid 2-related factor 2 (Nrf2), the master regulator of the antioxidant system, is weakly expressed in neurons [29,30]. Although unclear, this defect may result from the development of the neuronal cell. Maintaining an appropriate level of ROS is critical to avoid axonal degeneration due to a high level of oxidative stress (OS) and axonal growth inhibition induced by a low level of OS [31]. Astrocytes provide antioxidant support to neighboring neurons by releasing glutathione (GSH), a potent antioxidant, into the extracellular space [32,33]. Moreover, Nrf2 can play a role in translational regulation as RNA-binding proteins, such as FTH1 [34]. Meanwhile, Nrf2 can inhibit neuroinflammation by suppressing nuclear factor kappa-light-chain-enhancer of activated B cells (NF-κB) activation via hindering the degradation of nuclear factor of kappa light polypeptide gene enhancer in B-cell inhibitor α (IκB-α). IκB-α can prevent NF-κB translocation to the nucleus.

Neuroinflammation is a cellular defensive response against antigens in the central nervous system (CNS), primarily mediated by microglia, astrocytes, endothelial cells, and pericytes. Neuroinflammation enhances the immune system and increases the penetration of endothelial tissues by immune cells. In addition, neuroinflammation reduces antigens proliferation. Neuroinflammation is a crucial hallmark of neurodegenerative disease [35]. Cellular ROS or extracellular antigens initiate neuroinflammation. NF-κB promotes inflammatory cytokines, such as IL-6, TNF-α, and IL-1β [36], and forms inflammasomes to maintain normal conditions [37]. However, chronic inflammation induces apoptosis and neurodegenerative diseases, accompanied by increased OS. Increased OS can promote mitochondrial dysfunction and disease progress caused by chronic inflammation [38]. OS can cause aggregated forms of proteins, including amyloid-beta (Aβ), neurofibrillary tangles (NFTs), and alpha-synuclein (α-syn). Especially, 4-hydroxy-2,3-*trans*-nonenal (4-HNE), an end-product of lipid peroxidation, is a key molecule to form detrimental proteins during iron-mediated neuroinflammation. The Fenton reaction facilitates lipid peroxidation and forms 4-HNE as a final product [39]. 4-HNE has reactive bonds and can cause conformational change while producing aggregated forms of Aβ fibril, NFT, or α-syn [40,41]. Inflammation responses can be regulated by antioxidant function (e.g., Nrf2 or post-transcriptional modification (e.g., silent information regulator 2-related enzyme 1, SIRT1).

Considering that studies of ferroptosis have newly elucidated iron's role in cell death, the present review aims to describe the relationship between intracellular iron homeostasis and neuroinflammation based on recent studies and findings.

2. Intracellular Iron Homeostasis

The IRP/IRE system regulates intracellular iron homeostasis. IRPs consist of IRP1 and IRP2, possessing RNA-binding capability. IRPs bind to the IRE in the 5′-untranslated region (5′-UTR) or the 3′-untranslated region (3′-UTR) on mRNA and regulate the translation stage [42,43] (Figure 1).

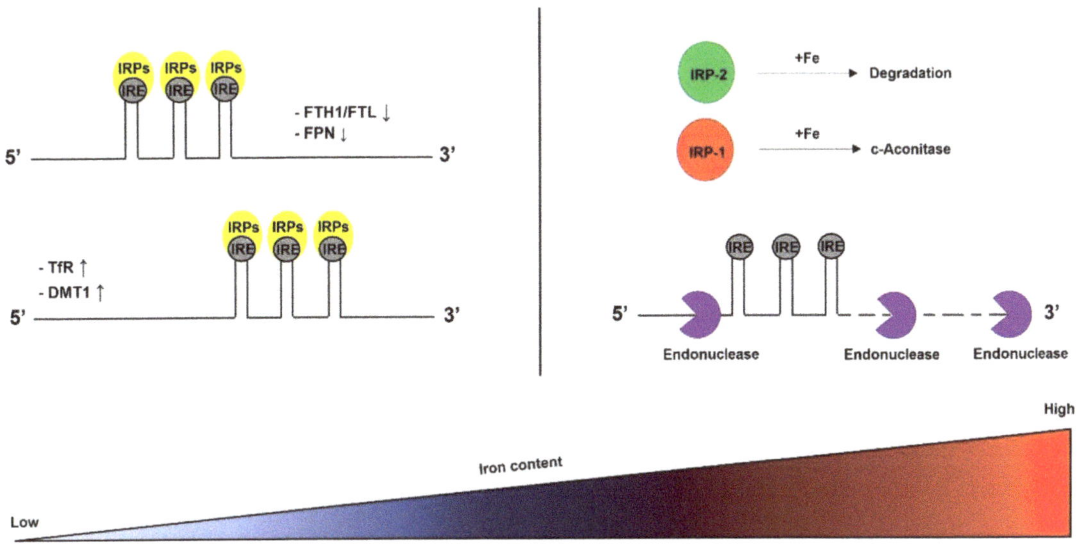

Figure 1. The IRP/IRE system. IRPs consist of two proteins, IRP1 and IRP2. Under iron-rich conditions, iron forms iron–sulfur clusters. Iron–sulfur clusters bind to IRP1. IRP1 acts as c-aconitase. Additionally, iron–sulfur clusters bind to FBXL5 (not described) and mediate IRP2 ubiquitination-dependent degradation. Eventually, inhibition of IRPs leads to the degradation of iron uptake-related mRNAs by the endonuclease. By contrast, under iron shortage conditions, IRPs bind to the IRE within mRNA. This stabilizes the mRNAs or prevents their translation in the nucleus. DMT1, divalent metal transporter 1; FPN1, ferroportin 1; FTH1, ferritin heavy chain; FTL, ferritin light chain; IRE, iron-responsive element; IRP, iron-regulatory protein; IRP1, iron-regulatory protein 1; IRP2, iron-regulatory protein 2; Tfr, transferrin receptor.

During an iron shortage, iron levels are increased by iron influx proteins, such as DMT1, Tf, TfR, and hepcidin. By contrast, iron-efflux-related proteins, such as FPN1, increase under iron-replete conditions. The IRP/IRE system finely regulates these opposed processes. Once iron enters the intracellular space, iron is trafficked by carrier proteins, such as PCBPs, to FTH1/FTL for storage and enzymes for activation. When cellular iron is lacking, FTH1/FTL vesicles release iron to the cytoplasm via NCOA4-mediated ferritinophagy to increase cellular iron contents (Figure 2).

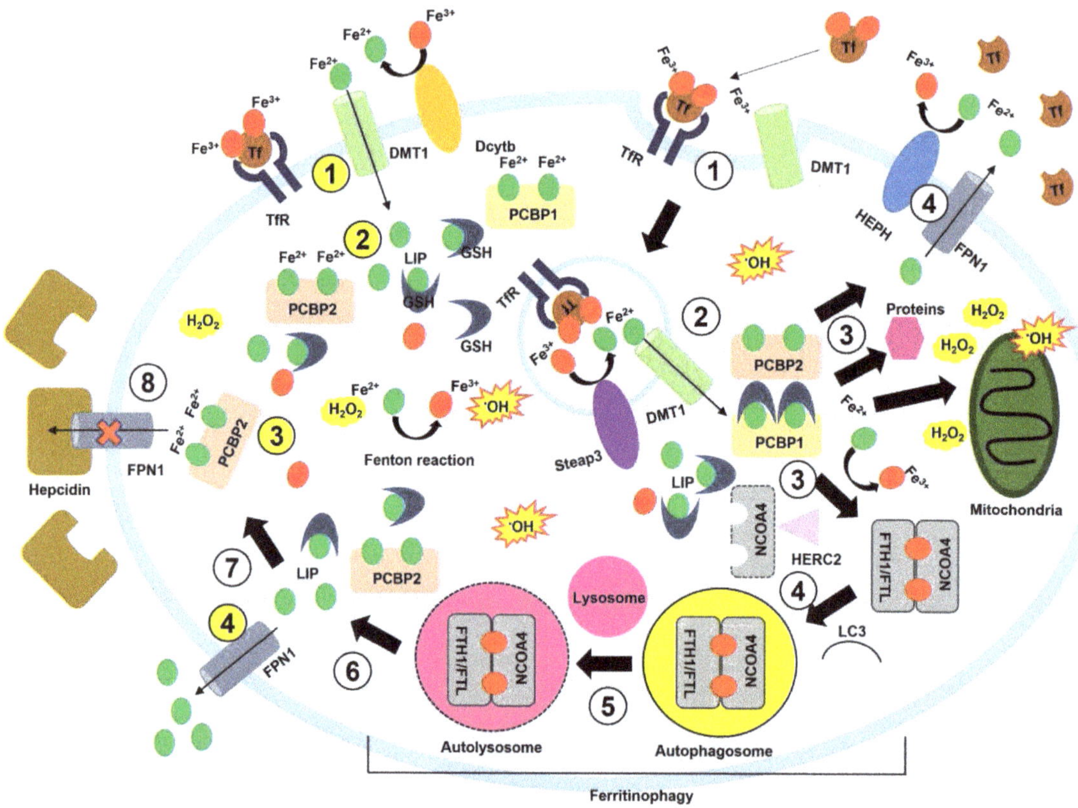

Figure 2. Cellular iron regulation in ferritinophagy. Fe^{3+} is reduced to Fe^{2+} via Dcytb, and Fe^{2+} is then transported into cells via Tf–Tfr or DMT1. Oxidized Fe^{3+} is encapsulated by vesicles called endosomes. Next, Steap3 in the vesicles reduces Fe^{3+} to Fe^{2+} and releases it into the cytoplasm. Fe^{2+} binds to PCBP1 or PCBP2 and is delivered to FTH1, the mitochondria, or FPN1. FTH1 interacts with NCOA4 to store iron. Meanwhile, the interaction between hepcidin and FPN1 blocks the leakage of intracellular iron. When iron is deficient, the FTH1–NCOA4 complex releases iron through ferritinophagy. When iron is repleted, FPN1 exports iron into the extracellular space. In the extracellular space, Fe^{2+} is oxidized to Fe^{3+} by HEPH. Intracellular iron responds to H_2O_2 and produces $^{\bullet}OH$. ROS damages organelles. A white circle with numbers means iron movement by endocytosis. A yellow circle with numbers shows iron movement through a channel. DMT1, divalent metal transporter 1; DcytB, duodenal cytochrome B; Fe^{2+}, ferrous iron; Fe^{3+}, ferric iron; FTH1, ferritin heavy chain; FTL, ferritin light chain; FPN1, ferroportin 1; GSH, glutathione; HEPH, hephaestin; HERC2, HECT domain and RCC1-like domain 2; $^{\bullet}OH$, hydroxyl radical; H_2O_2, hydrogen peroxide; LC3, microtubule-associated protein 1A/1B-light chain 3; LIP, labile iron pool; NCOA4, nuclear receptor coactivator 4; PCBP1, poly(rC)-binding protein 1; PCBP2, poly(rC)-binding protein 2; ROS, reactive oxygen species; Steap3, six-transmembrane epithelial antigen of prostate family member 3; Tf, transferrin; TfR, transferrin receptor.

A recent study showed that PCBP1 knockdown could promote ferritiniophagy and lipid peroxidation via binding to the 3′-UTR on beclin 1 (BECN1) mRNA and arachidonate 15-lipoxygenase (ALOX15) mRNA [44]. Although the process of intracellular iron homeostasis and related molecules are known, and new functions of the molecules have been discovered, more studies are needed about the interplay between iron redox homeostasis and neuroinflammation. Thus, this section describes the interaction between iron-related molecules and inflammation.

2.1. *Hepcidin*

Hepcidin is a peptide hormone produced by the liver in response to increased iron levels and inflammation. Hepcidin is involved in iron homeostasis, absorbing dietary iron, releasing recycled hemoglobin iron from macrophages, and transferring stored iron from hepatic cells [45,46]. Inflammation induces hepcidin release and reduces blood iron (i.e., hypoferremia). This increases host resistance to microbial infection and results in anemia. Hepcidin controls cellular iron efflux by interacting with FPN1. The hepcidin–FPN1 response promotes iron uptake [47] (Figure 2). The transcription of hepcidin is mainly regulated by the bone morphogenetic protein (BMP)/suppressor of mothers against the decapentaplegic (SMAD) pathway [48]. A high iron level stimulates BMP6 expression and leads to hepcidin expression by binding to a BMP-responsive element on the hepcidin gene promoter. An increase in hepcidin hinders iron efflux from the cell. Hepcidin levels are closely linked to IL-6 levels. IL-6 increases hepcidin and accumulates iron in the intracellular space while promoting the degradation of FPN1 by hepcidin [49,50]. Accumulated iron in the cell increases the Fenton reaction and ultimately produces excessive ROS, causing inflammation and cellular damage [51,52].

2.2. *NCOA4*

NCOA4 is a selective cargo receptor in ferritinophagy. NCOA4 finely regulates cellular iron homeostasis by anticipating the autophagic degradation of ferritin. Under iron-replete cellular conditions, HERC2-mediated ubiquitylation facilitates the turnover of NCOA4. However, under iron-deficient cellular conditions, NCOA4 is stabilized, thereby promoting ferritinophagy, a type of autophagy, by forming an autophagosome and directing it to the lysosome, which, in turn, increases cellular iron levels [20]. Thus, two selective processes occur according to whether NCOA4 binds to iron. In cells with excess iron, the direct binding of cytosolic iron to NCOA4 mediates its interaction with HERC2 and subsequent degradation, and ferritin is not degraded, thus retaining its stored iron. NCOA4-mediated iron homeostasis also facilitates ferroptosis by increasing cellular iron levels via ferritinophagy [19,53] (Figure 2).

2.3. *PCBPs*

PCBPs are multifunctional proteins that regulate gene expression and bind to iron to form delivery complexes [54]. These complexes deliver iron to other molecules requiring iron for activation. PCBP1 and PCBP2 are essential to maintain the LIP in cells. PCBP2 interacts with DMT1 and FPN1 and directly regulates Fe^{2+} trafficking in and out of the cytosol [55] (Figure 2), whereas PCBP1 plays various roles in the regulation of gene expression as a major iron chaperon [22,44,55,56]. A recent study showed that PCBP1 could regulate ferritinophagy via the interaction between BECN1, an autophagy regulator protein, and PCBP1. PCBP inhibited BECN1 translation by binding to the CU-rich elements in the 3′-UTR of BECN1 mRNA. This binding hampered microtubule-associated protein 1A/1B-light chain 3 (LC3) from forming autophagosomes [44]. In addition, inhibiting PCBPs leads to an iron shortage response because PCBPs cannot deliver iron to iron-related proteins using iron as a cofactor. Although the extracellular iron continuously enters cells, BECN1 promotes the formation of autophagosomes to release stored iron due to the absence of iron delivery proteins interacting with LC3 and NCOA4. In the last stage, autophagosomes fuse lysosomes, called autolysosomes, and release iron into the cytoplasm [19]. Increased iron can expedite the Fenton reaction, and increased ROS damages mitochondria. This aggravates an iron famine because mitochondria can induce the iron starvation response [57–59]. Moreover, constitutive deletion of PCBP1 and PCBP2 genes results in early embryonic lethality in mice [60]. Especially, PCBP1 can form a PCBP1–GSH–Fe^{2+} complex and balance the level of cytosolic LIP while delivering Fe^{2+} to an enzyme or ferritin. This process decreases the production of cellular ROS by the Fenton reaction [61,62] and ultimately attenuates lipid peroxidation via NRf2 activation.

2.4. IRP/IRE System

The IRP/IRE system consists of IRP1, IRP2, and IRE. IRP1 and IRP2 are the core molecules responsible for iron homeostasis. IRP1 and IRP2 bind to the specific region of target mRNAs called IREs [42]. Under iron deficiency conditions, IRP1 and IRP2 bind to IREs in the UTRs of the iron homeostasis-related mRNAs: ferritin, FPN1, and TfR. The binding of IRPs to the 5′-UTR of IREs in ferritin and FPN1 blocks translation initiation by interfering with the recruitment of the small ribosomal subunit [43]. In contrast, IRPs work differently with TfR mRNA. IRPs protect TfR mRNA from nucleolytic degradation by binding to its 3′-UTR. These reciprocal effects boost iron uptake and repress iron efflux. Under iron-replete conditions, the lack of interaction between IRPs and IREs increases the synthesis of ferritin and FPN1. However, it does not decrease TfR synthesis because TfR mRNA is degraded by endonuclease [63] (Figure 1). As a result, the iron uptake decreases but the export of iron increases. Meanwhile, activation of the IRP/IRE system can be diminished by ROS. This results in iron deficiency in cells.

2.5. DMT1

DMT1 (SLC11A2) transports Fe^{2+} out of endosomes. Ferrireductases on the cell surface reduce most non-Tf-bound iron and then enter the cytosol by DMT1. Expression of DMT1 is elaborately managed in an iron-dependent manner. DMT1 mRNA has the IRE region in the 3′-UTR, and IRPs bind to IREs under iron deficiency [64]. The binding of IRPs to IREs stabilizes DMT1 mRNA and increases DMTI1 synthesis. There is also the non-IRE containing region on DMT1 mRNA. Alternative splicing determines DMT1 fates, such as DMT1-I with IRE or DMT1-II without the IRE. The DMT1-II isoform is unresponsive to post-transcriptional regulation by intracellular iron concentration because it does not include the IRE [65,66]. Most cells implement the Tf–TfR-mediated process to uptake iron. The Tf–TfR complex forms an endosome with DMT1 and six-transmembrane epithelial antigen of prostate family member 3 (Steap3), acidified to pH 5.5–6.0 via an ATP-dependent proton pump [67]. The Tf–Fe^{3+} complex is released from Tf due to low pH, and then Steap3 reduces Fe^{3+} to Fe^{2+}, transferring Fe^{2+} into the cytosol using DMT1 [68]. This process provides cells with Fe^{2+} associated with iron delivery proteins, such as PCBP1 and PCBP2, in the cytosol (Figure 2). DMT1 contributes to the pathogenesis of Parkinson's disease (PD). Julio et al. suggested that DMT1 expression was increased in PD model mice and patients with PD. In contrast, mutated DMT1 protected rodents from parkinsonism induced by treatment with 1-methyl-4-phenyl-1,2,3,6-tetrahydropyridine (MPTP) and 6-hydroxydopamine [69]. Given that inflammatory cytokines (e.g., TNF-α and IFN-γ) increased DMT1 expression [70], it is reasonable for DMT1 to correlate with inflammation associated with PD development. A study showed that glial cells, activated by inflammatory cytokines, promoted PD progress [71]. Pioglitazone (a peroxisome proliferator-activated receptor alpha [PPAR-α] agonist) effectively attenuated the loss of dopaminergic neurons in substantia nigra in mice by suppressing MPTP-induced microglial activation. Interestingly, caspase inhibitors could not inhibit the degenerative process when dopaminergic neurons were already engaged in apoptosis or autophagic degeneration [72]. Instead, it was efficient for dopaminergic neurons, yet arrived at the final stage [73]. This means that inhibition of DMT1-induced inflammation may impact cell stress during PD, and therapy mainly focuses on the preventive aspect by regulating inflammation.

2.6. Ferritin

Ferritin is the main iron storage protein consisting of 24 subunit shells. It has two distinct subunits with different amino acid sequences, designated as FTH1 and FTL. Ferritin synthesis is regulated at the post-translational level through the IRP/IRE system, α-syn, and amyloid precursor protein (APP) [74]. The efficiency of IRE binding to ferritin mRNA is determined by iron (IRP1) and the redox status (IRP2). When iron levels are high, IRP1 forms an iron–sulfur cluster and activates aconitase. However, IRP1 loses RNA-binding activity [75]. IRP2 does not have an iron–sulfur cluster and is regulated by the ubiquitin-proteosome system (UPS) by an E3 ubiquitin ligase complex [76] (Figure 1). Heme is also known to regulate ferritin synthesis. This occurs via BTB and CNC homology 1 (Bach1) binding and IRP2 [77]. FTH1 has a di-iron ferroxidase center that oxidizes Fe^{2+} to Fe^{3+}, whereas FTL is considered to form the nucleation site in the mineral iron core [16]. The ferritin complex (FTH1 and FTL) can contain a few hundred to five thousand iron atoms [78]. Fe^{2+} is oxidized to Fe^{3+} via the ferroxidase in FTH1, and subsequently, Fe^{3+} moves toward the nucleation site in FTL and is mineralized and stored. This process is important for efficiency because iron mineralization of ferritin (specifically, FTL) can foster iron oxidation and accelerate circulation between Fe^{2+} and Fe^{3+} in the ferritin complex. However, FTL cannot oxidize Fe^{2+} to Fe^{3+} [79,80] (Figure 2). Under iron starvation conditions, the ferritin complex releases stored iron by promoting autophagy (i.e., ferritinophagy) [19,53]. Increased iron levels help to maintain cellular iron levels and activate iron-dependent enzymes, but excessive iron can increase ROS generation through the Fenton reaction and ultimately induce cell death due to failure in redox control (i.e., ferroptosis) [5,81]. During the inflammation process, ferritin synthesis is indirectly promoted by the IL-6–signal transducer and activator of transcription 3 (STAT3) pathway via hepcidin [82,83] (Figure 3).

IL-1β, IL-6, and TNF-α induce ferritin synthesis by increasing hepcidin transcription [83,84]. Increased ferritin synthesis often leads to hyperferritinemia in serum [84]. The role of extracellular ferritin is still unclear, but several theories are suggested: an iron carrier [85–87], to promote angiogenesis [88], to regulate the immune response and inflammatory signaling [82,89–96]. In other words, ferritin helps to decrease stress originating from iron and to maintain a normal immune system during inflammation.

2.7. Ferroportin

FPN1 is the sole iron export protein. When iron is overloaded, FPN1 promotes iron efflux. Fe^{2+} binds to the PCBP2 protein and is then transported to FPN1. This balances cellular iron levels [47,97,98]. The degradation of FPN1 is closely related to hepcidin, as mentioned above [47,49]. A lack of FPN1 increases the amounts of intracellular iron and facilitates the Fenton reaction [99] (Figure 2). ROS generated by the Fenton reaction attack PUFAs and promote lipid peroxidation by producing lipid peroxyl radicals. Eventually, lipid peroxyl radicals lead to ferroptosis. Accordingly, the expression of FPN1 is tightly regulated in cells [100] (Figure 4).

2.8. Neuroinflammation

Neuroinflammation in the CNS depends on specific cell types: microglia, astrocytes, endothelial cells, and pericytes. Additionally, disruption of the blood–brain barrier leads to the inflammatory response via macrophages [101]. Iron accumulation is identified in many neurodegenerative diseases, including Alzheimer's disease (AD), PD, and amyotrophic lateral sclerosis (ALS). In these neurodegenerative diseases, inflammation is promoted in glial cells and neurons, but there is still a lack of understanding of the role of iron in neuroinflammation. Considering the active redox trait of iron, increased iron levels in the intracellular space can have detrimental effects because they can produce •OH through the Fenton reaction and subsequently damage biomolecules, causing cell death [102]. In ferroptosis, •OH induces lipid peroxidation and promotes inflammation by activating cyclooxygenase-2 (COX2) [103,104]. Recently, researchers have been studying the relationship between ferroptosis and neurodegenerative diseases [105,106]. However, a few studies have shown a relationship between iron and neuroinflammation.

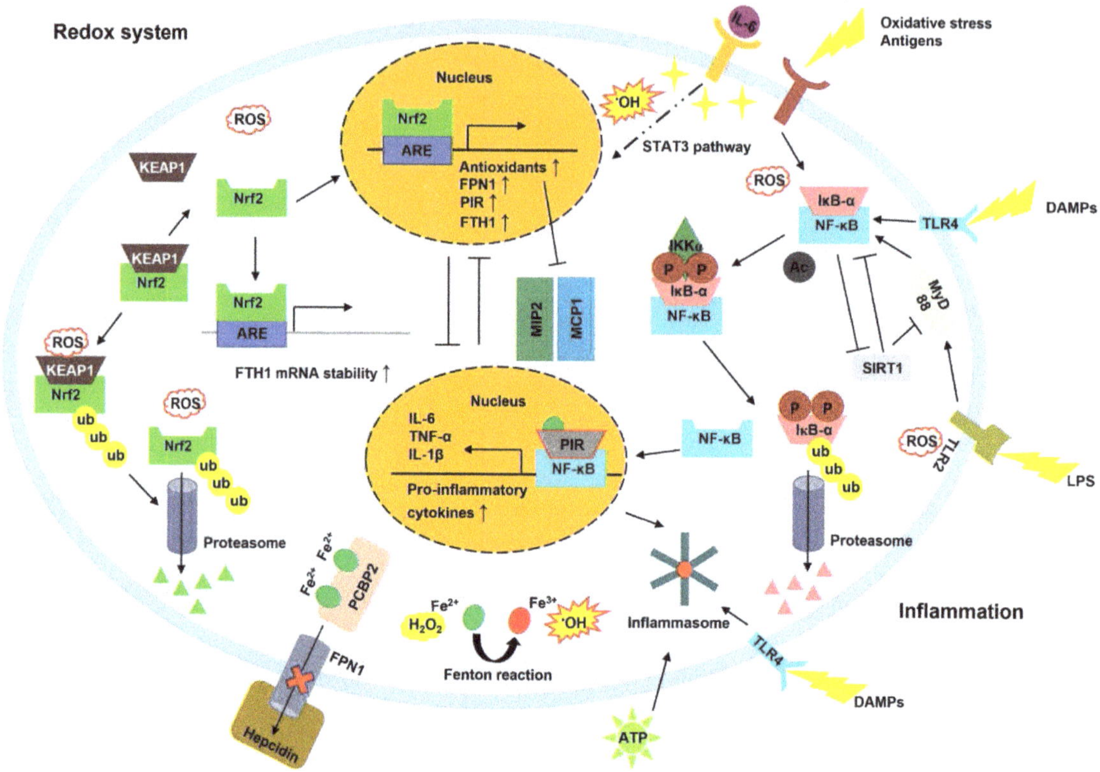

Figure 3. The regulation of cellular redox balance and inflammation. In redox regulation, ROS produced by IL-6 or the Fenton reaction promotes the dissociation of Nrf2 from Keap1 and activates Nrf2. Activated Nrf2 is translocated to the nucleus and initiates the transcription of antioxidant enzymes and proteins requiring iron. This process protects cells from ROS. During inflammation, ROS, DAMPs, or LPS activate NF-κB signal transduction by eliminating IκB-α via ubiquitination. NF-κB moves to the nucleus and induces the transcription of pro-inflammatory cytokines. In this process, inflammasomes are activated, and inflammation is increased. To prevent excessive inflammation, the Nrf2 pathway is activated, which suppresses inflammation-related proteins, such as inflammasomes, MIP2, MCP1, and the NF-κB pathway. Additionally, SIRT1 acts as a regulator and inhibits the activation of NF-κB. NF-κB also regulates the activation of the uncontrolled redox system by inhibiting Nrf2 activation. ARE, antioxidant response element; ATP, adenosine triphosphate; DAMP, damage-associated molecular pattern; FPN1, ferroportin 1; FTH1, ferritin heavy chain; •OH, hydroxyl radical; H_2O_2, hydrogen peroxide; IκB-α, nuclear factor of kappa light polypeptide gene enhancer in B-cells inhibitor, alpha; IKKα, IκB kinase alpha; IL-1β, interleukin-1β; IL-6, interleukin-6; KEAP1, Kelch-like ECH-associated protein 1; LPS, lipopolysaccharide; MCP1, monocyte chemoattractant protein 1; MIP2, macrophage inflammatory protein 2; MyD88, myeloid differentiation primary response protein 88; NF-κB, nuclear factor-kappa B; Nrf2, nuclear factor erythroid 2-related factor 2; P, phosphorylation; PCBP2, poly(rC)-binding protein 2; PIR, pirin; ROS, reactive oxygen species; SIRT1, silent information regulator factor 2-related enzyme 1; STAT3, signal transducer and activator of transcription 3; TLR2, Toll-like receptor 2; TLR4, Toll-like receptor 4; TNF-α, tumor necrosis factor-alpha; Ub, ubiquitin.

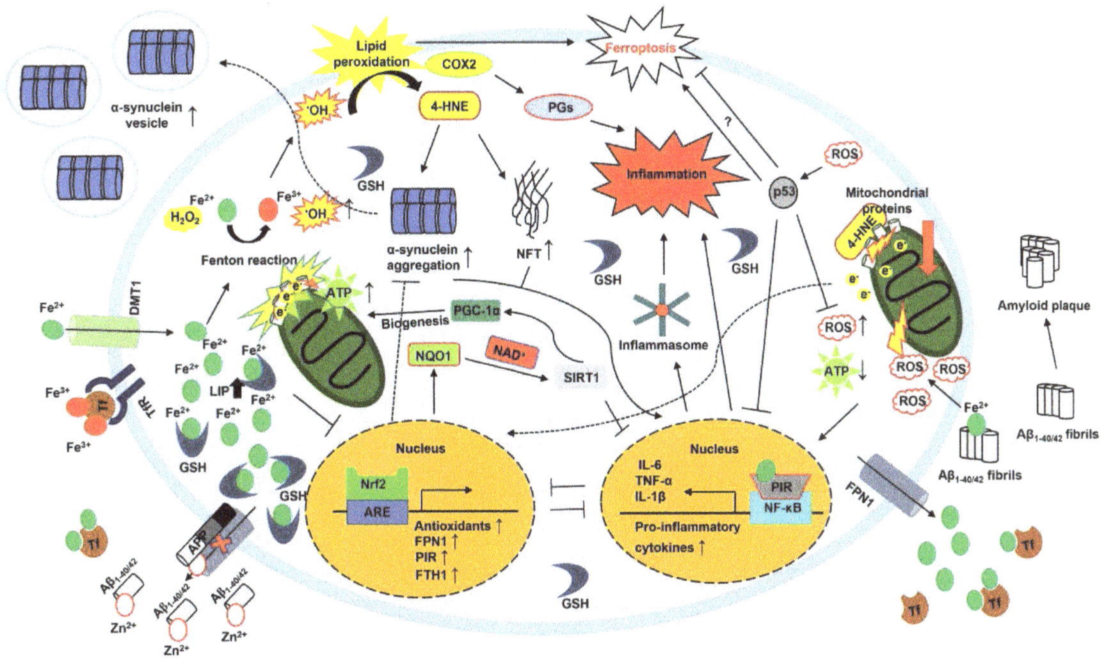

Figure 4. The interplay between iron homeostasis and inflammation in AD and PD. Increased cellular iron accelerates $^{\bullet}OH$ production via the Fenton reaction. Excessive $^{\bullet}OH$ increases lipid peroxidation, producing 4-HNE and activating COX2. 4-HNE promotes α-syn aggregation and continuously generates NFTs. These products are released to the extracellular space with the vesicles or activate the inflammatory response in cells. Additionally, 4-HNE can induce mitochondrial dysfunction by conjugating with mitochondrial proteins, causing electron leakage and enhancing ROS production. Ultimately, this response leads to the activation of the Nrf2-mediated antioxidant response. ROS can stimulate p53. p53 inhibits the NK-κB pathway and reduces ROS. However, there is still controversy about whether p53 prevents ferroptosis. COX2 is a pro-inflammatory enzyme. In addition, iron-binding to $A\beta_{1-40/42}$ fibrils drastically promotes the production of ROS. This damages the mitochondria and boosts ROS production. Ultimately, this cascade leads to the NF-κB-mediated inflammatory response and inflammasome formation. Meanwhile, NQO1 facilitates SIRT1 activation by providing more NAD^{+}. SIRT1 promotes mitochondrial biogenesis by activating PGC1-α. Interaction between $A\beta_{1-40/42}$ and Zn^{2+} increases cellular iron content by blocking FPN. α-syn, alpha-synuclein; Aβ, amyloid β; AD, Alzheimer's disease; APP, amyloid precursor protein; ARE, antioxidant response element; ATP, adenosine triphosphate; COX2, cyclooxygenase 2; DMT1, divalent metal transporter 1; e^{-}, electron; Fe^{2+}, ferrous iron; Fe^{3+}, ferric iron; FPN1, ferroportin 1; FTH1, ferritin heavy chain; 4-HNE, hydroxy-2,3-*trans*-nonenal; H_2O_2, hydrogen peroxide; $^{\bullet}OH$, hydroxyl radical; IL-1β, interleukin-1β; IL-6, interleukin-6; LIP, labile iron pool; NAD^{+}, nicotinamide adenine dinucleotide; NQO1, NAD(P)H quinone dehydrogenase 1; NF-κB, nuclear factor-κB; NFT, neurofibrillary tangle; Nrf2, nuclear factor erythroid 2-related factor 2; p53, tumor protein P53; PD, Parkinson's disease; PGC-1α, peroxisome proliferator-activated receptor gamma coactivator 1-alpha; PIR, pirin; PGs, prostaglandins; ROS, reactive oxygen species; SIRT1, silent information regulator factor 2-related enzyme 1; Tf, transferrin; Tfr, transferrin receptor; TNF-α, tumor necrosis factor-alpha; Zn^{2+}, zinc ion.

2.9. *NF-κB*

NF-κB consists of five transcription factors; NF-κB1 (p105/p50), NF-κB2 (p100/p52), RelA (p65), RelB, and c-Rel. Activated NF-κB participates in the inflammatory response by promoting pro-inflammatory genes. Activation of NF-κB leads to two distinct pathways: canonical and noncanonical. These two distinct pathways have different stimuli. In the canonical pathway, inflammatory stimuli, such as cytokines, antigens, and damage-associated molecular patterns (DAMPs), release p65/p50 dimers from IκBα, phosphorylating IκBα and degrading it through the UPS. Free p65/p50 dimers are translocated to the nucleus, activating the transcription of NF-κB target genes [107]. The trigger is a subset of tumor necrosis factor receptor (TNFR) superfamily members in the noncanonical pathway. They activate NF-κB-inducing kinase (NIK), and NIK phosphorylates IκB kinase alpha (IKKα). Following the phosphorylation cascade, p52/RelB enters the nucleus and promotes the expression of NF-κB target genes. NF-κB signaling is important for immune cell development [108] (Figure 3). Given that Toll-like receptors (TLRs, an inducer of inflammatory response) of microglia are highly expressed in AD [109], it is reasonable for NF-κB to be involved in AD progression. TLRs promote the canonical NF-κB signal transduction, which leads to chronic inflammation in AD due to stimuli, such as cytokines and Aβ plaques [110]. Patients with PD showed increased levels of OS. Immunohistochemical analyses of brain sections with PD showed increased activation of NF-κB, consistent with elevated levels of OS and decreased Nrf2 activation [111]. Interestingly, Fe^{2+} is related to excessive abnormal ROS generation in neuroblastoma. Fe^{2+} inhibits the Nrf2 signal pathway, exacerbates mitochondrial dysfunction, and promotes α-syn aggregation [112] (Figure 4). Recent studies revealed that severe OS could promote α-syn proteostasis [41,113], indicating that OS increased by Fe^{2+}-induced inhibition of Nrf2 may promote neuroinflammation by interfering with the Nrf2 countereffect on NF-κB activation in PD. In contrast, NF-κB is also known to induce FTH1 expression. Increased FTH1 can indirectly inhibit ROS accumulation by sequestrating iron and reducing the Fenton reaction, leading to the attenuation of apoptosis [114]. This process can oppose the detrimental role. The final effect of these two opposing roles may be determined by the antioxidant level.

2.10. SIRT1

Sirtuins are class III (NAD^+-dependent) histone deacetylases. In mammals, the sirtuin family is comprised of seven members, SIRT1–SIRT7 [115,116]. The sirtuins regulate diverse genes through epigenetic modification. This regulation mainly involves genomic stabilization, stress response, apoptosis, metabolism, senescence, proliferation, and inflammation [117–119]. Especially, SIRT1 is well studied because of its various physiological functions. SIRT1 promotes the epithelial–mesenchymal transition (EMT) process in cancer while endowing more aggressive traits to cancer but decreasing the antioxidant system [120–123]. SIRT1 promotes neuronal fortification during neuroinflammation and neurodegenerative diseases [124,125]. Once SIRT1 is activated, for example, by using NAD^+ produced by the enzymatic action of NAD(P)H quinone dehydrogenase 1 (NQO1), it can inhibit NF-κB by deacetylating the p65 subunit of NF-κB and vice versa (Figure 3). Antagonistic crosstalk between SIRT1 and NF-κB is finely regulated to maintain cellular homeostasis [126]. Moreover, several studies showed that SIRT1 weakened neuroinflammation by inhibiting the TLR pathway. Resveratrol, a SIRT1 activator, decreased neuroinflammatory cytokines such as IL1β and TNF-α, and improved spatial reference memory through repression of TLR2–myeloid differentiation primary response protein 88 (MyD88)–NF-κB signal transduction [127] (Figure 3). Recent evidence demonstrated the protective effects of SIRT on inflammation in AD and PD [128,129]. In AD, resveratrol decreases the expression of Aβ, promotes deacetylation of the tau protein, and represses apoptosis [130–135]. Overexpression of SIRT1 in the hippocampus enhanced learning and memory by reducing Aβ and tau in the triple-transgenic (3xTg) AD mouse model [136]. However, considering that resveratrol is not a SIRT1-specific activator, further study is needed to show the effects of SIRT1 on AD using a SIRT1-specific activator, SRT1720 [120]. In PD, resveratrol decreases apoptosis by inhibiting NF-κB and degrading α-syn via deacetylation of LC3 [137–140]. Additionally, a recent study showed that SIRT1 could promote mitochondrial biogenesis by activating peroxisome proliferator-activated receptor-gamma coactivator 1 (PGC-1) [141] (Figure 4).

2.11. Inflammasome

Inflammasomes are cytosolic molecular complexes that promote inflammatory responses to activate immune defenses. Inflammasomes are classified as nucleotide-binding oligomerization-like receptor (NLR) domain and leucine-rich repeat and pyrin domain-containing protein 1 (NLRP1), NLRP3, NLR family CARD domain-containing 4 (NLRC4), AIM2, and pyrin inflammasomes [142]. Inflammasomes consist of the NLR protein or AIM2-like receptor, apoptosis-associated speck-like protein containing a CARD (ASC), and pro-caspase-1. The NLR protein can sense an intracellular signal that promotes the formation of inflammasomes. Once inflammasomes are formed, activated caspase-1 mediates the catalytic cleavage and release of the pro-forms of pro-inflammatory cytokines, such as IL-1β and IL-18 [143]. In the CNS, inflammasome formation occurs in microglia, neurons, and astrocytes. Especially, NLRP3 inflammasome plays a crucial role in the neuroinflammation response [144]. NLRP3 inflammasome and NLRP3-dependent inflammatory cytokines are found in the periphery plasma of patients with PD [145]. Aggregated α-syn released from neurons can interact with TLRs in microglia, which activates NLRP3 inflammasome in microglia. In turn, NF-κB is translocated to the nucleus, leading to an increase in pro-inflammatory cytokines. Furthermore, pathological α-syn impairs mitochondrial homeostasis, interfering with protein transport via the translocase of the outer membrane (TOM) receptor, such as TOM20, and inhibiting SIRT3 activation in the mitochondria of microglia [146]. Meanwhile, mitochondrial ROS activates nicotinamide adenine dinucleotide phosphate oxidase 2 (NOX2) in microglia, resulting in microglial activation and neurotoxicity [147], ultimately leading to neuroinflammation and neuronal dysfunction [148–150]. However, another study reported that macrophages could regulate the inflammatory response via the NF-κB–p62-mitophagy pathway (a type of autophagy). NF-κB promotes p62 activation, an adaptor that binds polyubiquitinated proteins and helps to form autophagosomes [37]. Mitophagy eliminates damaged mitochondria, restrains NLRP3 activation, and, ultimately, attenuates the inflammatory response [151]. In AD, there are two main inflammasome activation pathways: the MYD88-dependent pathway (signal 1) and the ATP-dependent pathway (signal 2). The MYD88-dependent pathway utilizes DAMPs as a trigger. DAMPs stimulate NF-κB activation via TLRs in microglia (Figure 3). This increases the production of pro-inflammatory cytokines and facilitates the formation of inflammasomes. Activated inflammasomes trim pro-inflammatory cytokines into active forms. IL-1β is intimately linked to the pathogenesis of AD. Among other pro-inflammatory cytokines, IL-1β levels are increased in patients with AD. In signal 2, P2X purinergic receptor 7 (P2X7R), a trimeric ATP-gated cation channel, is a protagonist in forming inflammasomes. A study reported that P2X7R is related to chronic inflammatory neurological disorders [152]. P2X7R was highly expressed in immune cells, such as macrophages, mast cells, microglia, and oligodendrocytes, but to a lesser extent in astrocytes and neurons. In high-ATP conditions, P2X7R was activated, promoting the activation of inflammasomes [153].

2.12. NRF2

Nrf2 is known as a master regulator of cytoprotection against oxidative and xenobiotic stresses [154]. Nrf2 is a ubiquitously expressed redox-sensitive transcription factor with an important role in redox homeostasis and cell inflammation. Nrf2 promotes the expression of antioxidant enzymes and anti-inflammatory molecules [155–157]. Under normal conditions, Nrf2 is maintained at low basal levels in the cytoplasm because of its degradation by the UPS. In a normal state, Kelch-like ECH-associated protein 1 (Keap1), an adaptor protein for a cullin 3 (Cul3)-based ubiquitin E3 ligase, tightly binds to Nrf2, targeting Nrf2 for degradation by the proteasome [158–160]. However, OS and Nrf2-inducing chemicals reduce the E3 ligase activity of the Keap1–Cul3 complex and liberate Nrf2 from the Nrf2–Keap1 complex. This stabilizes Nrf2 against degradation, and Nrf2 is translocated to the nucleus. Continuously, Nrf2 binds to the antioxidant response element (ARE) that has the promoter for transcription of phase II detoxifying antioxidant enzymes. Once Nrf2 binds to the ARE motif, antioxidant enzymes are transcribed, and cellular antioxidant systems are simultaneously activated to protect cells from harmful molecules [161,162] (Figure 3). Activation of antioxidants is intertwined with inflammation. They block inflammatory mediators, including IL-6, TNF-α, monocyte chemoattractant protein 1 (MCP1), and macrophage inflammatory protein 2 (MIP2) [163]. This process is important in the progression of neurodegenerative diseases. A study showed that inflammatory markers, such as inducible nitric oxide synthase (iNOS), TNF-α, and IL-6, were increased in the hippocampus of Nrf2-knockout mice [164]. Despite its anti-inflammatory role, Nrf2 has Janus-like roles. On the one hand, Nrf2 inhibits NLRP3 inflammasome by increasing the expression of NQO1, one of the antioxidant enzymes induced by Nrf2, in macrophages [165,166]. On the other hand, Nrf2 has been shown to activate NLRP3 and AIM2 inflammasomes [167]. However, many studies demonstrated that Nrf2 negatively regulated NF-κB and vice versa. Nrf2 negatively influenced NF-κB-induced inflammation in three aspects: degradation of IKKβ by Keap1 [168], inhibition of OS by activation of Nrf2 induced by the cyclopentenone prostaglandin 15d-PGJ2 [169], and forming a complex with the competitive Nrf2 transcriptional coactivator CREB-binding protein (CBP) [170,171]. The result of three aspects ends in the inactivation of NF-κB. Furthermore, Nrf2-induced heme oxygenase 1 (HO-1) prohibited the translocation of NF-κB to the nucleus [172]. The disease phase affects the Nrf2 response. In the frontal cortex of patients with AD, NQO1 activity was increased during the initial stages of AD but reduced or maintained in the latter stage of AD [173]. This inducible cellular defense system helps cells resist unfavorable environments. In PD, Nrf2 can effectively reduce α-syn aggregation [174], whereas Nrf2 deficiency leads to increased α-syn aggregation, loss of neurons, and enhanced inflammation [175] (Figure 4).

Nrf2 is also closely associated with iron metabolism [176–178]. Nrf2 coordinates iron homeostasis within LIPs. Especially, Nrf2 promotes ferritin expression. Nrf2-deficient mice showed lower basal FTH1 and FTL levels than wild-type mice [179,180]. The regulation mechanism was uncovered by Pietsch et al. They proved that Nrf2 is directly bound to the ARE on FTH1 mRNA [181], suggesting that Nrf2 activation promotes iron storage and reduces labile iron levels by boosting ferritin expression. Meanwhile, Nrf2 is also involved in FPN1 expression. Nrf2 activation may displace Bach1 and inhibit the transcription of HO-1 and FPN1 genes through direct DNA binding [182]. Other studies suggested that Nrf2 activators (e.g., diethyl malate, sulforaphane) could increase FPN1 mRNA in murine macrophages in an iron-independent manner. Interaction between Nrf2 and FPN1 helped macrophages to offset the suppression of FPN1 mRNA expression following lipopolysaccharide (LPS) treatment [183]. Furthermore, Nrf2 increases pirin (PIR) transcription. PIR is known to regulate NF-κB transcriptional signaling and has an enzymatic redox function. Activation of PIR requires iron as a cofactor to form a PIR–iron complex. The PIR–iron complex alters the allosteric capability of NF-κB to bind to DNA [184,185]. Ultimately, the PIR–iron–NF-κB complex increases the NF-κB transcription of target genes (Figure 4). Nrf2 knockdown in HeLa cells reduced PIR expression, whereas Nfr2 overexpression increased the PIR mRNA level by 30% compared to the control [186]. Overall, Nrf2 activation plays a key role in cellular iron homeostasis and helps protect cells from oxidative damage.

3. Conclusions and Perspectives

Iron homeostasis is critical for the functioning of cells and organisms. Impairment of iron homeostasis can have devastating effects on human health. Ferroptosis induced by an imbalanced iron level emphasizes the importance of iron homeostasis. ROS generated by the Fenton reaction stimulate cellular antioxidant systems. However, cell damage occurs when the ROS burden exceeds the capacity of the antioxidant systems. Increased IL-6 in the immune response promotes the interaction between hepcidin and FPN1. This response inhibits the utilization of iron, an essential element of antigens. However, this process accelerates detrimental effects by promoting iron uptake instead of enhancing the immune system in extracellular space. Cellular iron shortage can also facilitate iron uptake through the DMT1–Tf–TfR complex and stimulates ferritinophagy via NCOA4. Increased intracellular iron is transferred to iron-dependent enzymes and inhibits ferritin (FTH1/FTL) turnover through PCBPs. Nevertheless, excessive iron can accelerate the Fenton reaction and lead to excessive ROS generation, boosting inflammation and cellular damage. Cells initiate the transcription of antioxidants using the Nrf2–ARE pathway to hinder severe injury. In this respect, the IRP/IRE system has a crucial role in the relationship between iron homeostasis and inflammation. Activation of Nrf2 inhibits the NF-κB pathway by preventing the degradation of IκB-α. This hinders the translocation of NF-κB to the nucleus and the transcription of pro-inflammatory cytokines. Prolonged activation of NF-κB promotes chronic inflammation and OS. In AD, $A\beta_{1–40/42}$ binds to redox-active metal ions (Cu^{2+}, Zn^{2+}, and Fe^{2+}) to form Aβ oligomers and, ultimately, Aβ fibrils (components of amyloid plaques). In forming Aβ–metal ions complex, OS and APP increase the cellular iron influx. Interestingly, AD progression is related to ferroptosis. In ferroptosis, iron promotes iron-based lipid peroxidation and ultimately produces 4-HNE. Continuously, 4-HNE induces tau protein aggregation, producing NFTs through modifying tau conformation. Moreover, 4-HNE can conjugate with mitochondrial proteins involved in energy production. This conjugation results in a conformational change and increases electron leakage from the electron transport chain, causing ROS generation. Consequently, this decreases ATP production and increases the level of OS due to mitochondrial dysfunction. In addition, COX2 is activated during ferroptosis and promotes inflammation. In the initial stage of PD, 4-HNE promotes α-syn aggregation. Suppression of the Nrf2 pathway by Fe^{2+} may promote OS and α-syn aggregation due to increased OS in PD. Iron-associated ROS production also facilitates inflammasome formation via NF-κB or P2X purinoceptor 7 (P2XR7) activation. Considering the importance of the antioxidant system, NAD(P)H-dependent enzymes may also be involved in regulating iron-induced inflammation. Enzymes requiring NAD(P)H possess antioxidant properties and a role as an energy provider. As an energy provider, a representative enzyme is NQO1. NQO1 increases NAD^+ and activates SIRT1. Activated SIRT1 can inhibit NF-κB via deacetylation of p65. This process may decrease OS and inflammation. In addition, PGC-1 activation by SIRT1 may compensate for the loss of mitochondria by promoting mitochondrial biogenesis. This may offer a practical benefit for patients with mitochondria dysfunction.

The relationship between iron and cell death has been known for over 30 years, but advanced research on the mechanism of iron-dependent cell death has recently been achieved in the cancer field. New findings will help to understand iron and diseases. Thus, the interplay between iron, cell death, and inflammation in neurobiology needs to be re-examined considering recent findings. The imbalance of iron homeostasis and excessive inflammation can cause detrimental effects on cells, highlighting the importance of their regulation. Many studies mainly focus on inflammation or the relationship between iron homeostasis and OS because iron-dependent cell death has actively been studied. Iron homeostasis is intimately associated with inflammation. However, the interaction of each molecule will need further study to understand the exact connection between them. Furthermore, considering that many molecules require energy for activation, further examination of iron homeostasis and inflammation is needed from the viewpoint of energy metabolism. This will improve the understanding of neurodegenerative diseases.

Author Contributions: Conceptualization, J.L. and D.-H.H.; Writing–Original Draft Preparation, J.L. and D.-H.H.; Review & Editing, J.L. and D.-H.H.; Funding Acquisition, D.-H.H. All authors have read and agreed to the published version of the manuscript.

Funding: This research was supported by the National Research Foundation of Korea (NRF) of the Korean government, grant number NRF-2021R1F1A1051212.

Conflicts of Interest: The authors declare no conflict of interest.

Abbreviations

α-syn	alpha-synuclein
Aβ	amyloid-beta
AD	Alzheimer's disease
ALOX15	arachidonate 15-lipoxygenase
ALS	amyotrophic lateral sclerosis
APP	amyloid precursor protein
ARE	antioxidant response element
ASC	apoptosis-associated speck-like protein containing a CARD
ATP	adenosine triphosphate
Bach1	BTB and CNC homology 1
BECN1	beclin 1
BMP	bone morphogenetic protein
CNS	central nervous system
COX2	cyclooxygenase-2
CRB	CREB-binding protein
DAMP	damage-associated molecular pattern
DcytB	duodenal cytochrome B
DMT1	divalent metal transporter 1
e^-	electron
EMT	epithelial–mesenchymal transition
Fe^{2+}	ferrous iron
Fe^{3+}	ferric iron
FPN1	ferroportin 1
FTH1	ferritin heavy chain
FTL	ferritin light chain
GSH	glutathione
4-HNE	4-hydroxy-2,3-trans-nonenal
HEPH	hephaestin
HERC2	HECT domain and RCC1-like domain 2
HMOX-1	heme oxygenase
H_2O_2	hydrogen peroxide
HO-1	heme oxygenase-1
IκB-α	nuclear factor of kappa light polypeptide gene enhancer in B-cells inhibitor alpha
IKKα	IκB kinase alpha
IL	interleukin
iNOS	inducible nitric oxide synthase
IRE	iron-responsive element
IRP	iron-regulatory protein
KEAP1	Kelch-like ECH-associated protein 1
LC3	microtubule-associated protein 1A/1B-light chain 3
LIP	labile iron pool
LPS	lipopolysaccharide
MCP1	monocyte chemoattractant protein 1
MIP2	macrophage inflammatory protein 2
MPTP	1-methyl-4-phenyl-1,2,3,6-tetrahydropyridine
MyD88	myeloid differentiation primary response protein 88
NAD^+	nicotinamide adenine dinucleotide
NCOA4	nuclear receptor coactivator 4
NF-κB	nuclear factor kappa-light-chain-enhancer of activated B cells
NFT	neurofibrillary tangle
NIK	NF-κB-inducing kinase
NLRP	nucleotide-binding domain and leucine-rich repeat and pyrin domain-containing prot
NOX2	nicotinamide adenine dinucleotide phosphate oxidase 2

NQO1	NAD(P)H quinone dehydrogenase 1
Nrf2	nuclear factor erythroid 2-related factor 2
•OH	hydroxyl radical
OS	oxidative stress
P	phosphorylation
P2X7R	P2X purinergic receptor 7
p53	tumor protein P53
PCBP	poly(rC)-binding protein
PD	Parkinson's disease
PGC-1α	peroxisome proliferator-activated receptor gamma coactivator 1-alpha
PIR	pirin
PPAR	peroxisome proliferator-activated receptor
PGs	prostaglandins
PUFA	polyunsaturated fatty acid
ROS	reactive oxygen species
SIRT1	silent information regulator factor 2-related enzyme 1
STAT3	signal transducer and activator of transcription 3
Steap3	six-transmembrane epithelial antigen of prostate family member 3
SMAD	suppressor of mothers against the decapentaplegic
Tf	transferrin
TfR	transferrin receptor
TLR	Toll-like receptor
TNF-α	tumor necrosis factor-α
TNFR	tumor necrosis factor receptor
TOM20	translocase of the outer membrane 20
Ub	ubiquitin
UPS	ubiquitin–proteosome system
UTR	untranslated region
Zn^{2+}	zinc ion

References

1. Drayer, B.; Burger, P.; Darwin, R.; Riederer, S.; Herfkens, R.; Johnson, G.A. MRI of brain iron. *AJR Am. J. Roentgenol.* **1986**, *147*, 103–110. [CrossRef] [PubMed]
2. Bartzokis, G.; Tishler, T.A.; Lu, P.H.; Villablanca, P.; Altshuler, L.L.; Carter, M.; Huang, D.; Edwards, N.; Mintz, J. Brain ferritin iron may influence age- and gender-related risks of neurodegeneration. *Neurobiol. Aging* **2007**, *28*, 414–423. [CrossRef] [PubMed]
3. Cornelissen, A.; Guo, L.; Sakamoto, A.; Virmani, R.; Finn, A.V. New insights into the role of iron in inflammation and atherosclerosis. *EBioMedicine* **2019**, *47*, 598–606. [CrossRef]
4. Kernan, K.F.; Carcillo, J.A. Hyperferritinemia and inflammation. *Int. Immunol.* **2017**, *29*, 401–409. [CrossRef] [PubMed]
5. Dixon, S.J.; Lemberg, K.M.; Lamprecht, M.R.; Skouta, R.; Zaitsev, E.M.; Gleason, C.E.; Patel, D.N.; Bauer, A.J.; Cantley, A.M.; Yang, W.S.; et al. Ferroptosis: An iron-dependent form of nonapoptotic cell death. *Cell* **2012**, *149*, 1060–1072. [CrossRef]
6. Chen, X.; Li, J.; Kang, R.; Klionsky, D.J.; Tang, D. Ferroptosis: Machinery and regulation. *Autophagy* **2021**, *17*, 2054–2081. [CrossRef]
7. Wolozin, B.; Golts, N. Iron and Parkinson's disease. *Neuroscientist* **2002**, *8*, 22–32. [CrossRef]
8. Liu, J.L.; Fan, Y.G.; Yang, Z.S.; Wang, Z.Y.; Guo, C. Iron and Alzheimer's Disease: From Pathogenesis to Therapeutic Implications. *Front. Neurosci.* **2018**, *12*, 632. [CrossRef]
9. Rouault, T.A. Iron metabolism in the CNS: Implications for neurodegenerative diseases. *Nat. Rev. Neurosci.* **2013**, *14*, 551–564. [CrossRef]
10. Meyer, E.; Kurian, M.A.; Hayflick, S.J. Neurodegeneration with Brain Iron Accumulation: Genetic Diversity and Pathophysiological Mechanisms. *Annu. Rev. Genomics Hum. Genet.* **2015**, *16*, 257–279. [CrossRef]
11. Reinert, A.; Morawski, M.; Seeger, J.; Arendt, T.; Reinert, T. Iron concentrations in neurons and glial cells with estimates on ferritin concentrations. *BMC Neurosci.* **2019**, *20*, 25. [CrossRef] [PubMed]
12. Zucca, F.A.; Basso, E.; Cupaioli, F.A.; Ferrari, E.; Sulzer, D.; Casella, L.; Zecca, L. Neuromelanin of the human substantia nigra: An update. *Neurotox. Res.* **2014**, *25*, 13–23. [CrossRef] [PubMed]
13. Daher, R.; Manceau, H.; Karim, Z. Iron metabolism and the role of the iron-regulating hormone hepcidin in health and disease. *Presse. Med.* **2017**, *46*, e272–e278. [CrossRef] [PubMed]
14. Finch, C.A.; Huebers, H. Perspectives in iron metabolism. *N. Engl. J. Med.* **1982**, *306*, 1520–1528. [CrossRef] [PubMed]
15. Vogt, A.S.; Arsiwala, T.; Mohsen, M.; Vogel, M.; Manolova, V.; Bachmann, M.F. On Iron Metabolism and Its Regulation. *Int. J. Mol. Sci.* **2021**, *22*, 4591. [CrossRef]
16. Crichton, R. *Iron Metabolism-From Molecular Mechanisms to Clinical Consequences*; Wiley and Sons, Ltd.: Chichester, UK, 2009.

17. Theil, E.C. Iron, ferritin, and nutrition. *Annu. Rev. Nutr.* **2004**, *24*, 327–343. [CrossRef]
18. Macara, I.G.; Hoy, T.G.; Harrison, P.M. The formation of ferritin from apoferritin. Kinetics and mechanism of iron uptake. *Biochem. J.* **1972**, *126*, 151–162. [CrossRef]
19. Hou, W.; Xie, Y.; Song, X.; Sun, X.; Lotze, M.T.; Zeh, H.J., 3rd; Kang, R.; Tang, D. Autophagy promotes ferroptosis by degradation of ferritin. *Autophagy* **2016**, *12*, 1425–1428. [CrossRef]
20. Mancias, J.D.; Pontano Vaites, L.; Nissim, S.; Biancur, D.E.; Kim, A.J.; Wang, X.; Liu, Y.; Goessling, W.; Kimmelman, A.C.; Harper, J.W. Ferritinophagy via NCOA4 is required for erythropoiesis and is regulated by iron dependent HERC2-mediated proteolysis. *Elife* **2015**, *4*, e10308. [CrossRef]
21. Leidgens, S.; Bullough, K.Z.; Shi, H.; Li, F.; Shakoury-Elizeh, M.; Yabe, T.; Subramanian, P.; Hsu, E.; Natarajan, N.; Nandal, A.; et al. Each member of the poly-r(C)-binding protein 1 (PCBP) family exhibits iron chaperone activity toward ferritin. *J. Biol. Chem.* **2013**, *288*, 17791–17802. [CrossRef]
22. Shi, H.; Bencze, K.Z.; Stemmler, T.L.; Philpott, C.C. A cytosolic iron chaperone that delivers iron to ferritin. *Science* **2008**, *320*, 1207–1210. [CrossRef]
23. Philpott, C.C.; Ryu, M.S.; Frey, A.; Patel, S. Cytosolic iron chaperones: Proteins delivering iron cofactors in the cytosol of mammalian cells. *J. Biol. Chem.* **2017**, *292*, 12764–12771. [CrossRef] [PubMed]
24. Johnson, D.C.; Dean, D.R.; Smith, A.D.; Johnson, M.K. Structure, function, and formation of biological iron-sulfur clusters. *Annu. Rev. Biochem.* **2005**, *74*, 247–281. [CrossRef] [PubMed]
25. Koppenol, W.H.; Hider, R.H. Iron and redox cycling. Do's and don'ts. *Free. Radic. Biol. Med.* **2019**, *133*, 3–10. [CrossRef] [PubMed]
26. Torti, S.V.; Torti, F.M. Iron: The cancer connection. *Mol. Asp. Med.* **2020**, *75*, 100860. [CrossRef]
27. Yang, W.S.; SriRamaratnam, R.; Welsch, M.E.; Shimada, K.; Skouta, R.; Viswanathan, V.S.; Cheah, J.H.; Clemons, P.A.; Shamji, A.F.; Clish, C.B.; et al. Regulation of ferroptotic cancer cell death by GPX4. *Cell* **2014**, *156*, 317–331. [CrossRef] [PubMed]
28. Ahlgren-Beckendorf, J.A.; Reising, A.M.; Schander, M.A.; Herdler, J.W.; Johnson, J.A. Coordinate regulation of NAD(P)H:quinone oxidoreductase and glutathione-S-transferases in primary cultures of rat neurons and glia: Role of the antioxidant/electrophile responsive element. *Glia* **1999**, *25*, 131–142. [CrossRef]
29. Kennedy, K.A.; Sandiford, S.D.; Skerjanc, I.S.; Li, S.S. Reactive oxygen species and the neuronal fate. *Cell. Mol. Life Sci.* **2012**, *69*, 215–221. [CrossRef]
30. Vieira, H.L.; Alves, P.M.; Vercelli, A. Modulation of neuronal stem cell differentiation by hypoxia and reactive oxygen species. *Prog. Neurobiol.* **2011**, *93*, 444–455. [CrossRef]
31. Wilson, C.; Munoz-Palma, E.; Gonzalez-Billault, C. From birth to death: A role for reactive oxygen species in neuronal development. *Semin. Cell Dev. Biol.* **2018**, *80*, 43–49. [CrossRef]
32. Dringen, R.; Pawlowski, P.G.; Hirrlinger, J. Peroxide detoxification by brain cells. *J. Neurosci. Res.* **2005**, *79*, 157–165. [CrossRef] [PubMed]
33. Hirrlinger, J.; Schulz, J.B.; Dringen, R. Glutathione release from cultured brain cells: Multidrug resistance protein 1 mediates the release of GSH from rat astroglial cells. *J. Neurosci. Res.* **2002**, *69*, 318–326. [CrossRef] [PubMed]
34. Sun, X.; Ou, Z.; Chen, R.; Niu, X.; Chen, D.; Kang, R.; Tang, D. Activation of the p62-Keap1-NRF2 pathway protects against ferroptosis in hepatocellular carcinoma cells. *Hepatology* **2016**, *63*, 173–184. [CrossRef] [PubMed]
35. Stephenson, J.; Nutma, E.; van der Valk, P.; Amor, S. Inflammation in CNS neurodegenerative diseases. *Immunology* **2018**, *154*, 204–219. [CrossRef] [PubMed]
36. Karin, M.; Yamamoto, Y.; Wang, Q.M. The IKK NF-kappa B system: A treasure trove for drug development. *Nat. Rev. Drug Discov.* **2004**, *3*, 17–26. [CrossRef]
37. Komatsu, M.; Kageyama, S.; Ichimura, Y. p62/SQSTM1/A170: Physiology and pathology. *Pharmacol. Res.* **2012**, *66*, 457–462. [CrossRef]
38. Di Filippo, M.; Chiasserini, D.; Tozzi, A.; Picconi, B.; Calabresi, P. Mitochondria and the link between neuroinflammation and neurodegeneration. *J. Alzheimers Dis.* **2010**, *20* (Suppl. 2), S369–S379. [CrossRef]
39. Gow, A.J.; Duran, D.; Malcolm, S.; Ischiropoulos, H. Effects of peroxynitrite-induced protein modifications on tyrosine phosphorylation and degradation. *FEBS Lett.* **1996**, *385*, 63–66. [CrossRef]
40. Liu, Q.; Smith, M.A.; Avila, J.; DeBernardis, J.; Kansal, M.; Takeda, A.; Zhu, X.; Nunomura, A.; Honda, K.; Moreira, P.I.; et al. Alzheimer-specific epitopes of tau represent lipid peroxidation-induced conformations. *Free Radic. Biol. Med.* **2005**, *38*, 746–754. [CrossRef]
41. Bae, E.J.; Ho, D.H.; Park, E.; Jung, J.W.; Cho, K.; Hong, J.H.; Lee, H.J.; Kim, K.P.; Lee, S.J. Lipid peroxidation product 4-hydroxy-2-nonenal promotes seeding-capable oligomer formation and cell-to-cell transfer of alpha-synuclein. *Antioxid. Redox Signal.* **2013**, *18*, 770–783. [CrossRef]
42. Kuhn, L.C. Iron regulatory proteins and their role in controlling iron metabolism. *Metallomics* **2015**, *7*, 232–243. [CrossRef] [PubMed]
43. Muckenthaler, M.; Gray, N.K.; Hentze, M.W. IRP-1 binding to ferritin mRNA prevents the recruitment of the small ribosomal subunit by the cap-binding complex eIF4F. *Mol. Cell* **1998**, *2*, 383–388. [CrossRef]
44. Lee, J.; You, J.H.; Roh, J.L. Poly(rC)-binding protein 1 represses ferritinophagy-mediated ferroptosis in head and neck cancer. *Redox Biol.* **2022**, *51*, 102276. [CrossRef] [PubMed]

5. Ganz, T. Hepcidin, a key regulator of iron metabolism and mediator of anemia of inflammation. *Blood* **2003**, *102*, 783–788. [CrossRef] [PubMed]
6. Nicolas, G.; Viatte, L.; Bennoun, M.; Beaumont, C.; Kahn, A.; Vaulont, S. Hepcidin, a new iron regulatory peptide. *Blood Cells Mol. Dis.* **2002**, *29*, 327–335. [CrossRef] [PubMed]
7. Nemeth, E.; Tuttle, M.S.; Powelson, J.; Vaughn, M.B.; Donovan, A.; Ward, D.M.; Ganz, T.; Kaplan, J. Hepcidin regulates cellular iron efflux by binding to ferroportin and inducing its internalization. *Science* **2004**, *306*, 2090–2093. [CrossRef]
8. Hawula, Z.J.; Wallace, D.F.; Subramaniam, V.N.; Rishi, G. Therapeutic Advances in Regulating the Hepcidin/Ferroportin Axis. *Pharmaceuticals* **2019**, *12*, 170. [CrossRef]
9. Ganz, T.; Nemeth, E. Iron sequestration and anemia of inflammation. *Semin. Hematol.* **2009**, *46*, 387–393. [CrossRef]
0. Nemeth, E.; Rivera, S.; Gabayan, V.; Keller, C.; Taudorf, S.; Pedersen, B.K.; Ganz, T. IL-6 mediates hypoferremia of inflammation by inducing the synthesis of the iron regulatory hormone hepcidin. *J. Clin. Investig.* **2004**, *113*, 1271–1276. [CrossRef]
1. Nakamura, T.; Naguro, I.; Ichijo, H. Iron homeostasis and iron-regulated ROS in cell death, senescence and human diseases. *Biochim. Biophys. Acta Gen. Subj.* **2019**, *1863*, 1398–1409. [CrossRef]
2. Wincup, C.; Sawford, N.; Rahman, A. Pathological mechanisms of abnormal iron metabolism and mitochondrial dysfunction in systemic lupus erythematosus. *Expert Rev. Clin. Immunol.* **2021**, *17*, 957–967. [CrossRef]
3. Gao, M.; Monian, P.; Pan, Q.; Zhang, W.; Xiang, J.; Jiang, X. Ferroptosis is an autophagic cell death process. *Cell Res.* **2016**, *26*, 1021–1032. [CrossRef] [PubMed]
4. Yanatori, I.; Richardson, D.R.; Toyokuni, S.; Kishi, F. The new role of poly (rC)-binding proteins as iron transport chaperones: Proteins that could couple with inter-organelle interactions to safely traffic iron. *Biochim. Biophys. Acta Gen. Subj.* **2020**, *1864*, 129685. [CrossRef] [PubMed]
5. Philpott, C.C.; Jadhav, S. The ins and outs of iron: Escorting iron through the mammalian cytosol. *Free Radic. Biol. Med.* **2019**, *133*, 112–117. [CrossRef] [PubMed]
6. Patel, S.J.; Protchenko, O.; Shakoury-Elizeh, M.; Baratz, E.; Jadhav, S.; Philpott, C.C. The iron chaperone and nucleic acid-binding activities of poly(rC)-binding protein 1 are separable and independently essential. *Proc. Natl. Acad. Sci. USA* **2021**, *118*, e2104666118. [CrossRef]
7. Lee, J.; You, J.H.; Shin, D.; Roh, J.L. Inhibition of Glutaredoxin 5 predisposes Cisplatin-resistant Head and Neck Cancer Cells to Ferroptosis. *Theranostics* **2020**, *10*, 7775–7786. [CrossRef]
8. Yuan, H.; Li, X.; Zhang, X.; Kang, R.; Tang, D. CISD1 inhibits ferroptosis by protection against mitochondrial lipid peroxidation. *Biochem. Biophys. Res. Commun.* **2016**, *478*, 838–844. [CrossRef]
9. Kim, E.H.; Shin, D.; Lee, J.; Jung, A.R.; Roh, J.L. CISD2 inhibition overcomes resistance to sulfasalazine-induced ferroptotic cell death in head and neck cancer. *Cancer Lett.* **2018**, *432*, 180–190. [CrossRef]
0. Ghanem, L.R.; Kromer, A.; Silverman, I.M.; Chatterji, P.; Traxler, E.; Penzo-Mendez, A.; Weiss, M.J.; Stanger, B.Z.; Liebhaber, S.A. The Poly(C) Binding Protein Pcbp2 and Its Retrotransposed Derivative Pcbp1 Are Independently Essential to Mouse Development. *Mol. Cell. Biol.* **2016**, *36*, 304–319. [CrossRef]
1. Winterbourn, C.C. Toxicity of iron and hydrogen peroxide: The Fenton reaction. *Toxicol. Lett.* **1995**, *82*, 969–974. [CrossRef]
2. Hoes, M.F.; Grote Beverborg, N.; Kijlstra, J.D.; Kuipers, J.; Swinkels, D.W.; Giepmans, B.N.G.; Rodenburg, R.J.; van Veldhuisen, D.J.; de Boer, R.A.; van der Meer, P. Iron deficiency impairs contractility of human cardiomyocytes through decreased mitochondrial function. *Eur. J. Heart Fail.* **2018**, *20*, 910–919. [CrossRef] [PubMed]
3. Yoshinaga, M.; Nakatsuka, Y.; Vandenbon, A.; Ori, D.; Uehata, T.; Tsujimura, T.; Suzuki, Y.; Mino, T.; Takeuchi, O. Regnase-1 Maintains Iron Homeostasis via the Degradation of Transferrin Receptor 1 and Prolyl-Hydroxylase-Domain-Containing Protein 3 mRNAs. *Cell Rep.* **2017**, *19*, 1614–1630. [CrossRef]
4. Leibold, E.A.; Munro, H.N. Cytoplasmic protein binds in vitro to a highly conserved sequence in the 5′ untranslated region of ferritin heavy- and light-subunit mRNAs. *Proc. Natl. Acad. Sci. USA* **1988**, *85*, 2171–2175. [CrossRef] [PubMed]
5. Tabuchi, M.; Yoshimori, T.; Yamaguchi, K.; Yoshida, T.; Kishi, F. Human NRAMP2/DMT1, which mediates iron transport across endosomal membranes, is localized to late endosomes and lysosomes in HEp-2 cells. *J. Biol. Chem.* **2000**, *275*, 22220–22228. [CrossRef]
6. Lee, P.L.; Gelbart, T.; West, C.; Halloran, C.; Beutler, E. The human Nramp2 gene: Characterization of the gene structure, alternative splicing, promoter region and polymorphisms. *Blood Cells Mol. Dis.* **1998**, *24*, 199–215. [CrossRef]
7. Van Weert, A.W.; Dunn, K.W.; Geuze, H.J.; Maxfield, F.R.; Stoorvogel, W. Transport from late endosomes to lysosomes, but not sorting of integral membrane proteins in endosomes, depends on the vacuolar proton pump. *J. Cell Biol.* **1995**, *130*, 821–834. [CrossRef]
8. Ohgami, R.S.; Campagna, D.R.; Greer, E.L.; Antiochos, B.; McDonald, A.; Chen, J.; Sharp, J.J.; Fujiwara, Y.; Barker, J.E.; Fleming, M.D. Identification of a ferrireductase required for efficient transferrin-dependent iron uptake in erythroid cells. *Nat. Genet.* **2005**, *37*, 1264–1269. [CrossRef] [PubMed]
9. Salazar, J.; Mena, N.; Hunot, S.; Prigent, A.; Alvarez-Fischer, D.; Arredondo, M.; Duyckaerts, C.; Sazdovitch, V.; Zhao, L.; Garrick, L.M.; et al. Divalent metal transporter 1 (DMT1) contributes to neurodegeneration in animal models of Parkinson's disease. *Proc. Natl. Acad. Sci. USA* **2008**, *105*, 18578–18583. [CrossRef]
0. Wang, X.; Garrick, M.D.; Yang, F.; Dailey, L.A.; Piantadosi, C.A.; Ghio, A.J. TNF, IFN-gamma, and endotoxin increase expression of DMT1 in bronchial epithelial cells. *Am. J. Physiol. Lung Cell. Mol. Physiol.* **2005**, *289*, L24–L33. [CrossRef]

71. Hirsch, E.C.; Breidert, T.; Rousselet, E.; Hunot, S.; Hartmann, A.; Michel, P.P. The role of glial reaction and inflammation ir Parkinson's disease. *Ann. N.Y. Acad. Sci.* **2003**, *991*, 214–228. [CrossRef]
72. Hartmann, A.; Troadec, J.D.; Hunot, S.; Kikly, K.; Faucheux, B.A.; Mouatt-Prigent, A.; Ruberg, M.; Agid, Y.; Hirsch, E.C. Caspase- is an effector in apoptotic death of dopaminergic neurons in Parkinson's disease, but pathway inhibition results in neurona necrosis. *J. Neurosci.* **2001**, *21*, 2247–2255. [CrossRef] [PubMed]
73. Forno, L.S. Neuropathology of Parkinson's disease. *J. Neuropathol. Exp. Neurol.* **1996**, *55*, 259–272. [CrossRef] [PubMed]
74. Zhou, Z.D.; Tan, E.K. Iron regulatory protein (IRP)-iron responsive element (IRE) signaling pathway in human neurodegenerativ diseases. *Mol. Neurodegener.* **2017**, *12*, 75. [CrossRef] [PubMed]
75. Muckenthaler, M.U.; Galy, B.; Hentze, M.W. Systemic iron homeostasis and the iron-responsive element/iron-regulatory protei (IRE/IRP) regulatory network. *Annu. Rev. Nutr.* **2008**, *28*, 197–213. [CrossRef] [PubMed]
76. Salahudeen, A.A.; Thompson, J.W.; Ruiz, J.C.; Ma, H.W.; Kinch, L.N.; Li, Q.; Grishin, N.V.; Bruick, R.K. An E3 ligase possessing a iron-responsive hemerythrin domain is a regulator of iron homeostasis. *Science* **2009**, *326*, 722–726. [CrossRef]
77. Hintze, K.J.; Katoh, Y.; Igarashi, K.; Theil, E.C. Bach1 repression of ferritin and thioredoxin reductase1 is heme-sensitive in cell and in vitro and coordinates expression with heme oxygenase1, beta-globin, and NADP(H) quinone (oxido) reductase1. *J. Bio Chem.* **2007**, *282*, 34365–34371. [CrossRef]
78. Jian, N.; Dowle, M.; Horniblow, R.D.; Tselepis, C.; Palmer, R.E. Morphology of the ferritin iron core by aberration correctec scanning transmission electron microscopy. *Nanotechnology* **2016**, *27*, 46LT02. [CrossRef]
79. Mehlenbacher, M.; Poli, M.; Arosio, P.; Santambrogio, P.; Levi, S.; Chasteen, N.D.; Bou-Abdallah, F. Iron Oxidation and Cor Formation in Recombinant Heteropolymeric Human Ferritins. *Biochemistry* **2017**, *56*, 3900–3912. [CrossRef]
80. Bou-Abdallah, F.; Zhao, G.; Biasiotto, G.; Poli, M.; Arosio, P.; Chasteen, N.D. Facilitated diffusion of iron(II) and dioxyger substrates into human H-chain ferritin. A fluorescence and absorbance study employing the ferroxidase center substitution Y34V *J. Am. Chem. Soc.* **2008**, *130*, 17801–17811. [CrossRef]
81. Conrad, M.; Pratt, D.A. The chemical basis of ferroptosis. *Nat. Chem. Biol.* **2019**, *15*, 1137–1147. [CrossRef]
82. Nairz, M.; Weiss, G. Iron in infection and immunity. *Mol. Aspects Med.* **2020**, *75*, 100864. [CrossRef] [PubMed]
83. Weiss, G.; Ganz, T.; Goodnough, L.T. Anemia of inflammation. *Blood* **2019**, *133*, 40–50. [CrossRef] [PubMed]
84. Rosario, C.; Zandman-Goddard, G.; Meyron-Holtz, E.G.; D'Cruz, D.P.; Shoenfeld, Y. The hyperferritinemic syndrome Macrophage activation syndrome, Still's disease, septic shock and catastrophic antiphospholipid syndrome. *BMC Med.* **2013**, *11* 185. [CrossRef] [PubMed]
85. Sibille, J.C.; Kondo, H.; Aisen, P. Interactions between isolated hepatocytes and Kupffer cells in iron metabolism: A possible rol for ferritin as an iron carrier protein. *Hepatology* **1988**, *8*, 296–301. [CrossRef]
86. Leimberg, M.J.; Prus, E.; Konijn, A.M.; Fibach, E. Macrophages function as a ferritin iron source for cultured human erythroi precursors. *J. Cell. Biochem.* **2008**, *103*, 1211–1218. [CrossRef]
87. Li, L.; Fang, C.J.; Ryan, J.C.; Niemi, E.C.; Lebron, J.A.; Bjorkman, P.J.; Arase, H.; Torti, F.M.; Torti, S.V.; Nakamura, M.C.; et a Binding and uptake of H-ferritin are mediated by human transferrin receptor-1. *Proc. Natl. Acad. Sci. USA* **2010**, *107*, 3505–351 [CrossRef]
88. Coffman, L.G.; Parsonage, D.; D'Agostino, R., Jr.; Torti, F.M.; Torti, S.V. Regulatory effects of ferritin on angiogenesis. *Proc. Nat Acad. Sci. USA* **2009**, *106*, 570–575. [CrossRef]
89. Wang, W.; Knovich, M.A.; Coffman, L.G.; Torti, F.M.; Torti, S.V. Serum ferritin: Past, present and future. *Biochim. Biophys. Act* **2010**, *1800*, 760–769. [CrossRef]
90. Ruddell, R.G.; Hoang-Le, D.; Barwood, J.M.; Rutherford, P.S.; Piva, T.J.; Watters, D.J.; Santambrogio, P.; Arosio, P.; Ramm, G.A Ferritin functions as a proinflammatory cytokine via iron-independent protein kinase C zeta/nuclear factor kappaB-regulate signaling in rat hepatic stellate cells. *Hepatology* **2009**, *49*, 887–900. [CrossRef]
91. Matzner, Y.; Hershko, C.; Polliack, A.; Konijn, A.M.; Izak, G. Suppressive effect of ferritin on in vitro lymphocyte function. *Br. Haematol.* **1979**, *42*, 345–353. [CrossRef]
92. Broxmeyer, H.E.; Williams, D.E.; Geissler, K.; Hangoc, G.; Cooper, S.; Bicknell, D.C.; Levi, S.; Arosio, P. Suppressive effects in viv of purified recombinant human H-subunit (acidic) ferritin on murine myelopoiesis. *Blood* **1989**, *73*, 74–79. [CrossRef] [PubMed]
93. Li, R.; Luo, C.; Mines, M.; Zhang, J.; Fan, G.H. Chemokine CXCL12 induces binding of ferritin heavy chain to the chemokin receptor CXCR4, alters CXCR4 signaling, and induces phosphorylation and nuclear translocation of ferritin heavy chain. *J. Bio Chem.* **2006**, *281*, 37616–37627. [CrossRef] [PubMed]
94. Mesquita, G.; Silva, T.; Gomes, A.C.; Oliveira, P.F.; Alves, M.G.; Fernandes, R.; Almeida, A.A.; Moreira, A.C.; Gomes, M.S H-Ferritin is essential for macrophages' capacity to store or detoxify exogenously added iron. *Sci. Rep.* **2020**, *10*, 3061. [CrossRef] [PubMed]
95. Recalcati, S.; Invernizzi, P.; Arosio, P.; Cairo, G. New functions for an iron storage protein: The role of ferritin in immunity and autoimmunity. *J. Autoimmun.* **2008**, *30*, 84–89. [CrossRef] [PubMed]
96. Zandman-Goddard, G.; Shoenfeld, Y. Ferritin in autoimmune diseases. *Autoimmun. Rev.* **2007**, *6*, 457–463. [CrossRef] [PubMed]
97. Liu, X.B.; Yang, F.; Haile, D.J. Functional consequences of ferroportin 1 mutations. *Blood Cells Mol. Dis.* **2005**, *35*, 33–46. [CrossRef] [PubMed]
98. Donovan, A.; Lima, C.A.; Pinkus, J.L.; Pinkus, G.S.; Zon, L.I.; Robine, S.; Andrews, N.C. The iron exporter ferroportin/Slc40a1 is essential for iron homeostasis. *Cell Metab.* **2005**, *1*, 191–200. [CrossRef]

99. Ma, S.; Dielschneider, R.F.; Henson, E.S.; Xiao, W.; Choquette, T.R.; Blankstein, A.R.; Chen, Y.; Gibson, S.B. Ferroptosis and autophagy induced cell death occur independently after siramesine and lapatinib treatment in breast cancer cells. *PLoS ONE* **2017**, *12*, e0182921. [CrossRef]
100. Bogdan, A.R.; Miyazawa, M.; Hashimoto, K.; Tsuji, Y. Regulators of Iron Homeostasis: New Players in Metabolism, Cell Death, and Disease. *Trends Biochem. Sci.* **2016**, *41*, 274–286. [CrossRef]
101. Abe, N.; Nishihara, T.; Yorozuya, T.; Tanaka, J. Microglia and Macrophages in the Pathological Central and Peripheral Nervous Systems. *Cells* **2020**, *9*, 2132. [CrossRef]
102. Ward, R.J.; Zucca, F.A.; Duyn, J.H.; Crichton, R.R.; Zecca, L. The role of iron in brain ageing and neurodegenerative disorders. *Lancet. Neurol.* **2014**, *13*, 1045–1060. [CrossRef] [PubMed]
103. Ayala, A.; Munoz, M.F.; Arguelles, S. Lipid peroxidation: Production, metabolism, and signaling mechanisms of malondialdehyde and 4-hydroxy-2-nonenal. *Oxid. Med. Cell. Longev.* **2014**, *2014*, 360438. [CrossRef] [PubMed]
104. Nassar, A.; Radhakrishnan, A.; Cabrero, I.A.; Cotsonis, G.; Cohen, C. COX-2 expression in invasive breast cancer: Correlation with prognostic parameters and outcome. *Appl. Immunohistochem. Mol. Morphol.* **2007**, *15*, 255–259. [CrossRef] [PubMed]
105. Chen, L.; Hambright, W.S.; Na, R.; Ran, Q. Ablation of the Ferroptosis Inhibitor Glutathione Peroxidase 4 in Neurons Results in Rapid Motor Neuron Degeneration and Paralysis. *J. Biol. Chem.* **2015**, *290*, 28097–28106. [CrossRef] [PubMed]
106. Ryan, S.K.; Zelic, M.; Han, Y.; Teeple, E.; Chen, L.; Sadeghi, M.; Shankara, S.; Guo, L.; Li, C.; Pontarelli, F.; et al. Microglia ferroptosis is regulated by SEC24B and contributes to neurodegeneration. *Nat. Neurosci.* **2023**, *26*, 12–26. [CrossRef]
107. Martin, M.; Sun, M.; Motolani, A.; Lu, T. The Pivotal Player: Components of NF-kappaB Pathway as Promising Biomarkers in Colorectal Cancer. *Int. J. Mol. Sci.* **2021**, *22*, 7429. [CrossRef]
108. Sun, S.C. Non-canonical NF-kappaB signaling pathway. *Cell Res.* **2011**, *21*, 71–85. [CrossRef]
109. Chiarini, A.; Armato, U.; Hu, P.; Dal Pra, I. Danger-Sensing/Patten Recognition Receptors and Neuroinflammation in Alzheimer's Disease. *Int. J. Mol. Sci.* **2020**, *21*, 9036. [CrossRef]
110. Thawkar, B.S.; Kaur, G. Inhibitors of NF-kappaB and P2X7/NLRP3/Caspase 1 pathway in microglia: Novel therapeutic opportunities in neuroinflammation induced early-stage Alzheimer's disease. *J. Neuroimmunol.* **2019**, *326*, 62–74. [CrossRef]
111. Mattson, M.P.; Camandola, S. NF-kappaB in neuronal plasticity and neurodegenerative disorders. *J. Clin. Investig.* **2001**, *107*, 247–254. [CrossRef]
112. Gan, L.; Johnson, J.A. Oxidative damage and the Nrf2-ARE pathway in neurodegenerative diseases. *Biochim. Biophys. Acta* **2014**, *1842*, 1208–1218. [CrossRef] [PubMed]
113. Nasstrom, T.; Fagerqvist, T.; Barbu, M.; Karlsson, M.; Nikolajeff, F.; Kasrayan, A.; Ekberg, M.; Lannfelt, L.; Ingelsson, M.; Bergstrom, J. The lipid peroxidation products 4-oxo-2-nonenal and 4-hydroxy-2-nonenal promote the formation of alpha-synuclein oligomers with distinct biochemical, morphological, and functional properties. *Free Radic. Biol. Med.* **2011**, *50*, 428–437. [CrossRef] [PubMed]
114. Pham, C.G.; Bubici, C.; Zazzeroni, F.; Papa, S.; Jones, J.; Alvarez, K.; Jayawardena, S.; De Smaele, E.; Cong, R.; Beaumont, C.; et al. Ferritin heavy chain upregulation by NF-kappaB inhibits TNFalpha-induced apoptosis by suppressing reactive oxygen species. *Cell* **2004**, *119*, 529–542. [CrossRef]
115. Kitada, M.; Kume, S.; Takeda-Watanabe, A.; Kanasaki, K.; Koya, D. Sirtuins and renal diseases: Relationship with aging and diabetic nephropathy. *Clin. Sci. (Lond.)* **2013**, *124*, 153–164. [CrossRef] [PubMed]
116. Michan, S.; Sinclair, D. Sirtuins in mammals: Insights into their biological function. *Biochem. J.* **2007**, *404*, 1–13. [CrossRef] [PubMed]
117. Finkel, T.; Deng, C.X.; Mostoslavsky, R. Recent progress in the biology and physiology of sirtuins. *Nature* **2009**, *460*, 587–591. [CrossRef]
118. Carafa, V.; Rotili, D.; Forgione, M.; Cuomo, F.; Serretiello, E.; Hailu, G.S.; Jarho, E.; Lahtela-Kakkonen, M.; Mai, A.; Altucci, L. Sirtuin functions and modulation: From chemistry to the clinic. *Clin. Epigenetics* **2016**, *8*, 61. [CrossRef]
119. Mendes, K.L.; Lelis, D.F.; Santos, S.H.S. Nuclear sirtuins and inflammatory signaling pathways. *Cytokine Growth Factor Rev.* **2017**, *38*, 98–105. [CrossRef]
120. Lee, J.; You, J.H.; Kim, M.S.; Roh, J.L. Epigenetic reprogramming of epithelial-mesenchymal transition promotes ferroptosis of head and neck cancer. *Redox Biol.* **2020**, *37*, 101697. [CrossRef]
121. Hao, C.; Zhu, P.X.; Yang, X.; Han, Z.P.; Jiang, J.H.; Zong, C.; Zhang, X.G.; Liu, W.T.; Zhao, Q.D.; Fan, T.T.; et al. Overexpression of SIRT1 promotes metastasis through epithelial-mesenchymal transition in hepatocellular carcinoma. *BMC Cancer* **2014**, *14*, 978. [CrossRef]
122. Byles, V.; Zhu, L.; Lovaas, J.D.; Chmilewski, L.K.; Wang, J.; Faller, D.V.; Dai, Y. SIRT1 induces EMT by cooperating with EMT transcription factors and enhances prostate cancer cell migration and metastasis. *Oncogene* **2012**, *31*, 4619–4629. [CrossRef] [PubMed]
123. Sun, L.; Kokura, K.; Izumi, V.; Koomen, J.M.; Seto, E.; Chen, J.; Fang, J. MPP8 and SIRT1 crosstalk in E-cadherin gene silencing and epithelial-mesenchymal transition. *EMBO Rep.* **2015**, *16*, 689–699. [CrossRef] [PubMed]
124. Zhang, Y.; Anoopkumar-Dukie, S.; Arora, D.; Davey, A.K. Review of the anti-inflammatory effect of SIRT1 and SIRT2 modulators on neurodegenerative diseases. *Eur. J. Pharmacol.* **2020**, *867*, 172847. [CrossRef] [PubMed]
125. Donmez, G. The neurobiology of sirtuins and their role in neurodegeneration. *Trends Pharmacol. Sci.* **2012**, *33*, 494–501. [CrossRef] [PubMed]

126. Kauppinen, A.; Suuronen, T.; Ojala, J.; Kaarniranta, K.; Salminen, A. Antagonistic crosstalk between NF-kappaB and SIRT1 in the regulation of inflammation and metabolic disorders. *Cell. Signal.* **2013**, *25*, 1939–1948. [CrossRef] [PubMed]
127. Qi, B.; Shi, C.; Meng, J.; Xu, S.; Liu, J. Resveratrol alleviates ethanol-induced neuroinflammation in vivo and in vitro: Involvement of TLR2-MyD88-NF-kappaB pathway. *Int. J. Biochem. Cell Biol.* **2018**, *103*, 56–64. [CrossRef]
128. Heppner, F.L.; Ransohoff, R.M.; Becher, B. Immune attack: The role of inflammation in Alzheimer disease. *Nat. Rev. Neurosci.* **2015**, *16*, 358–372. [CrossRef]
129. Kalampokini, S.; Becker, A.; Fassbender, K.; Lyros, E.; Unger, M.M. Nonpharmacological Modulation of Chronic Inflammation in Parkinson's Disease: Role of Diet Interventions. *Parkinsons Dis.* **2019**, *2019*, 7535472. [CrossRef]
130. Zhao, H.F.; Li, N.; Wang, Q.; Cheng, X.J.; Li, X.M.; Liu, T.T. Resveratrol decreases the insoluble Abeta1-42 level in hippocampus and protects the integrity of the blood-brain barrier in AD rats. *Neuroscience* **2015**, *310*, 641–649. [CrossRef]
131. Qi, Y.; Shang, L.; Liao, Z.; Su, H.; Jing, H.; Wu, B.; Bi, K.; Jia, Y. Intracerebroventricular injection of resveratrol ameliorated Abeta-induced learning and cognitive decline in mice. *Metab. Brain Dis.* **2019**, *34*, 257–266. [CrossRef]
132. Feng, X.; Liang, N.; Zhu, D.; Gao, Q.; Peng, L.; Dong, H.; Yue, Q.; Liu, H.; Bao, L.; Zhang, J.; et al. Resveratrol inhibits beta-amyloid-induced neuronal apoptosis through regulation of SIRT1-ROCK1 signaling pathway. *PLoS ONE* **2013**, *8*, e59888.
133. Ai, Z.; Li, C.; Li, L.; He, G. Resveratrol inhibits beta-amyloid-induced neuronal apoptosis via regulation of p53 acetylation in PC12 cells. *Mol. Med. Rep.* **2015**, *11*, 2429–2434. [CrossRef]
134. Zhang, J.; Feng, X.; Wu, J.; Xu, H.; Li, G.; Zhu, D.; Yue, Q.; Liu, H.; Zhang, Y.; Sun, D.; et al. Neuroprotective effects of resveratrol on damages of mouse cortical neurons induced by beta-amyloid through activation of SIRT1/Akt1 pathway. *Biofactors* **2014**, *40*, 258–267. [CrossRef] [PubMed]
135. Li, M.Z.; Zheng, L.J.; Shen, J.; Li, X.Y.; Zhang, Q.; Bai, X.; Wang, Q.S.; Ji, J.G. SIRT1 facilitates amyloid beta peptide degradation by upregulating lysosome number in primary astrocytes. *Neural. Regen. Res.* **2018**, *13*, 2005–2013. [CrossRef] [PubMed]
136. Corpas, R.; Revilla, S.; Ursulet, S.; Castro-Freire, M.; Kaliman, P.; Petegnief, V.; Gimenez-Llort, L.; Sarkis, C.; Pallas, M.; Sanfeliu, C. SIRT1 Overexpression in Mouse Hippocampus Induces Cognitive Enhancement Through Proteostatic and Neurotrophic Mechanisms. *Mol. Neurobiol.* **2017**, *54*, 5604–5619. [CrossRef] [PubMed]
137. Feng, Y.; Liu, T.; Dong, S.Y.; Guo, Y.J.; Jankovic, J.; Xu, H.; Wu, Y.C. Rotenone affects p53 transcriptional activity and apoptosis via targeting SIRT1 and H3K9 acetylation in SH-SY5Y cells. *J. Neurochem.* **2015**, *134*, 668–676. [CrossRef]
138. Wu, Y.; Li, X.; Zhu, J.X.; Xie, W.; Le, W.; Fan, Z.; Jankovic, J.; Pan, T. Resveratrol-activated AMPK/SIRT1/autophagy in cellular models of Parkinson's disease. *Neurosignals* **2011**, *19*, 163–174. [CrossRef]
139. Singh, P.; Hanson, P.S.; Morris, C.M. SIRT1 ameliorates oxidative stress induced neural cell death and is down-regulated in Parkinson's disease. *BMC Neurosci.* **2017**, *18*, 46. [CrossRef]
140. Guo, Y.J.; Dong, S.Y.; Cui, X.X.; Feng, Y.; Liu, T.; Yin, M.; Kuo, S.H.; Tan, E.K.; Zhao, W.J.; Wu, Y.C. Resveratrol alleviates MPTP-induced motor impairments and pathological changes by autophagic degradation of alpha-synuclein via SIRT1-deacetylated LC3. *Mol. Nutr. Food Res.* **2016**, *60*, 2161–2175. [CrossRef]
141. Majeed, Y.; Halabi, N.; Madani, A.Y.; Engelke, R.; Bhagwat, A.M.; Abdesselem, H.; Agha, M.V.; Vakayil, M.; Courjaret, R.; Goswami, N.; et al. SIRT1 promotes lipid metabolism and mitochondrial biogenesis in adipocytes and coordinates adipogenesis by targeting key enzymatic pathways. *Sci. Rep.* **2021**, *11*, 8177. [CrossRef]
142. Kanneganti, T.D. The inflammasome: Firing up innate immunity. *Immunol. Rev.* **2015**, *265*, 1–5. [CrossRef]
143. Rathinam, V.A.; Fitzgerald, K.A. Inflammasome Complexes: Emerging Mechanisms and Effector Functions. *Cell* **2016**, *165*, 792–800. [CrossRef] [PubMed]
144. Albornoz, E.A.; Woodruff, T.M.; Gordon, R. Inflammasomes in CNS Diseases. *Exp. Suppl.* **2018**, *108*, 41–60. [PubMed]
145. Fan, Z.; Pan, Y.T.; Zhang, Z.Y.; Yang, H.; Yu, S.Y.; Zheng, Y.; Ma, J.H.; Wang, X.M. Systemic activation of NLRP3 inflammasome and plasma alpha-synuclein levels are correlated with motor severity and progression in Parkinson's disease. *J. Neuroinflammation* **2020**, *17*, 11. [CrossRef] [PubMed]
146. Di Maio, R.; Barrett, P.J.; Hoffman, E.K.; Barrett, C.W.; Zharikov, A.; Borah, A.; Hu, X.; McCoy, J.; Chu, C.T.; Burton, E.A.; et al. alpha-Synuclein binds to TOM20 and inhibits mitochondrial protein import in Parkinson's disease. *Sci. Transl. Med.* **2016**, *8*, 342ra378. [CrossRef]
147. Hou, L.; Bao, X.; Zang, C.; Yang, H.; Sun, F.; Che, Y.; Wu, X.; Li, S.; Zhang, D.; Wang, Q. Integrin CD11b mediates alpha-synuclein-induced activation of NADPH oxidase through a Rho-dependent pathway. *Redox Biol.* **2018**, *14*, 600–608. [CrossRef]
148. Gustot, A.; Gallea, J.I.; Sarroukh, R.; Celej, M.S.; Ruysschaert, J.M.; Raussens, V. Amyloid fibrils are the molecular trigger of inflammation in Parkinson's disease. *Biochem. J.* **2015**, *471*, 323–333. [CrossRef]
149. Codolo, G.; Plotegher, N.; Pozzobon, T.; Brucale, M.; Tessari, I.; Bubacco, L.; de Bernard, M. Triggering of inflammasome by aggregated alpha-synuclein, an inflammatory response in synucleinopathies. *PLoS ONE* **2013**, *8*, e55375. [CrossRef]
150. Harms, A.S.; Thome, A.D.; Yan, Z.; Schonhoff, A.M.; Williams, G.P.; Li, X.; Liu, Y.; Qin, H.; Benveniste, E.N.; Standaert, D.G. Peripheral monocyte entry is required for alpha-Synuclein induced inflammation and Neurodegeneration in a model of Parkinson disease. *Exp. Neurol.* **2018**, *300*, 179–187. [CrossRef]
151. Zhong, Z.; Umemura, A.; Sanchez-Lopez, E.; Liang, S.; Shalapour, S.; Wong, J.; He, F.; Boassa, D.; Perkins, G.; Ali, S.R.; et al. NF-kappaB Restricts Inflammasome Activation via Elimination of Damaged Mitochondria. *Cell* **2016**, *164*, 896–910. [CrossRef]
152. Adinolfi, E.; Giuliani, A.L.; De Marchi, E.; Pegoraro, A.; Orioli, E.; Di Virgilio, F. The P2X7 receptor: A main player in inflammation. *Biochem. Pharmacol.* **2018**, *151*, 234–244. [CrossRef] [PubMed]

53. Wiley, J.S.; Sluyter, R.; Gu, B.J.; Stokes, L.; Fuller, S.J. The human P2X7 receptor and its role in innate immunity. *Tissue Antigens* **2011**, *78*, 321–332. [CrossRef] [PubMed]
54. Suzuki, T.; Motohashi, H.; Yamamoto, M. Toward clinical application of the Keap1-Nrf2 pathway. *Trends Pharmacol. Sci.* **2013**, *34*, 340–346. [CrossRef]
55. Sandberg, M.; Patil, J.; D'Angelo, B.; Weber, S.G.; Mallard, C. NRF2-regulation in brain health and disease: Implication of cerebral inflammation. *Neuropharmacology* **2014**, *79*, 298–306. [CrossRef]
56. Sajja, R.K.; Green, K.N.; Cucullo, L. Altered Nrf2 signaling mediates hypoglycemia-induced blood-brain barrier endothelial dysfunction in vitro. *PLoS ONE* **2015**, *10*, e0122358. [CrossRef] [PubMed]
57. Dinkova-Kostova, A.T.; Abramov, A.Y. The emerging role of Nrf2 in mitochondrial function. *Free Radic. Biol. Med.* **2015**, *88*, 179–188. [CrossRef]
58. Itoh, K.; Wakabayashi, N.; Katoh, Y.; Ishii, T.; Igarashi, K.; Engel, J.D.; Yamamoto, M. Keap1 represses nuclear activation of antioxidant responsive elements by Nrf2 through binding to the amino-terminal Neh2 domain. *Genes. Dev.* **1999**, *13*, 76–86. [CrossRef]
59. Kobayashi, A.; Kang, M.I.; Okawa, H.; Ohtsuji, M.; Zenke, Y.; Chiba, T.; Igarashi, K.; Yamamoto, M. Oxidative stress sensor Keap1 functions as an adaptor for Cul3-based E3 ligase to regulate proteasomal degradation of Nrf2. *Mol. Cell. Biol.* **2004**, *24*, 7130–7139. [CrossRef]
60. Aseervatham G, S.B.; Choi, S.; Krishnan, J.; Ruckmani, K. Cigarette smoke and related risk factors in neurological disorders: An update. *Biomed. Pharmacother.* **2017**, *85*, 79–86.
61. Vomhof-Dekrey, E.E.; Picklo, M.J., Sr. The Nrf2-antioxidant response element pathway: A target for regulating energy metabolism. *J. Nutr. Biochem.* **2012**, *23*, 1201–1206. [CrossRef]
62. Sivandzade, F.; Prasad, S.; Bhalerao, A.; Cucullo, L. NRF2 and NF-κB interplay in cerebrovascular and neurodegenerative disorders: Molecular mechanisms and possible therapeutic approaches. *Redox Biol.* **2019**, *21*, 101059. [CrossRef] [PubMed]
63. Thimmulappa, R.K.; Scollick, C.; Traore, K.; Yates, M.; Trush, M.A.; Liby, K.T.; Sporn, M.B.; Yamamoto, M.; Kensler, T.W.; Biswal, S. Nrf2-dependent protection from LPS induced inflammatory response and mortality by CDDO-Imidazolide. *Biochem. Biophys. Res. Commun.* **2006**, *351*, 883–889. [CrossRef] [PubMed]
64. Innamorato, N.G.; Rojo, A.I.; Garcia-Yague, A.J.; Yamamoto, M.; de Ceballos, M.L.; Cuadrado, A. The transcription factor Nrf2 is a therapeutic target against brain inflammation. *J. Immunol.* **2008**, *181*, 680–689. [CrossRef] [PubMed]
65. Liu, X.; Zhang, X.; Ding, Y.; Zhou, W.; Tao, L.; Lu, P.; Wang, Y.; Hu, R. Nuclear Factor E2-Related Factor-2 Negatively Regulates NLRP3 Inflammasome Activity by Inhibiting Reactive Oxygen Species-Induced NLRP3 Priming. *Antioxid. Redox Signal.* **2017**, *26*, 28–43. [CrossRef]
66. Asher, G.; Shaul, Y. Ubiquitin-independent degradation: Lessons from the p53 model. *Isr. Med. Assoc. J.* **2006**, *8*, 229–232. [PubMed]
67. Zhao, C.; Gillette, D.D.; Li, X.; Zhang, Z.; Wen, H. Nuclear factor E2-related factor-2 (Nrf2) is required for NLRP3 and AIM2 inflammasome activation. *J. Biol. Chem.* **2014**, *289*, 17020–17029. [CrossRef]
68. Lee, D.F.; Kuo, H.P.; Liu, M.; Chou, C.K.; Xia, W.; Du, Y.; Shen, J.; Chen, C.T.; Huo, L.; Hsu, M.C.; et al. KEAP1 E3 ligase-mediated downregulation of NF-kappaB signaling by targeting IKKbeta. *Mol. Cell* **2009**, *36*, 131–140. [CrossRef]
69. Brigelius-Flohe, R.; Flohe, L. Basic principles and emerging concepts in the redox control of transcription factors. *Antioxid. Redox Signal.* **2011**, *15*, 2335–2381. [CrossRef]
70. Yang, H.; Magilnick, N.; Ou, X.; Lu, S.C. Tumour necrosis factor alpha induces co-ordinated activation of rat GSH synthetic enzymes via nuclear factor kappaB and activator protein-1. *Biochem. J.* **2005**, *391*, 399–408. [CrossRef]
71. Liu, G.H.; Qu, J.; Shen, X. NF-kappaB/p65 antagonizes Nrf2-ARE pathway by depriving CBP from Nrf2 and facilitating recruitment of HDAC3 to MafK. *Biochim. Biophys. Acta* **2008**, *1783*, 713–727. [CrossRef]
72. Bellezza, I.; Tucci, A.; Galli, F.; Grottelli, S.; Mierla, A.L.; Pilolli, F.; Minelli, A. Inhibition of NF-kappaB nuclear translocation via HO-1 activation underlies alpha-tocopheryl succinate toxicity. *J. Nutr. Biochem.* **2012**, *23*, 1583–1591. [CrossRef] [PubMed]
73. Ansari, M.A.; Scheff, S.W. Oxidative stress in the progression of Alzheimer disease in the frontal cortex. *J. Neuropathol. Exp. Neurol.* **2010**, *69*, 155–167. [CrossRef] [PubMed]
74. Lastres-Becker, I.; Garcia-Yague, A.J.; Scannevin, R.H.; Casarejos, M.J.; Kugler, S.; Rabano, A.; Cuadrado, A. Repurposing the NRF2 Activator Dimethyl Fumarate as Therapy Against Synucleinopathy in Parkinson's Disease. *Antioxid. Redox Signal.* **2016**, *25*, 61–77. [CrossRef]
75. Lastres-Becker, I.; Ulusoy, A.; Innamorato, N.G.; Sahin, G.; Rabano, A.; Kirik, D.; Cuadrado, A. alpha-Synuclein expression and Nrf2 deficiency cooperate to aggravate protein aggregation, neuronal death and inflammation in early-stage Parkinson's disease. *Hum. Mol. Genet.* **2012**, *21*, 3173–3192. [CrossRef] [PubMed]
76. Malhotra, D.; Portales-Casamar, E.; Singh, A.; Srivastava, S.; Arenillas, D.; Happel, C.; Shyr, C.; Wakabayashi, N.; Kensler, T.W.; Wasserman, W.W.; et al. Global mapping of binding sites for Nrf2 identifies novel targets in cell survival response through ChIP-Seq profiling and network analysis. *Nucleic Acids Res.* **2010**, *38*, 5718–5734. [CrossRef] [PubMed]
77. Hirotsu, Y.; Katsuoka, F.; Funayama, R.; Nagashima, T.; Nishida, Y.; Nakayama, K.; Engel, J.D.; Yamamoto, M. Nrf2-MafG heterodimers contribute globally to antioxidant and metabolic networks. *Nucleic Acids Res.* **2012**, *40*, 10228–10239. [CrossRef] [PubMed]

178. Chorley, B.N.; Campbell, M.R.; Wang, X.; Karaca, M.; Sambandan, D.; Bangura, F.; Xue, P.; Pi, J.; Kleeberger, S.R.; Bell, D.A. Identification of novel NRF2-regulated genes by ChIP-Seq: Influence on retinoid X receptor alpha. *Nucleic Acids Res.* **2012**, 40, 7416–7429. [CrossRef]
179. Kwak, M.K.; Itoh, K.; Yamamoto, M.; Sutter, T.R.; Kensler, T.W. Role of transcription factor Nrf2 in the induction of hepatic phase 2 and antioxidative enzymes in vivo by the cancer chemoprotective agent, 3H-1, 2-dimethiole-3-thione. *Mol. Med.* **2001**, 7, 135–145. [CrossRef]
180. Thimmulappa, R.K.; Mai, K.H.; Srisuma, S.; Kensler, T.W.; Yamamoto, M.; Biswal, S. Identification of Nrf2-regulated genes induced by the chemopreventive agent sulforaphane by oligonucleotide microarray. *Cancer Res.* **2002**, *62*, 5196–5203.
181. Pietsch, E.C.; Chan, J.Y.; Torti, F.M.; Torti, S.V. Nrf2 mediates the induction of ferritin H in response to xenobiotics and cancer chemopreventive dithiolethiones. *J. Biol. Chem.* **2003**, *278*, 2361–2369. [CrossRef] [PubMed]
182. Ogawa, K.; Sun, J.; Taketani, S.; Nakajima, O.; Nishitani, C.; Sassa, S.; Hayashi, N.; Yamamoto, M.; Shibahara, S.; Fujita, H.; et al. Heme mediates derepression of Maf recognition element through direct binding to transcription repressor Bach1. *EMBO J.* **2001**, *20*, 2835–2843. [CrossRef]
183. Harada, N.; Kanayama, M.; Maruyama, A.; Yoshida, A.; Tazumi, K.; Hosoya, T.; Mimura, J.; Toki, T.; Maher, J.M.; Yamamoto, M.; et al. Nrf2 regulates ferroportin 1-mediated iron efflux and counteracts lipopolysaccharide-induced ferroportin 1 mRNA suppression in macrophages. *Arch. Biochem. Biophys.* **2011**, *508*, 101–109. [CrossRef] [PubMed]
184. Liu, F.; Rehmani, I.; Esaki, S.; Fu, R.; Chen, L.; de Serrano, V.; Liu, A. Pirin is an iron-dependent redox regulator of NF-kappaB. *Proc. Natl. Acad. Sci. USA* **2013**, *110*, 9722–9727. [CrossRef] [PubMed]
185. Pang, H.; Bartlam, M.; Zeng, Q.; Miyatake, H.; Hisano, T.; Miki, K.; Wong, L.L.; Gao, G.F.; Rao, Z. Crystal structure of human pirin: An iron-binding nuclear protein and transcription cofactor. *J. Biol. Chem.* **2004**, *279*, 1491–1498. [CrossRef] [PubMed]
186. Brzoska, K.; Stepkowski, T.M.; Kruszewski, M. Basal PIR expression in HeLa cells is driven by NRF2 via evolutionary conserved antioxidant response element. *Mol. Cell. Biochem.* **2014**, *389*, 99–111. [CrossRef]

antioxidants

Review

Ferroptosis and Its Potential Role in Glioma: From Molecular Mechanisms to Therapeutic Opportunities

Yusong Luo [1,2,3], Guopeng Tian [1,2,3], Xiang Fang [1,2,3], Shengwei Bai [1,2,3], Guoqiang Yuan [1,2,3,*] and Yawen Pan [1,2,3,*]

1 Department of Neurosurgery, Lanzhou University Second Hospital, Lanzhou 730030, China
2 Key Laboratory of Neurology of Gansu Province, Lanzhou 730030, China
3 The Second Clinical Medical School, Lanzhou University, Lanzhou 730030, China
* Correspondence: yuangq08@lzu.edu.cn (G.Y.); pyw@lzu.edu.cn (Y.P.)

Abstract: Glioma is the most common intracranial malignant tumor, and the current main standard treatment option is a combination of tumor surgical resection, chemotherapy and radiotherapy. Due to the terribly poor five-year survival rate of patients with gliomas and the high recurrence rate of gliomas, some new and efficient therapeutic strategies are expected. Recently, ferroptosis, as a new form of cell death, has played a significant role in the treatment of gliomas. Specifically, studies have revealed key processes of ferroptosis, including iron overload in cells, occurrence of lipid peroxidation, inactivation of cysteine/glutathione antiporter system Xc^- (xCT) and glutathione peroxidase 4 (GPX4). In the present review, we summarized the molecular mechanisms of ferroptosis and introduced the application and challenges of ferroptosis in the development and treatment of gliomas. Moreover, we highlighted the therapeutic opportunities of manipulating ferroptosis to improve glioma treatments, which may improve the clinical outcome.

Keywords: glioma; ferroptosis; lipid peroxidation; molecular mechanisms; treatment

Citation: Luo, Y.; Tian, G.; Fang, X.; Bai, S.; Yuan, G.; Pan, Y. Ferroptosis and Its Potential Role in Glioma: From Molecular Mechanisms to Therapeutic Opportunities. *Antioxidants* **2022**, *11*, 2123. https://doi.org/10.3390/antiox11112123

Academic Editors: Yan-Zhong Chang and Stanley Omaye

Received: 13 September 2022
Accepted: 26 October 2022
Published: 28 October 2022

1. Introduction

Glioma is the most common primary malignant tumor of the brain, accounting for approximately 50–60% of the central nervous system (CNS) tumors [1] and approximately 81% of intracranial malignancies [2,3]. Patients with gliomas have significantly higher recurrence rates than those with other tumors of the CNS [4]. Gliomas have been classified by the World Health Organization (WHO) grading system into four grades, where gliomas of grade 1 and grade 2 indicate low-grade gliomas, and gliomas of grade 3 and grade 4 reveal high-grade gliomas [5]. The median overall survival (OS) time of low-grade glioma patients is approximately 11.6 years [6]. However, patients with grade 3 glioma have a median OS time of approximately three years, and the median OS time of grade 4 glioma patients is approximately 15 months [7]. While current clinical treatments for glioma consist of surgical resection, radiotherapy, chemotherapy, novel molecular targeted therapy and immunotherapy [8], these treatments have not brought desirable benefits to patients, and the prognosis of patients remains extremely poor [9,10]. Therefore, there is a great need to develop new therapeutic strategies, including novel therapeutic targets inhibiting glioma cells, to improve OS time and the quality of life for these patients. The common deaths of different cells in the body include necrosis, apoptosis, autophagy and pyroptosis [11]. Recently, ferroptosis, as a new nonapoptotic cell death pattern resulting from iron-dependent lipid peroxidation injury, has attracted more attention [12–14].

Ferroptosis is a new type of programmed cell death triggered by cell membrane damage arising from these processes, including intracellular iron accumulation, production of reactive oxygen species (ROS), lipid peroxidation, failure activity of glutathione peroxidase (GPX) and xCT [15,16]. Cells undergoing ferroptosis not only have changes in cell composition, but also in cell morphology. When cells undergo ferroptosis, although the morphology of the nucleus does not change significantly, the morphology of mitochondria shows increased bilayer membrane density, reduction or disappearance of cristae, and reduced volume [16,17]. Many recent studies have shown that significant progress has been made on the impacts of ferroptosis on glioma. Ferroptosis inducers, as compounds from plants and others, have certain effects on the treatment of glioma by affecting ferroptosis processes.

Herein, we will mainly focused on the essential molecular mechanisms of ferroptosis, as well as the potential impact of ferroptosis on glioma growth and treatment. We also provided an overview of the challenges of ferroptosis in glioma therapy and discussed the therapeutic opportunities of manipulating ferroptosis to improve treatment.

2. Molecular Mechanisms of Ferroptosis

J.M. Gutteridge found in 1984 that iron salts could induce lipid peroxidation by breaking down lipid peroxides into alkoxyl and peroxyl radicals, and iron complexed with ethylene diamine tetraacetic acid (EDTA) could also initiate lipid peroxidation by reacting with hydrogen peroxide (H_2O_2) to form hydroxyl radicals ($^{\bullet}OH$), which may lay the foundation for iron-dependent cell death [18]. In addition, it has been reported that exogenous glutamate could induce cell death by inhibiting cystine uptake through xCT to lead to decreased glutathione production, and a unique programmed cell death pathway called oxytosis, which was dependent on oxidative stress and ROS production and was introduced [19]. This laid the prior groundwork for the discovery and proposal of ferroptosis. Ferroptosis was defined as a new form of programmed cell death by Brent R. Stockwell in 2012 that expresses the process of iron-dependent cell death in cancer cells [13]. Molecular mechanisms of ferroptosis differ from other major forms of regulated cell death (RCD) (Table 1). The main biochemical processes of ferroptosis consist of excess iron and accumulation of ROS in cells, lipid peroxidation, inactivation of xCT and depletion of glutathione and lipid repair enzyme [20–22].

Table 1. The features of different forms of RCD.

	Morphological Features		Biochemical Features	Common Inspection Indicators
Ferroptosis	cell membrane	plasma membrane integrity	Iron accumulation and lipid peroxidation	GSH, GPX4, MDA, SLC7A11, NRF2, ACSL4, FSP1,LPO
	Cell cytoplasm	small mitochondria and increased mitochondrial membrane densities		
	Cell nucleus	no obvious alteration		
Apoptosis	Cell membrane	plasma membrane disruption,	DNA fragmentation	Caspase, Bcl-2, TUNEL,Annexin-V JC-1
	Cell cytoplasm	cell volume reduction		
	Cell nucleus	nuclear volume reduction chromatin agglutination		
Necroptosis	Cell membrane	plasma membrane disruption,	Drop in ATP levels	RIP1, RIP3 Calcein-AM
	Cell cytoplasm	generalized swelling of the cytoplasm and organelles		
	Cell nucleus	oderate chromatin condensation and leakage of cellular constituents		
Autophagy	Cell membrane	no obvious alteration	Increased lysosomal activity	LC3, ATG family proteins(ATG5, ATG7)
	Cell cytoplasm	formation of double-membraned autolysosomes		
	Cell nucleus	no chromatin agglutination		

2.1. Iron Metabolism

Iron is a basic trace element for various cells to carry out various biological functions. Dietary iron comes in many forms but is typically classified as either non-heme or heme iron (Fe^{2+} complexed with protoporphyrin IX). Non-heme dietary iron exists largely as ferric salts, which are reduced back to Fe^{2+} by iron reductase in the intestine. Fe^{2+} enters intestinal epithelial cells (IECs) via the brush-border transporter divalent metal transporter 1 (DMT1) and exits through ferroportin 1 (FPN1) in the basolateral membranes [23,24]. Fe^{2+} is oxidized to Fe^{3+} by ceruloplasmin (CP) and hephaestin (HP), and Fe^{3+} combines with transferrin (Tf) to be transported in the blood [25]. Tf-Fe^{3+} attaches to the transferrin receptor (TfR) on the cell membrane and then internalizes to the cell as endosomes [26,27]. Fe^{3+} is released and reduced to Fe^{2+} by 6-transmembrane epithelial antigen of the prostate 3 (STEAP3) in a variety of cells; then, Fe^{2+} enters the cytoplasm via DMT1 on the endosomal membrane [28,29]. Heme iron is a part of the hemoproteins hemoglobin and myoglobin. However, the molecular mechanism of heme iron absorption is still unclear. There is some evidence that ingested heme iron is decomposed by heme oxygenase in intestinal cells, thus releasing free ferric iron. A large amount of Fe^{2+} accumulates in the cytoplasm to form a labile iron pool, and the metabolic activity of Fe^{2+} has a vital impact on various biological functions, such as ferroptosis [30,31]. Intracellular iron overload, with H_2O_2, triggers the Fenton reaction, inducing the formation of ROS, such as $^{\bullet}OH$, which cause lipid peroxidation to provoke ferroptosis [32–34] (Figure 1). Iron responsive element binding protein 2 (IREB2), as a significant regulator of iron metabolism, may develop sensitivity to ferroptosis [35–37]. Meanwhile, autophagy can regulate the iron pool by affecting the recruitment of ferritin to autophagosomes for lysosomal degradation to release free iron [38–40]. For example, ferritinophagy directly recognizes the ferritin heavy chain 1 (FTH1) by the cargo receptor nuclear receptor coactivator 4 (NCOA4) and then releases iron by transporting the ferritin complex to autophagosomes for lysosomal degradation [41]. Conversely, reduced intracellular Fe^{2+} levels could impede the process of ferroptosis. For instance, erastin-induced ferroptosis was weakened by decreased intracellular Fe^{2+} because of the knockout of autophagy-related 5 (ATG5) or autophagy-related 7 (ATG7) [42]. Thus, the metabolism of iron plays a vital role in ferroptosis.

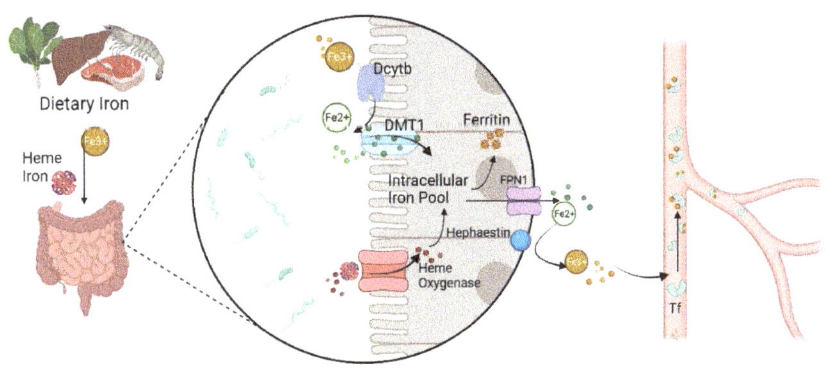

Figure 1. Iron absorption and metabolism in the body. Fe^{2+}, ferrous cation; Fe^{3+}, ferric cation; Dcytb, duodenal cytochrome b; DMT1, divalent metal transporter 1; FPN1, ferroportin 1.

2.2. Lipid Metabolism

Lipid peroxidation, a hallmark feature of ferroptosis, is the ultimate executor of ferroptosis. ROS generated by the Fenton reaction interact with polyunsaturated fatty acids (PUFAs) on cellular or organelle membranes to generate toxic phospholipid hydroperoxides (PLOOHs), thereby inducing ferroptosis [43–45]. Research has shown that some factors, such as acyl–coenzyme A synthetase long-chain family 4 (ACSL4), lysophosphatidylcholine acyltransferase 3 (LPCAT3) and lipoxygenases (LOXs), participate in the production of lipid peroxidation [46–48]. ACSL4 (as a required lipid metabolism enzyme) and LPCAT3 (as a class of key enzymes catalyzing the reacylation of lysophospholipids to phospholipids) activate free long-chain polyunsaturated fatty acids, promote lysophosphatidylcholine (LPC) conversion into lecithin, mediate the synthesis of oxidized cell membrane phospholipids and subsequently regulate ferroptosis development [49,50]. Meanwhile, ACSL4 esterifies arachidonic acid (AA) into acyl-coenzyme A (acyl-CoA) for the biosynthesis of PUFAs which plays a key role in lipid peroxidation and ferroptosis [51]. LOXs, pivotal regulators of ferroptosis, may have a vital effect on the initiation of ferroptosis by promoting lipid autoxidation and predicting ferroptosis sensitivity [47,52]. Therefore, lipid peroxidation in ferroptosis executes cell death by the destruction of the lipid bilayer on cellular or organelle membranes (Figure 2a).

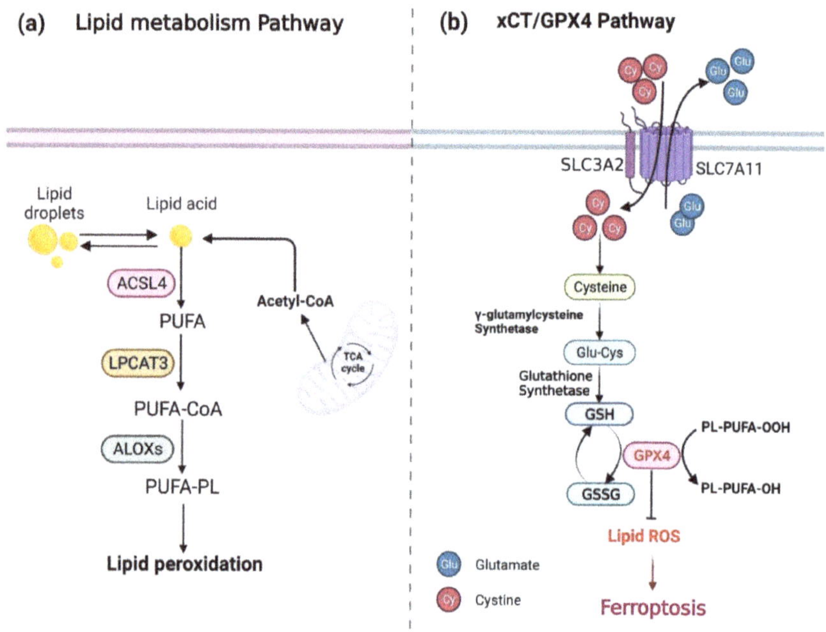

Figure 2. Molecular mechanisms of ferroptosis. (**a**) The lipid metabolism pathway; (**b**) the xCT/GPX4 pathway. ACSL4, long-chain fatty acid CoA ligase 4; PUFA, polyunsaturated fatty acid; LPCAT3, lysophosphatidylcholine acyltransferase 3; PUFA-CoA, polyunsaturated fatty acid coenzyme A; ALOXs, lipoxygenases; PUFA-PL, polyunsaturated fatty acid-containing phospholipid; TCA, tricarboxylic acid cycle; GSH, glutathione; GSSG, glutathione disulfide; GPX4, glutathione peroxidase 4.

2.3. *The xCT and GPX4*

Environmental pressure (such as high temperature and hypoxia) can cause iron reaction, so the cell also needs to establish an appropriate mechanism of defense ferroptosis. The most classic defense way of ferroptosis is the antioxidant axis formed by xCT, glutathione (GSH), and GPX4. The xCT, as a transmembrane protein, consists of light-chain solute carrier family 7 member 11 (SLC7A11) and heavy-chain solute carrier family 3 member 2 (SLC3A2, CD98hc or 4F2hc). SLC7A11, which is a main functional subunit of xCT, aims to regulate extracellular cysteine (Cys) into cells and intracellular glutamic acid (Glu) out of cells, and SLC3A2, as an important subunit, maintains the stability of xCT by anchoring and stabilizing SLC7A11 [53,54]. Then, Cys generates reduced GSH with Glu and glycine (Gly) under the catalysis of glutamate cysteine ligase (GCL) and glutamylcysteine synthetase (GCS) [55,56]. Beclin 1 suppresses xCT activity to promote ferroptosis by adhering to SLC7A11 directly [57]. GPX4, as a key enzyme in ferroptosis reduces PLOOH to nontoxic phospholipid alcohols (PLOHs) in membranes with GSH to prevent ferroptosis [58]. The inhibitor 2-cyano-3,12-dioxooleana-1,9 (11)-dien-28-oic acid (CDDO) prevents the specific degradation of GPX4 via chaperone-mediated autophagy (CMA) by affecting the interaction between heat shock protein 90 (HSP90) and lysosomes, which inhibits the ferroptosis of cells [59]. However, the inhibition of CMA is relieved by inhibiting the mammalian target of rapamycin (MTOR) pathway [60], which may involve the degradation of GPX4 to promote ferroptosis. Therefore, xCT, GPX4 and GSH could be regulated and have an important effect on ferroptosis (Figure 2b).

2.4. FSP1 and DHODH

In addition to the classic GPX4 defense pathway, recent studies have identified ferroptosis suppressor protein 1 (FSP1) and dihydroorotate dehydrogenase (DHODH) independent of the GPX4 signaling pathway. They all involve ubiquinone, a metabolite molecule that exists in both chemically reduced and oxidized states. Ubiquinone (or CoQ) is a lipid that functions on cell membranes and mitochondrial membranes. FSP1 in the cell membrane inhibits ferroptosis by reducing ubiquitin to ubiquinol (CoQH2), which acts as a free radical trapping antioxidant to prevent lipid peroxidation at the cell membrane [61,62]. Similar to the FSP1 system mechanism, DHODH-mediated regulation of panthenol production is an effective system that is specifically designed to alleviate lipid peroxidation in the mitochondria. Mitochondria produce a large amount of ROS in the electron transport chain located in the inner membrane during oxidative phosphorylation. Lipid peroxidation occurs when the mitochondrial antioxidant system is damaged and unable to remove ROS. DHODH is a flavin-dependent enzyme located in the mitochondrial inner membrane and its main function is to catalyze the fourth step of the pyrimidine biosynthesis pathway [63]. The oxidation of dihydroorotate (DHO) to orotate (OA) simultaneously transfers electrons to the ubiquitin in the inner mitochondrial membrane for its reduction to ubiquinol. Gan's team found that when cells were treated with GPX4 inhibitors, such as RSL3, metabolomic analysis revealed a significant decrease in N-aminoformyl-aspartate (C-Asp), increase in Uridine and increase in the synthesis of uridine 5′-monophosphate (UMP) [64]. This suggests a possible relationship between ferroptosis and pyrimidine nucleotide synthesis. By supplementing intermediate metabolites for pyrimidine synthesis, the authors found that dihydroorotic acid inhibited RSL3-induced ferroptosis, whereas orotic acid made cells more sensitive to RSL3. Since DHO and OA are substrates and products of DHODH, respectively, this further confirms that DHODH may be involved in the regulation of ferroptosis. Interestingly, further studies have revealed the use of the DHODH inhibitor Brequinar (BQR) to induce ferroptosis in GPX4 low-expression cells, while the high expression of GPX4, BQR treatment significantly increased the sensitivity of cells to ferroptosis inducers. As is known, most GBM cell lines have higher expression levels of DHODH and GPX4 compared to normal human astrocyte cytoplasm (NHA) [65]. The research also confirms that, in the solid tumor with a high expression of GPX4, the combination of ferroptosis inducer sulfasalazine and DHODH inhibitor can have good therapeutic effect. This finding provides a new strategy for how to target ferroptosis in glioma therapy.

In general, there are at least three types of iron-death defense systems in cells based on different subcellular localization: GPX4 in the cytosol and mitochondria, FSP1 in the cell membrane, and DHODH in mitochondria.

3. Targeting Ferroptosis to Treat Glioma

3.1. Metabolic Pathway

Iron in the brain plays a crucial role in maintaining proper functioning of the central nervous system through its participation in many cellular activities, such as myelination, neurotransmitter synthesis, and energy production. The maintenance of this homeostasis depends on the function of the blood–brain barrier (BBB), which is composed of brain microvascular endothelial cells (BMVEC), astrocytes, microglia and pericytes. It has long been thought that the development of the BBB leads to a reduction in iron absorption in infancy. However, some evidence has revealed that iron levels in the brain increase with ageing [66,67]. Additionally, the mechanism of iron uptake into BMVEC is thought to be primarily via the Tf/TfR1 pathway. Iron was first uptake into BMVEC from blood by TfR1-mediated endocytosis. Then, the action of an H+-ATPase, the membrane of endocytosomes reduced the pH of the endocytosomes, resulting in the dissociation of Fe^{3+} from Tf and their reduction to Fe^{2+}, which will cross the endosomal membrane by a process mediated by DMT1. When the PH rises to 7.4, the non-iron-bound Tf (apo-Tf) and TfR1 return to the luminal membrane where apo-Tf-TfR1 is released into the blood for the next round of iron uptake [68]. Interestingly, after iron was released into the brain interstitial fluid, it was unevenly distributed among different cell types in different regions of the brain [69]. Almost all iron transport-related proteins are expressed in glial cells, but not in neurons.

As is known, iron is a cofactor for many enzymes, including ribonucleic acid reductase (RR), which is an enzyme involved in DNA synthesis. To maintain proliferation, GBM cells need to increase iron uptake, thereby regulating the expression of proteins involved in iron uptake. Recently, studies have reported that there are higher free iron levels in glioma than in other brain tumors, such as meningioma cells and glioblastoma cancer stem cells [70]. Iron-related gene expression in gliomas, such as TfR1 and TfR2, is different from that in other brain tumors and normal human brain tissue [71,72]. TfR levels in glioma sample tissues appear higher than those in meningiomas and other brain tumors in general, which may be correlated with the high levels of iron in gliomas [73]. In addition, the high expression of TfR2 not only promotes glioma cell proliferation, but also contributes to the better sensitivity to temozolomide. The proliferation of glioma cells is attributed to TfR2, which could be localized in lipid rafts and stimulate the ERK1/ERK2 phosphorylation by combining with Tf, but the mechanism of TfR2-induced glioma hypersensitivity to temozolomide remains unclear [74]. Therefore, the effect of TfR2 on glioma is still controversial, and more reports are needed to verify this. Several recent studies have indicated that iron homeostasis and ferroptosis are also affected by iron-sulfur cluster (ISC) proteins. Loss of ISC could induce ferroptosis by initiating iron-starvation responses to lead to iron overload in tumor cells. This mechanism is that ISC synthesis inhibition could activate the iron regulatory protein (IRP); IRP increases TfR levels and reduces FPN1 production by binding to target mRNAs, which promotes intracellular excessive iron accumulation [75]. However, whether ISC has a similar role in glioma requires more evidence.

DMT1 may be related to increased iron levels in glioma cells and is currently a molecular marker in neurodegenerative diseases [76]. In an experimental rat model with C6 glioma cells, propofol inhibits DMT1 expression, tumor cell proliferation and eventually decreased glioma weight [77]. Additionally, this tumor suppressive effect was further found to be associated with a significant reduction in the GSH and ROS. However, a study showed that temozolomide (TMZ) may suppress tumor growth by inducing ferroptosis by targeting DMT1 expression in glioblastoma cells [78]. These results suggest that DMT1 may affect glioma proliferation by regulating ferroptosis and ROS levels and has been investigated as a potential therapeutic target. While the STEAP3 protein plays a vital role in other processes, such as affecting the inflammatory response by regulating the Toll-like receptor 4-mediated macrophage production of chemoattractant protein-5, interferon-beta and interferon-induced protein-10 [79,80], it could also have an essential impact on ferroptosis by reducing Fe^{3+} to Fe^{2+} [81]. The expression of STEAP3 in glioma cells is higher than that in normal brain tissues, which could be regarded as a potential prognostic marker and reduce the overall survival of patients with glioma [82,83]. In addition, STEAP3 not only regulates ferroptosis by enhancing TfR expression and inducing mesenchymal transition, but also has a direct influence on glioma cell proliferation, invasion, and sphere formation in vitro and on glioma growth in vivo [84]. Poly(C)-binding protein 2 (PCBP2), as a significant factor in iron metabolism and posttranscriptional and translational regulation, possibly affects the process of ferroptosis. While PCBP2 is upregulated in glioma tissues and cell lines, the development and proliferation of glioma are suppressed when it is knocked down or when its inhibitor microRNA-214 is applied [85]. The higher levels of ferritin detected in the serum of patients with tumors possibly predict that the prognosis of these patients will deteriorate more, which indicates that iron metabolism plays a necessary role in the progression and therapy of tumors. Ferritins in the serum and cerebrospinal fluid of patients with gliomas, which could come from glioma cells, were higher than those of patients without gliomas [86]. Significant evidence has suggested that nuclear factor erythroid-related factor 2 (NRF2) acts as a key regulator of antioxidant responses, which favors cancer cell growth and leads to increased drug resistance in tumor cells [87–89]. NRF2 mainly targets heme oxygenase-1 (HO-1) to reduce the levels of ROS and degrade prooxidants [90]. Research has shown that neuronal precursor cell-expressed developmentally downregulated 4-1 (NEDD4-1) induces resistance to TMZ treatment in gliomas via activating the AKT/NRF2/HO-1 axis [91]. TMZ also induces ferroptosis by inhibiting the NRF2/HO-1 signaling pathway in gliomas [78]. NRF2/HO-1axis appears to play an important role in glioma therapy. In addition, triptolide and brusatol, as NRF2 inhibitors, suppresses potently IDH1-mutated glioma cells by targeting the NRF2-driven glutathione synthesis pathway to induce lipid peroxidation [92,93]. The upregulated cystathionine-γ-lyase (CSE) in IDH1-mutant astrocytomas promotes cell survival by maintaining GSH to drive antioxidant defense, and whether it is related to the NRF2 needs further verification [94]. Therefore, ferroptosis in gliomas could be associated with the regulation of NRF2.

3.2. The xCT Pathway

Cysteine deprivation is an important inducer of ferroptosis and greatly contributes to the ferroptosis in GBM [95,96]. The study conducted by Takeuchi et al., including 40 patients with gliomas, concluded that high levels of xCT could predict a short progression-free survival and a low overall survival [97]. Specifically, the high levels of xCT possibly promote glioma cells to grow and survive by enhancing mitochondrial biogenesis and adenosine tri-phosphate (ATP) generation, as well as by reducing the accumulation of ROS [98]. These findings suggest that we can inhibit system Xc-induces ferroptosis in glioma. Radiation, chemotherapy (such as TMZ) and immunotherapy could lead to the activation of ferroptosis by downregulating the expression of xCT to induce the death of glioma cells [99–101]. Some widely used clinical drugs have also been applied to the treatment of gliomas by managing xCT and sequentially regulating ferroptosis. Gao et al. found that ibuprofen could enhance ferroptosis by depleting the expression of xCT and GPX4 to inhibit the growth of glioma cells [102]. Another study found that sulfasalazine could stimulate ferroptosis by inhibiting the activity of xCT and sequentially decreasing the formation of GSH strengthened the effect of radiation therapy to increase the overall survival of mice [103]. In addition, tumor suppressor P53 is a frequently mutated gene in various cancers, including glioma. P53 suppresses glioma growth by the induction of ferroptosis [104]. Notably, P53 possibly inhibits the activity of xCT by directly depleting the level of SLC7A11, thus promoting the ability of ferroptosis to suppress the growth of glioma cells [105]. However, an interesting phenomenon is that, in glucose deprivation environments, the treatment of epidermal growth factor will upregulate xCT in glioma cell lines, leading to tumor death [106–109].

In summary, xCT could play a dual role in the development and treatment of gliomas, so further studies are needed to express the practical effect of xCT in gliomas.

3.3. GPX4 Expression

Recently, a study revealed that when GPX4 is knocked down or reacts with its inhibitors, ferroptosis is activated to induce the death of glioma cells by accumulating lipid peroxides to damage the cell membrane and organelle membrane [110]. Some ferroptosis inducers, such as plumbagin, triggers ferroptosis by inducing GPX4 degradation via the lysosome pathway and inhibiting glioma growth [111]. In addition, it has been demonstrated that certain traditional Chinese herbs induce ferroptosis in glioma. For instance, capsaicin, as a potential anticancer ferroptosis inducer, suppresses the proliferative effects of glioma cells by increasing ACSL4 levels and decreasing GPX4 levels to induce ferroptosis [112]. Dihydrotanshinone I (DHI), which boosted ferroptosis by decreasing the expression level of GPX4 and increasing that of ACSL4, inhibited the growth and proliferation of glioma cells [113]. The curcumin analog ALZ003 inhibited the development and growth of glioma cells and enhanced their sensitivity to temozolomide treatment by promoting the androgen receptor (AR) ubiquitination and downregulating GPX4 to highlight ferroptosis in vitro which improved the survival of experimental rodents in vivo [114]. Additionally, the natural compound artesunate (ART), as an antimalarial drug, was previously demonstrated and ART also exhibited an anti-tumor effect and was a specific inducer of ferroptosis in a number of different types of cancer, including glioma. It inhibits the proliferation of glioma cells in vitro and in vivo by promoting GSH depletion and low GPX4 expression to increase ferroptosis [115]. In another study, dihydroartemisinin (DHA), as an inhibitor could have the prospect of treating glioma because of its role in promoting apoptosis and autophagy and reducing the invasion ability of glioma cells [116,117]. Specifically, DHA could promote the development of ACSL4 and xCT, but significantly downregulated GPX4 levels, initiating the death of glioma cells by maintaining ferroptosis [118]. Beyond the above stated aspects, there are many new nanomaterials involved. It was reported that the biomimetic nanoparticles (PIOC@CM NPs) increased the level of ROS, depleted GSH upon ultrasonic irradiation and attenuated the activity of GPX4 to kill glioma C6 cells by activating ferroptosis [119]. Iron oxide nanoparticles loaded with paclitaxel (IONP@PTX) not only inhibited the migration and invasion of glioma cells by enhancing ions, ROS and lipid peroxidation, but also promoted the autophagy-dependent ferroptosis pathway by decreasing the levels of GPX4 in vitro [120]. Above all, these findings provide some new drug treatment options for glioma and demonstrate that GPX4 degradation promotes ferroptosis in glioma.

3.4. *Tumor Immune Microenvironment*

As key regulatory components, immune cells and immune-related molecules have been shown to play pivotal roles in the development and treatment of glioma cells. Accumulating evidence indicates that ferroptosis not only promotes tumor cell death, but also affects the tumor immune microenvironment (TIME) [121,122] (Figure 3). As key regulatory components, immune cells and immune-related molecules have been shown to play pivotal roles in the development and treatment of glioma cells. The main tumor-associated macrophages (TAMs) are the M1-polarized subtype with proinflammatory and antitumoral functions and the M2-polarized subtype with anti-inflammatory and protumoral effects [123–125]. The M2 subtype plays a vital role in the TIME by excreting extracellular matrix (ECM) components, promoting T-cell anergy and stimulating angiogenesis [126,127]. In addition, TAMs may increase the tumorigenicity and chemoresistance of glioma cells by revising the stromal and blood vessel architecture [128]. Myeloid-derived suppressor cells (MDSCs), as myeloid-derived progenitor cells that accumulate in the TIME of glioma suppress the proliferation and activity of T cells by releasing inducible nitric oxide synthase ROS, cyclooxygenase-2 and transforming growth factor-β [129,130]. The regulatory T (Treg) cells accumulating in the TIME of glioma could inhibit immune surveillance and attack by excreting IL-10, IL-35 and TGF-β, which possibly predicts the poor prognosis of patients with gliomas [131–133]. Neutrophils are able to not only inhibit the antiangiogenic therapy of other tumors, but also predict the poor prognosis of patients with gliomas [134].

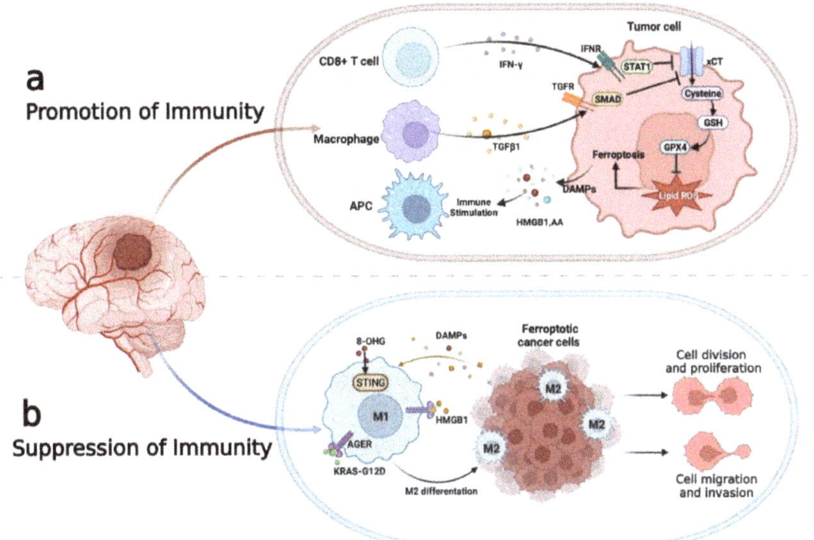

Figure 3. Role of ferroptosis in glioma immunity. (**a**) $CD8^+$ T cells release IFNγ to activate INFR (which inhibits SLC7A11 transcription through STAT1) to promote tumour cell ferroptosis. TGFβ1 released by macrophages induces the downregulation of system xCT mediated by SMAD proteins, thereby triggering lipid ROS-mediated ferroptosis via the GSH-GPX4 axis. In turn, ferroptotic glioma cells release DAMPs (such as HMGB1 and AA) to promote the recruitment and activation of immune cells. (**b**) In contrast, DAMPs, such as HMGB1, KRAS-G12D and 8-OHG, could affect the function of macrophages in the tumour microenvironment. In particular, KRAS-G12D binds AGER on the cell surface of macrophages to trigger M2 macrophage polarization, which might limit antitumour immunity. IFN-γ, interferon-γ; INFR, interferon receptor; STAT1, signal transducer and activator of transcription 1; GSH, glutathione; GPX4, glutathione peroxidase 4; DAMPs, damage-associated molecular patterns; HMGB1, high mobility group protein B1; AA, arachidonic acid; 8-OHG, 8-hydroxyguanosine; AGER, advanced glycosylation end productspecific receptor.

However, the regulation of ferroptosis in the TIME appears to show the treatment limitations of glioma. Due to the reducing supply of oxygen and nutrients, as well as promoting acidosis in the TIME [135,136], glioma cells will survive due to their own tremendous plasticity [137–139], but immune cells cannot adapt and lose the effect [140,141]. In addition, the function of immune cells could be inhibited due to the cytokines secreted by glioma cells and immune cells, which may express ferroptosis [142,143]. The Treg cells that impede immune surveillance of tumors, an immunosuppressive subset of CD4+ T cells, are resistant to ferroptosis, which is likely due to GPX4 induction in activated Treg cells [144]. A study has shown that KRAS-G12D can be released into exosomes by pancreatic cancer cells upon ferroptosis and taken up by macrophages via advanced glycation end products (AGEs), which ultimately stimulate tumor growth through the polarization of macrophages to the M2 phenotype. In addition, conditional deletion of Gpx4 in the pancreas of mice promoted mutant Kras-driven tumorigenesis through ferroptotic injury-induced DNA release and subsequent STING-activated inflammation in macrophages [12]. Previous experiments and database analysis showed that the infiltration of Treg cells, neutrophils, and M2-polarized subtype macrophages in the TIME was significantly increased at high levels of ferroptosis [121]. Similarly, MDSCs with immunosuppressive functions exhibit resistance to ferroptosis due to an inhibitory drive on the p53-heme oxygenase 1 (HMOX1) axis mediated by N-acylsphingosine aminohydrolase 2 (ASAH2). In addition, large amounts of lipid peroxides were detected in tumor-derived $CD8^+$ T cells, but not in lymph node-derived $CD8^+$ T cells, which suggested that ferroptosis may be the metabolic vulnerability of tumor-specific $CD8^+$ T cells. In addition, ferroptosis inducers (especially GPX4 inhibitors) may reduce antitumour immunity and promote tumor development by impairing $CD8^+$ T cells and follicular helper T cells (Tfhs) [144]. Under these conditions, ferroptosis may promote tumor growth by suppressing antitumour immunity. These advances indicate that ferroptosis has great potential to enhance the immunotherapy in cancer treatment [145,146].

4. Challenges of Ferroptosis in Glioma

While ferroptosis has a significant effect on the growth and treatment of gliomas, the mechanism needs to be further explored [147]. Most RCD effector molecules are proteases or porins, for example, caspases as well as mixed lineage kinase domain-like protein (MLKL) are involved in apoptosis and necrosis, respectively, and GasderminD participates in pyroptosis [148–150]. PLOOH is currently regarded as the ultimate executor of ferroptosis; however, whether effector molecules still exist downstream of PLOOH requires more exploration. [145]. Cytoplasmic GPX4 detoxifying lipid peroxides accumulated at the plasma membrane were unable to inhibit mitochondrial lipid peroxidation-induced ferroptosis in GPX4 knockout cells treated with DHODH inhibitors, which indicated that additional ferroptosis enforcement mechanisms may exist downstream of cytoplasmic lipid peroxidation [144]. The interaction between ferroptosis and other RCDs is also unclear. Some features of ferroptosis compared to other RCDs are not unique; for example, ferroptosis signals such as lipid peroxidation and regulators such as GPX4 and SLC7A11 can also regulate other types of RCDs. Ferroptosis may also modulate the TIME by interacting with other RCDs, thereby affecting the development and treatment of gliomas. In addition, ferroptosis currently has limitations in the diagnosis and treatment of gliomas. Currently, some targets, especially GPX4, ACSL4, P53 and FTH1 [151,152], are regarded as biomarkers, but they are not still a gold standard. Some significant biomarkers that can accurately predict the tumor response to ferroptosis induction are urgently needed, especially those that can be detected directly in patient blood, urine, feces and tumor tissue. It is also not clear what types of patients with gliomas are more sensitive to ferroptosis treatments. Three criteria, including iron levels, gene expression and mutations, can be combined to assess which patients are most likely to benefit from ferroptosis; for example, SLC7A11 inhibitors may be particularly effective against certain types of gliomas that overexpress this target. Finally, ferroptosis plays a dual role in the development and treatment of tumors. Ferroptosis promoted glioma cell death, but also reduced the treatment effect of gliomas by enhancing the levels of Treg cells, neutrophils, and M2-polarized subtype macrophages in the TIME to suppress antitumour immunity [121]. Tumor cells sacrificing themselves could also get the surrounding tumor cells to be in a stress state and finally avoid ferroptosis by secreting cytokines [145]. What substances are released by tumor cells after ferroptosis and the effects of these signals need to be further studied [153,154]. More evidence is needed to confirm whether the cytokines released by cells after ferroptosis enable surrounding glioma cells to evade immune surveillance by regulating TIME [155,156].

5. Conclusions and Future Perspectives

Ferroptosis, as a new programmed cell death mode, is different from other RCDs and is the result of iron-dependent lipid peroxidation accumulation. In this review, we focused on the regulatory mechanism of ferroptosis and found that ferroptosis plays multiple roles in the occurrence and development of glioma. Ferroptosis not only induces glioma cell death but also promotes glioma cell growth, invasion, migration, and resistance by regulating TAMs, MDSCs, Treg cells, neutrophils and $CD8^+$ T cells in the TIME. Notably, studies have shown that some new compounds (such as strychnine, dihydroartemisinin and ibuprofen) are capable of inducing ferroptosis in gliomas, and ferroptosis-induced chemosensitizers, including erastin, can be used in combination with various drugs (such as cisplatin, temozolomide and cytarabine), which may provide new therapeutic opportunities for glioma treatment. However, poor BBB penetration reduces targeting tumor ability, and potential compensatory mechanisms hinder the effectiveness of ferroptosis agents in glioma therapy. It has been suggested that designing nanoengineered systems to improve the targeted delivery of drugs can overcome these issues, which further enhances the effectiveness of glioma treatment [157–159] (Table 2). Overall, although ferroptosis has great advantages in glioma treatment, we still need multidisciplinary cooperation to further explore the pros and cons of targeting ferroptosis and to evaluate its potential value in clinical applications.

Table 2. Various NP-based drug delivery systems for the potential treatment of glioma.

Nanocarrier	Coating	Outcome	Reference
Cisplatin-Fe_3O_4 /Gd_2O_3	LF + RGD dimer	Increased accumulation in tumor Released Fe^{2+} and Fe^{3+}	[160]
Iron oxide	NIR-fluorescent silica	Visualized tumor-associated macrophage populations	[161]
PEG	Doxorubicin	Increased drug diffusion across BBB	[162]
PEtOz-SS-PCL micelle	Doxorubicin	Increased drug diffusion across BBB	[163]
Liposome	Temozolomide	Enhanced antitumor activity	[164]
OX26-PLGA	Temozolomide	Enhanced permeability Improved cellular uptake	[165]
Fa-PEG-PCL	Luteolin	Prolonged survival time Enhanced antitumor activity	[166]
Anti-miR-21-PLA	Temozolomide	Increased apoptotic cell death	[167]
Transferrin-PEG-PLA	Resveratrol	Improved drug accumulation	[168]
Albumin	Paclitaxel and fenretinide	Increased drug diffusion across BBB Increased survival rate	[169]

Author Contributions: Conceptualization, Y.P., G.Y. and Y.L.; investigation, Y.L. and G.T.; writing-original draft preparation, Y.L.; writing-review and editing, Y.L. and X.F.; visualization, Y.L. and S.B.; supervision, Y.P. and G.Y. All authors have read and agreed to the published version of the manuscript.

Funding: This work was supported by grants from the National Natural Science Foundation of China (81960541/82060455).

Conflicts of Interest: The authors declare no conflict of interest.

Abbreviations

•OH	Hydroxyl radicals
15-LOX	15-lipoxygenase
15-LOX-1	15-lipoxygenase-1
AA	Arachidonic acid
ACSL4	Acyl–coenzyme A synthetase long-chain family 4
acyl-CoA	Acyl-coenzyme A
AGEs	Advanced glycation end products
ALOXE3	Arachidonate lipoxygenase 3
AR	Androgen receptor
ART	Artesunate
ASAH2	N-acylsphingosine aminohydrolase 2
ATG5	Autophagy related 5
ATG7	Autophagy related 7
ATP	Adenosine triphosphate
BBB	Blood-brain barrier
CDDO	2-cyano-3,12-dioxooleana-1,9 (11)-dien-28-oic acid
CMA	Chaperone-mediated autophagy
CNS	Central nervous system
CP	Ceruloplasmin
Cys	Cysteine
DHA	Dihydroartemisinin

DHI	Dihydrotanshinone I
DHODH	Dihydroorotate dehydrogenase
DMT1	Divalent metal transporter 1
ECM	Extracellular matrix
Fe^{2+}	Ferrous cations
Fe^{3+}	Ferric cations
FTH1	Ferritin heavy chain 1
FXR1	Fragile-X mental retardation autosomal 1
GCL	Glutamate cysteine ligase
GCS	Glutamylcysteine synthetase
Glu	Glutamic acid
Gly	Glycine
GPX	Glutathione peroxidase
GPX4	Glutathione peroxidase 4
GSH	Glutathione
GSSG	Glutathione disulfide
H_2O_2	Hydrogen per-oxide
HMOX1	P53-heme oxygenase 1
HP	Hephaestin
HSP90	Heat shock protein 90
HSPA5	Heat shock protein family A member 5
IECs	Intestinal epithelium cells
IL-13Rα2	Interleukin receptor-13alpha2
IONPs	Iron oxide nanoparticles
IREB2	Iron responsive element binding protein 2
LOXs	Lipoxygenases
LPC	Lysophosphatidylcholine
LPCAT3	Lysophosphatidylcholine acyltransferase 3
MDSCs)	Myeloid-derived suppressor cells
MLKL	Mixed lineage kinase-like
MTOR	Mammalian target of rapamycin
NADPH	Nicotinamide adenine dinucleotide phosphate
NCOA4	Nuclear receptor coactivator 4
OS	Overall survival
PCBP2	Poly(C)-binding protein 2
PE-PUFA	Phosphatidylethanolamine polyunsaturated fatty acid
PLOHs	Phospholipid alcohols
PLOOHs	Phospholipid hydroperoxides
PUFAs	Polyunsaturated fatty acids
RCD	Regulated cell death
ROS	Reactive oxygen species
SLC3A2	Solute carrier family 3 member 2
SLC7A11	Solute carrier family 7 member 11
STEAP3	6-transmembrane epithelial antigen of the prostate 3
STING	Stimulator of interferon gene
TAMs	Tumor-associated macrophages
Tf	Transferrin
Tfh	Follicular helper T cell
TfR	Transferrin receptor
TIME	Tumor immune microenvironment
TMZ	Temozolomide
Treg	Regulatory T
VEGF	Vascular endothelial-derived growth factor
WHO	World Health Organization
xCT	Cysteine/glutathione antiporter system Xc^-

References

1. Fang, Y.; Zhang, Z. Arsenic trioxide as a novel anti-glioma drug: A review. *Cell. Mol. Biol. Lett.* **2020**, *25*, 44. [CrossRef]
2. Ostrom, Q.T.; Gittleman, H.; Truitt, G.; Boscia, A.; Kruchko, C.; Barnholtz-Sloan, J.S. CBTRUS Statistical Report: Primary Brain and Other Central Nervous System Tumors Diagnosed in the United States in 2011–2015. *Neuro-Oncology* **2018**, *20*, iv1–iv86. [CrossRef] [PubMed]
3. Xu, S.; Tang, L.; Li, X.; Fan, F.; Liu, Z. Immunotherapy for glioma: Current management and future application. *Cancer Lett.* **2020**, *476*, 1–12. [CrossRef] [PubMed]
4. Price, R.L.; Chiocca, E.A. Evolution of malignant glioma treatment: From chemotherapy to vaccines to viruses. *Neurosurgery* **2014**, *61* (Suppl. 1), 74–83. [CrossRef]
5. Louis, D.N.; Perry, A.; Reifenberger, G.; von Deimling, A.; Figarella-Branger, D.; Cavenee, W.K.; Ohgaki, H.; Wiestler, O.D.; Kleihues, P.; Ellison, D.W. The 2016 World Health Organization Classification of Tumors of the Central Nervous System: A summary. *Acta Neuropathol.* **2016**, *131*, 803–820. [CrossRef] [PubMed]
6. Ohgaki, H.; Kleihues, P. Population-based studies on incidence, survival rates, and genetic alterations in astrocytic and oligodendroglial gliomas. *J. Neuropathol. Exp. Neurol.* **2005**, *64*, 479–489. [CrossRef]
7. Bleeker, F.E.; Molenaar, R.J.; Leenstra, S. Recent advances in the molecular understanding of glioblastoma. *J. Neurooncol.* **2012**, *108*, 11–27. [CrossRef]
8. Jin, L.; Guo, S.; Zhang, X.; Mo, Y.; Ke, S.; Duan, C. Optimal treatment strategy for adult patients with newly diagnosed glioblastoma: A systematic review and network meta-analysis. *Neurosurg. Rev.* **2021**, *44*, 1943–1955. [CrossRef]
9. Milano, M.T.; Johnson, M.D.; Sul, J.; Mohile, N.A.; Korones, D.N.; Okunieff, P.; Walter, K.A. Primary spinal cord glioma: A Surveillance, Epidemiology, and End Results database study. *J. Neurooncol.* **2010**, *98*, 83–92. [CrossRef] [PubMed]
10. Miller, J.J.; Shih, H.A.; Andronesi, O.C.; Cahill, D.P. Isocitrate dehydrogenase-mutant glioma: Evolving clinical and therapeutic implications. *Cancer* **2017**, *123*, 4535–4546. [CrossRef] [PubMed]
11. Aguirre, J.I.; Castillo, E.J.; Kimmel, D.B. Biologic and pathologic aspects of osteocytes in the setting of medication-related osteonecrosis of the jaw (MRONJ). *Bone* **2021**, *153*, 116168. [CrossRef]
12. Chen, X.; Kang, R.; Kroemer, G.; Tang, D. Broadening horizons: The role of ferroptosis in cancer. *Nat. Rev. Clin. Oncol.* **2021**, *18*, 280–296. [CrossRef] [PubMed]
13. Dixon, S.J.; Lemberg, K.M.; Lamprecht, M.R.; Skouta, R.; Zaitsev, E.M.; Gleason, C.E.; Patel, D.N.; Bauer, A.J.; Cantley, A.M.; Yang, W.S.; et al. Ferroptosis: An iron-dependent form of nonapoptotic cell death. *Cell* **2012**, *149*, 1060–1072. [CrossRef] [PubMed]
14. Xiong, R.; He, R.; Liu, B.; Jiang, W.; Wang, B.; Li, N.; Geng, Q. Ferroptosis: A New Promising Target for Lung Cancer Therapy. *Oxid. Med. Cell. Longev.* **2021**, *2021*, 8457521. [CrossRef] [PubMed]
15. Kinowaki, Y.; Taguchi, T.; Onishi, I.; Kirimura, S.; Kitagawa, M.; Yamamoto, K. Overview of Ferroptosis and Synthetic Lethality Strategies. *Int. J. Mol. Sci.* **2021**, *22*, 9271. [CrossRef]
16. Chen, K.; Zhang, S.; Jiao, J.; Zhao, S. Ferroptosis and Its Potential Role in Lung Cancer: Updated Evidence from Pathogenesis to Therapy. *J. Inflamm. Res.* **2021**, *14*, 7079–7090. [CrossRef] [PubMed]
17. Li, J.; Cao, F.; Yin, H.L.; Huang, Z.J.; Lin, Z.T.; Mao, N.; Sun, B.; Wang, G. Ferroptosis: Past, present and future. *Cell Death Dis.* **2020**, *11*, 88. [CrossRef]
18. Gutteridge, J.M. Lipid peroxidation initiated by superoxide-dependent hydroxyl radicals using complexed iron and hydrogen peroxide. *FEBS Lett.* **1984**, *172*, 245–249. [CrossRef]
19. Tan, S.; Schubert, D.; Maher, P. Oxytosis: A novel form of programmed cell death. *Curr. Top. Med. Chem.* **2001**, *1*, 497–506. [CrossRef]
20. Cao, J.Y.; Dixon, S.J. Mechanisms of ferroptosis. *Cell. Mol. Life Sci.* **2016**, *73*, 2195–2209. [CrossRef]
21. Latunde-Dada, G.O. Ferroptosis: Role of lipid peroxidation, iron and ferritinophagy. *Biochim. Biophys. Acta Gen. Subj.* **2017**, *1861*, 1893–1900. [CrossRef] [PubMed]
22. Kajarabille, N.; Latunde-Dada, G.O. Programmed Cell-Death by Ferroptosis: Antioxidants as Mitigators. *Int. J. Mol. Sci.* **2019**, *20*, 4968. [CrossRef] [PubMed]
23. Anderson, G.J.; Frazer, D.M.; McKie, A.T.; Wilkins, S.J.; Vulpe, C.D. The expression and regulation of the iron transport molecules hephaestin and IREG1: Implications for the control of iron export from the small intestine. *Cell Biochem. Biophys.* **2002**, *36*, 137–146. [CrossRef]
24. Hirayama, T.; Inden, M.; Tsuboi, H.; Niwa, M.; Uchida, Y.; Naka, Y.; Hozumi, I.; Nagasawa, H. A Golgi-targeting fluorescent probe for labile Fe(ii) to reveal an abnormal cellular iron distribution induced by dysfunction of VPS35. *Chem. Sci.* **2019**, *10*, 1514–1521. [CrossRef]
25. Yang, Q.; Liu, W.; Zhang, S.; Liu, S. The cardinal roles of ferroportin and its partners in controlling cellular iron in and out. *Life Sci.* **2020**, *258*, 118135. [CrossRef]
26. Guan, W.; Xia, M.; Ji, M.; Chen, B.; Li, S.; Zhang, M.; Liang, S.; Chen, B.; Gong, W.; Dong, C.; et al. Iron induces two distinct Ca^{2+} signalling cascades in astrocytes. *Commun. Biol.* **2021**, *4*, 525. [CrossRef]
27. Wang, J.; Tian, S.; Petros, R.A.; Napier, M.E.; Desimone, J.M. The complex role of multivalency in nanoparticles targeting the transferrin receptor for cancer therapies. *J. Am. Chem. Soc.* **2010**, *132*, 11306–11313. [CrossRef]
28. Oosterheert, W.; van Bezouwen, L.S.; Rodenburg, R.N.P.; Granneman, J.; Forster, F.; Mattevi, A.; Gros, P. Cryo-EM structures of human STEAP4 reveal mechanism of iron(III) reduction. *Nat. Commun.* **2018**, *9*, 4337. [CrossRef]

29. Gao, G.; Li, J.; Zhang, Y.; Chang, Y.Z. Cellular Iron Metabolism and Regulation. *Adv. Exp. Med. Biol.* **2019**, *1173*, 21–32. [CrossRef]
30. Yan, H.F.; Zou, T.; Tuo, Q.Z.; Xu, S.; Li, H.; Belaidi, A.A.; Lei, P. Ferroptosis: Mechanisms and links with diseases. *Signal Transduct. Target. Ther.* **2021**, *6*, 49. [CrossRef]
31. Khalil, S.; Holy, M.; Grado, S.; Fleming, R.; Kurita, R.; Nakamura, Y.; Goldfarb, A. A specialized pathway for erythroid iron delivery through lysosomal trafficking of transferrin receptor 2. *Blood Adv.* **2017**, *1*, 1181–1194. [CrossRef] [PubMed]
32. Volani, C.; Doerrier, C.; Demetz, E.; Haschka, D.; Paglia, G.; Lavdas, A.A.; Gnaiger, E.; Weiss, G. Dietary iron loading negatively affects liver mitochondrial function. *Metallomics* **2017**, *9*, 1634–1644. [CrossRef] [PubMed]
33. Liu, J.; Kang, R.; Tang, D. Signaling pathways and defense mechanisms of ferroptosis. *FEBS J.* **2021**. [CrossRef] [PubMed]
34. Paul, B.D.; Snyder, S.H. The unusual amino acid L-ergothioneine is a physiologic cytoprotectant. *Cell Death Differ.* **2010**, *17*, 1134–1140. [CrossRef] [PubMed]
35. Kim, W.J.; Lee, S.D. Candidate genes for COPD: Current evidence and research. *Int. J. Chronic Obstruct. Pulmon. Dis.* **2015**, *10*, 2249–2255. [CrossRef]
36. Zumbrennen-Bullough, K.B.; Becker, L.; Garrett, L.; Holter, S.M.; Calzada-Wack, J.; Mossbrugger, I.; Quintanilla-Fend, L.; Racz, I.; Rathkolb, B.; Klopstock, T.; et al. Abnormal brain iron metabolism in Irp2 deficient mice is associated with mild neurological and behavioral impairments. *PLoS ONE* **2014**, *9*, e98072. [CrossRef]
37. Dobbyn, A.; Huckins, L.M.; Boocock, J.; Sloofman, L.G.; Glicksberg, B.S.; Giambartolomei, C.; Hoffman, G.E.; Perumal, T.M.; Girdhar, K.; Jiang, Y.; et al. Landscape of Conditional eQTL in Dorsolateral Prefrontal Cortex and Co-localization with Schizophrenia GWAS. *Am. J. Hum. Genet.* **2018**, *102*, 1169–1184. [CrossRef]
38. Mancias, J.D.; Wang, X.; Gygi, S.P.; Harper, J.W.; Kimmelman, A.C. Quantitative proteomics identifies NCOA4 as the cargo receptor mediating ferritinophagy. *Nature* **2014**, *509*, 105–109. [CrossRef]
39. Ryu, M.S.; Zhang, D.; Protchenko, O.; Shakoury-Elizeh, M.; Philpott, C.C. PCBP1 and NCOA4 regulate erythroid iron storage and heme biosynthesis. *J. Clin. Investig.* **2017**, *127*, 1786–1797. [CrossRef]
40. Park, E.; Chung, S.W. ROS-mediated autophagy increases intracellular iron levels and ferroptosis by ferritin and transferrin receptor regulation. *Cell Death Dis.* **2019**, *10*, 822. [CrossRef]
41. Fang, Y.; Chen, X.; Tan, Q.; Zhou, H.; Xu, J.; Gu, Q. Inhibiting Ferroptosis through Disrupting the NCOA4-FTH1 Interaction: A New Mechanism of Action. *ACS Cent. Sci.* **2021**, *7*, 980–989. [CrossRef] [PubMed]
42. Zhang, Z.; Yao, Z.; Wang, L.; Ding, H.; Shao, J.; Chen, A.; Zhang, F.; Zheng, S. Activation of ferritinophagy is required for the RNA-binding protein ELAVL1/HuR to regulate ferroptosis in hepatic stellate cells. *Autophagy* **2018**, *14*, 2083–2103. [CrossRef] [PubMed]
43. Florean, C.; Song, S.; Dicato, M.; Diederich, M. Redox biology of regulated cell death in cancer: A focus on necroptosis and ferroptosis. *Free Radic. Biol. Med.* **2019**, *134*, 177–189. [CrossRef] [PubMed]
44. Kong, H.; Reczek, C.R.; McElroy, G.S.; Steinert, E.M.; Wang, T.; Sabatini, D.M.; Chandel, N.S. Metabolic determinants of cellular fitness dependent on mitochondrial reactive oxygen species. *Sci. Adv.* **2020**, *6*, eabb7272. [CrossRef]
45. Chiu, C.F.; Saidi, W.A.; Kagan, V.E.; Star, A. Defect-Induced Near-Infrared Photoluminescence of Single-Walled Carbon Nanotubes Treated with Polyunsaturated Fatty Acids. *J. Am. Chem. Soc.* **2017**, *139*, 4859–4865. [CrossRef]
46. Sha, W.; Hu, F.; Xi, Y.; Chu, Y.; Bu, S. Mechanism of Ferroptosis and Its Role in Type 2 Diabetes Mellitus. *J. Diabetes Res.* **2021**, *2021*, 9999612. [CrossRef]
47. Shah, R.; Shchepinov, M.S.; Pratt, D.A. Resolving the Role of Lipoxygenases in the Initiation and Execution of Ferroptosis. *ACS Cent. Sci.* **2018**, *4*, 387–396. [CrossRef]
48. Chu, B.; Kon, N.; Chen, D.; Li, T.; Liu, T.; Jiang, L.; Song, S.; Tavana, O.; Gu, W. ALOX12 is required for p53-mediated tumour suppression through a distinct ferroptosis pathway. *Nat. Cell Biol.* **2019**, *21*, 579–591. [CrossRef]
49. Kagan, V.E.; Mao, G.; Qu, F.; Angeli, J.P.; Doll, S.; Croix, C.S.; Dar, H.H.; Liu, B.; Tyurin, V.A.; Ritov, V.B.; et al. Oxidized arachidonic and adrenic PEs navigate cells to ferroptosis. *Nat. Chem. Biol.* **2017**, *13*, 81–90. [CrossRef]
50. Tang, Y.; Zhou, J.; Hooi, S.C.; Jiang, Y.M.; Lu, G.D. Fatty acid activation in carcinogenesis and cancer development: Essential roles of long-chain acyl-CoA synthetases. *Oncol. Lett.* **2018**, *16*, 1390–1396. [CrossRef]
51. Orlando, U.D.; Garona, J.; Ripoll, G.V.; Maloberti, P.M.; Solano, A.R.; Avagnina, A.; Gomez, D.E.; Alonso, D.F.; Podesta, E.J. The functional interaction between Acyl-CoA synthetase 4, 5-lipooxygenase and cyclooxygenase-2 controls tumor growth: A novel therapeutic target. *PLoS ONE* **2012**, *7*, e40794. [CrossRef]
52. Li, Y.; Feng, D.; Wang, Z.; Zhao, Y.; Sun, R.; Tian, D.; Liu, D.; Zhang, F.; Ning, S.; Yao, J.; et al. Ischemia-induced ACSL4 activation contributes to ferroptosis-mediated tissue injury in intestinal ischemia/reperfusion. *Cell Death Differ.* **2019**, *26*, 2284–2299. [CrossRef]
53. Sakakura, Y.; Sato, H.; Shiiya, A.; Tamba, M.; Sagara, J.; Matsuda, M.; Okamura, N.; Makino, N.; Bannai, S. Expression and function of cystine/glutamate transporter in neutrophils. *J. Leukoc. Biol.* **2007**, *81*, 974–982. [CrossRef] [PubMed]
54. Miladinovic, T.; Ungard, R.G.; Linher-Melville, K.; Popovic, S.; Singh, G. Functional effects of TrkA inhibition on system X_C^--mediated glutamate release and cancer-induced bone pain. *Mol. Pain* **2018**, *14*, 1744806918776467. [CrossRef] [PubMed]
55. Yang, H.; Ramani, K.; Xia, M.; Ko, K.S.; Li, T.W.; Oh, P.; Li, J.; Lu, S.C. Dysregulation of glutathione synthesis during cholestasis in mice: Molecular mechanisms and therapeutic implications. *Hepatology* **2009**, *49*, 1982–1991. [CrossRef]
56. Desideri, E.; Filomeni, G.; Ciriolo, M.R. Glutathione participates in the modulation of starvation-induced autophagy in carcinoma cells. *Autophagy* **2012**, *8*, 1769–1781. [CrossRef]

7. Song, X.; Zhu, S.; Chen, P.; Hou, W.; Wen, Q.; Liu, J.; Xie, Y.; Liu, J.; Klionsky, D.J.; Kroemer, G.; et al. AMPK-Mediated BECN1 Phosphorylation Promotes Ferroptosis by Directly Blocking System X_C^- Activity. *Curr. Biol.* **2018**, *28*, 2388–2399.e2385. [CrossRef] [PubMed]

8. Chen, X.; Kang, R.; Kroemer, G.; Tang, D. Targeting ferroptosis in pancreatic cancer: A double-edged sword. *Trends Cancer* **2021**, *7*, 891–901. [CrossRef] [PubMed]

9. Wu, Z.; Geng, Y.; Lu, X.; Shi, Y.; Wu, G.; Zhang, M.; Shan, B.; Pan, H.; Yuan, J. Chaperone-mediated autophagy is involved in the execution of ferroptosis. *Proc. Natl. Acad. Sci. USA* **2019**, *116*, 2996–3005. [CrossRef]

0. Arias, E.; Koga, H.; Diaz, A.; Mocholi, E.; Patel, B.; Cuervo, A.M. Lysosomal mTORC2/PHLPP1/Akt Regulate Chaperone-Mediated Autophagy. *Mol. Cell* **2015**, *59*, 270–284. [CrossRef]

1. Doll, S.; Freitas, F.P.; Shah, R.; Aldrovandi, M.; da Silva, M.C.; Ingold, I.; Goya Grocin, A.; Xavier da Silva, T.N.; Panzilius, E.; Scheel, C.H.; et al. FSP1 is a glutathione-independent ferroptosis suppressor. *Nature* **2019**, *575*, 693–698. [CrossRef]

2. Bersuker, K.; Hendricks, J.M.; Li, Z.; Magtanong, L.; Ford, B.; Tang, P.H.; Roberts, M.A.; Tong, B.; Maimone, T.J.; Zoncu, R.; et al. The CoQ oxidoreductase FSP1 acts parallel to GPX4 to inhibit ferroptosis. *Nature* **2019**, *575*, 688–692. [CrossRef] [PubMed]

3. Zhou, Y.; Tao, L.; Zhou, X.; Zuo, Z.; Gong, J.; Liu, X.; Zhou, Y.; Liu, C.; Sang, N.; Liu, H.; et al. DHODH and cancer: Promising prospects to be explored. *Cancer Metab.* **2021**, *9*, 22. [CrossRef] [PubMed]

4. Mao, C.; Liu, X.; Zhang, Y.; Lei, G.; Yan, Y.; Lee, H.; Koppula, P.; Wu, S.; Zhuang, L.; Fang, B.; et al. Author Correction: DHODH-mediated ferroptosis defence is a targetable vulnerability in cancer. *Nature* **2021**, *596*, E13. [CrossRef] [PubMed]

5. Li, B.; Chen, X.; Qiu, W.; Zhao, R.; Duan, J.; Zhang, S.; Pan, Z.; Zhao, S.; Guo, Q.; Qi, Y.; et al. Synchronous Disintegration of Ferroptosis Defense Axis via Engineered Exosome-Conjugated Magnetic Nanoparticles for Glioblastoma Therapy. *Adv. Sci.* **2022**, *9*, e2105451. [CrossRef]

6. Wang, S.M.; Fu, L.J.; Duan, X.L.; Crooks, D.R.; Yu, P.; Qian, Z.M.; Di, X.J.; Li, J.; Rouault, T.A.; Chang, Y.Z. Role of hepcidin in murine brain iron metabolism. *Cell. Mol. Life Sci.* **2010**, *67*, 123–133. [CrossRef] [PubMed]

7. You, L.; Yu, P.P.; Dong, T.; Guo, W.; Chang, S.; Zheng, B.; Ci, Y.; Wang, F.; Yu, P.; Gao, G.; et al. Astrocyte-derived hepcidin controls iron traffic at the blood-brain-barrier via regulating ferroportin 1 of microvascular endothelial cells. *Cell Death Dis.* **2022**, *13*, 667. [CrossRef]

8. Yu, P.; Chang, Y.Z. Brain Iron Metabolism and Regulation. *Adv. Exp. Med. Biol.* **2019**, *1173*, 33–44. [CrossRef]

9. Rouault, T.A. Iron metabolism in the CNS: Implications for neurodegenerative diseases. *Nat. Rev. Neurosci.* **2013**, *14*, 551–564. [CrossRef]

0. Park, K.J.; Kim, J.; Testoff, T.; Adams, J.; Poklar, M.; Zborowski, M.; Venere, M.; Chalmers, J.J. Quantitative characterization of the regulation of iron metabolism in glioblastoma stem-like cells using magnetophoresis. *Biotechnol. Bioeng.* **2019**, *116*, 1644–1655. [CrossRef]

1. Hanninen, M.M.; Haapasalo, J.; Haapasalo, H.; Fleming, R.E.; Britton, R.S.; Bacon, B.R.; Parkkila, S. Expression of iron-related genes in human brain and brain tumors. *BMC Neurosci.* **2009**, *10*, 36. [CrossRef]

2. Voth, B.; Nagasawa, D.T.; Pelargos, P.E.; Chung, L.K.; Ung, N.; Gopen, Q.; Tenn, S.; Kamei, D.T.; Yang, I. Transferrin receptors and glioblastoma multiforme: Current findings and potential for treatment. *J. Clin. Neurosci.* **2015**, *22*, 1071–1076. [CrossRef] [PubMed]

3. Recht, L.; Torres, C.O.; Smith, T.W.; Raso, V.; Griffin, T.W. Transferrin receptor in normal and neoplastic brain tissue: Implications for brain-tumor immunotherapy. *J. Neurosurg.* **1990**, *72*, 941–945. [CrossRef] [PubMed]

4. Calzolari, A.; Larocca, L.M.; Deaglio, S.; Finisguerra, V.; Boe, A.; Raggi, C.; Ricci Vitani, L.; Pierconti, F.; Malavasi, F.; De Maria, R.; et al. Transferrin receptor 2 is frequently and highly expressed in glioblastomas. *Transl. Oncol.* **2010**, *3*, 123–134. [CrossRef] [PubMed]

5. Terzi, E.M.; Sviderskiy, V.O.; Alvarez, S.W.; Whiten, G.C.; Possemato, R. Iron-sulfur cluster deficiency can be sensed by IRP2 and regulates iron homeostasis and sensitivity to ferroptosis independent of IRP1 and FBXL5. *Sci. Adv.* **2021**, *7*, eabg4302. [CrossRef]

6. Ingrassia, R.; Garavaglia, B.; Memo, M. DMT1 Expression and Iron Levels at the Crossroads Between Aging and Neurodegeneration. *Front. Neurosci.* **2019**, *13*, 575. [CrossRef]

7. Yang, C.; Xia, Z.; Li, T.; Chen, Y.; Zhao, M.; Sun, Y.; Ma, J.; Wu, Y.; Wang, X.; Wang, P.; et al. Antioxidant Effect of Propofol in Gliomas and Its Association With Divalent Metal Transporter 1. *Front. Oncol.* **2020**, *10*, 590931. [CrossRef]

8. Song, Q.; Peng, S.; Sun, Z.; Heng, X.; Zhu, X. Temozolomide Drives Ferroptosis via a DMT1-Dependent Pathway in Glioblastoma Cells. *Yonsei Med. J.* **2021**, *62*, 843–849. [CrossRef]

9. Zhang, F.; Tao, Y.; Zhang, Z.; Guo, X.; An, P.; Shen, Y.; Wu, Q.; Yu, Y.; Wang, F. Metalloreductase Steap3 coordinates the regulation of iron homeostasis and inflammatory responses. *Haematologica* **2012**, *97*, 1826–1835. [CrossRef]

0. Li, P.L.; Liu, H.; Chen, G.P.; Li, L.; Shi, H.J.; Nie, H.Y.; Liu, Z.; Hu, Y.F.; Yang, J.; Zhang, P.; et al. STEAP3 (Six-Transmembrane Epithelial Antigen of Prostate 3) Inhibits Pathological Cardiac Hypertrophy. *Hypertension* **2020**, *76*, 1219–1230. [CrossRef]

1. Ohgami, R.S.; Campagna, D.R.; McDonald, A.; Fleming, M.D. The Steap proteins are metalloreductases. *Blood* **2006**, *108*, 1388–1394. [CrossRef] [PubMed]

2. Chen, H.; Xu, C.; Yu, Q.; Zhong, C.; Peng, Y.; Chen, J.; Chen, G. Comprehensive landscape of STEAP family functions and prognostic prediction value in glioblastoma. *J. Cell. Physiol.* **2021**, *236*, 2988–3000. [CrossRef]

83. Xiao, D.; Zhou, Y.; Wang, X.; Zhao, H.; Nie, C.; Jiang, X. A Ferroptosis-Related Prognostic Risk Score Model to Predict Clinica Significance and Immunogenic Characteristics in Glioblastoma Multiforme. *Oxid. Med. Cell. Longev.* **2021**, *2021*, 910785 [CrossRef] [PubMed]
84. Han, M.; Xu, R.; Wang, S.; Yang, N.; Ni, S.; Zhang, Q.; Xu, Y.; Zhang, X.; Zhang, C.; Wei, Y.; et al. Six-Transmembrane Epithelia Antigen of Prostate 3 Predicts Poor Prognosis and Promotes Glioblastoma Growth and Invasion. *Neoplasia* **2018**, *20*, 543–554 [CrossRef]
85. Han, W.; Xin, Z.; Zhao, Z.; Bao, W.; Lin, X.; Yin, B.; Zhao, J.; Yuan, J.; Qiang, B.; Peng, X. RNA-binding protein PCBP2 modulate glioma growth by regulating FHL3. *J. Clin. Investig.* **2013**, *123*, 2103–2118. [CrossRef] [PubMed]
86. Mitre, A.-O.; Florian, A.I.; Buruiana, A.; Boer, A.; Moldovan, I.; Soritau, O.; Florian, S.I.; Susman, S. Ferroptosis Involvement i Glioblastoma Treatment. *Medicina* **2022**, *58*, 319. [CrossRef]
87. Fan, Z.; Wirth, A.K.; Chen, D.; Wruck, C.J.; Rauh, M.; Buchfelder, M.; Savaskan, N. Nrf2-Keap1 pathway promotes cell proliferatior and diminishes ferroptosis. *Oncogenesis* **2017**, *6*, e371. [CrossRef]
88. Shin, D.; Kim, E.H.; Lee, J.; Roh, J.L. Nrf2 inhibition reverses resistance to GPX4 inhibitor-induced ferroptosis in head and nec cancer. *Free Radic. Biol. Med.* **2018**, *129*, 454–462. [CrossRef]
89. Wang, X.J.; Sun, Z.; Villeneuve, N.F.; Zhang, S.; Zhao, F.; Li, Y.; Chen, W.; Yi, X.; Zheng, W.; Wondrak, G.T.; et al. Nrf2 enhance resistance of cancer cells to chemotherapeutic drugs, the dark side of Nrf2. *Carcinogenesis* **2008**, *29*, 1235–1243. [CrossRef]
90. Jaramillo, M.C.; Zhang, D.D. The emerging role of the Nrf2-Keap1 signaling pathway in cancer. *Genes Dev.* **2013**, *27*, 2179–2191 [CrossRef]
91. Chuang, H.Y.; Hsu, L.Y.; Pan, C.M.; Pikatan, N.W.; Yadav, V.K.; Fong, I.H.; Chen, C.H.; Yeh, C.T.; Chiu, S.C. The E3 Ubiquiti Ligase NEDD4-1 Mediates Temozolomide-Resistant Glioblastoma through PTEN Attenuation and Redox Imbalance in Nrf2-HO- Axis. *Int. J. Mol. Sci.* **2021**, *22*, 10247. [CrossRef] [PubMed]
92. Yu, D.; Liu, Y.; Zhou, Y.; Ruiz-Rodado, V.; Larion, M.; Xu, G.; Yang, C. Triptolide suppresses IDH1-mutated malignancy vi Nrf2-driven glutathione metabolism. *Proc. Natl. Acad. Sci. USA* **2020**, *117*, 9964–9972. [CrossRef] [PubMed]
93. Tang, X.; Fu, X.; Liu, Y.; Yu, D.; Cai, S.J.; Yang, C. Blockade of Glutathione Metabolism in IDH1-Mutated Glioma. *Mol. Cancer The* **2020**, *19*, 221–230. [CrossRef]
94. Cano-Galiano, A.; Oudin, A.; Fack, F.; Allega, M.F.; Sumpton, D.; Martinez-Garcia, E.; Dittmar, G.; Hau, A.C.; De Falco, A Herold-Mende, C.; et al. Cystathionine-γ-lyase drives antioxidant defense in cysteine-restricted IDH1-mutant astrocytoma *Neuro-Oncol. Adv.* **2021**, *3*, vdab057. [CrossRef]
95. Koppula, P.; Zhuang, L.; Gan, B. Cystine transporter SLC7A11/xCT in cancer: Ferroptosis, nutrient dependency, and cance therapy. *Protein Cell* **2021**, *12*, 599–620. [CrossRef]
96. Hu, N.; Hu, W.H.; Zhou, S.L.; Yang, Z.; Liang, W.L.; Yang, R.Y.; Li, M.H.; Jing, Z.; Li, Z.A.; Fu, X.D.; et al. SLC7A11 negativel associates with mismatch repair gene expression and endows glioblastoma cells sensitive to radiation under low glucos conditions. *Neoplasma* **2021**, *68*, 1147–1156. [CrossRef] [PubMed]
97. Takeuchi, S.; Wada, K.; Toyooka, T.; Shinomiya, N.; Shimazaki, H.; Nakanishi, K.; Nagatani, K.; Otani, N.; Osada, H.; Uozumi, Y et al. Increased xCT expression correlates with tumor invasion and outcome in patients with glioblastomas. *Neurosurgery* **2013** *72*, 33–41; discussion 41. [CrossRef] [PubMed]
98. Polewski, M.D.; Reveron-Thornton, R.F.; Cherryholmes, G.A.; Marinov, G.K.; Cassady, K.; Aboody, K.S. Increased Expressior of System xc- in Glioblastoma Confers an Altered Metabolic State and Temozolomide Resistance. *Mol. Cancer Res.* **2016**, *14* 1229–1242. [CrossRef]
99. Lang, X.; Green, M.D.; Wang, W.; Yu, J.; Choi, J.E.; Jiang, L.; Liao, P.; Zhou, J.; Zhang, Q.; Dow, A.; et al. Radiotherapy and Immunotherapy Promote Tumoral Lipid Oxidation and Ferroptosis via Synergistic Repression of SLC7A11. *Cancer Discov.* **2019** *9*, 1673–1685. [CrossRef]
100. Lei, G.; Zhang, Y.; Koppula, P.; Liu, X.; Zhang, J.; Lin, S.H.; Ajani, J.A.; Xiao, Q.; Liao, Z.; Wang, H.; et al. The role of ferroptosis ir ionizing radiation-induced cell death and tumor suppression. *Cell Res.* **2020**, *30*, 146–162. [CrossRef]
101. Wang, W.; Green, M.; Choi, J.E.; Gijon, M.; Kennedy, P.D.; Johnson, J.K.; Liao, P.; Lang, X.; Kryczek, I.; Sell, A.; et al. $CD8^{+}$ T cell regulate tumour ferroptosis during cancer immunotherapy. *Nature* **2019**, *569*, 270–274. [CrossRef] [PubMed]
102. Gao, X.; Guo, N.; Xu, H.; Pan, T.; Lei, H.; Yan, A.; Mi, Y.; Xu, L. Ibuprofen induces ferroptosis of glioblastoma cells vi downregulation of nuclear factor erythroid 2-related factor 2 signaling pathway. *Anticancer Drugs* **2020**, *31*, 27–34. [CrossRef] [PubMed]
103. Sleire, L.; Skeie, B.S.; Netland, I.A.; Forde, H.E.; Dodoo, E.; Selheim, F.; Leiss, L.; Heggdal, J.I.; Pedersen, P.H.; Wang, J.; et a Drug repurposing: Sulfasalazine sensitizes gliomas to gamma knife radiosurgery by blocking cystine uptake through system Xc- leading to glutathione depletion. *Oncogene* **2015**, *34*, 5951–5959. [CrossRef] [PubMed]
104. Zhang, Y.; Dube, C.; Gibert, M., Jr.; Cruickshanks, N.; Wang, B.; Coughlan, M.; Yang, Y.; Setiady, I.; Deveau, C.; Saoud, K.; et a The p53 Pathway in Glioblastoma. *Cancers* **2018**, *10*, 297. [CrossRef] [PubMed]
105. Umans, R.A.; Martin, J.; Harrigan, M.E.; Patel, D.C.; Chaunsali, L.; Roshandel, A.; Iyer, K.; Powell, M.D.; Oestreich, K.; Sontheimer H. Transcriptional Regulation of Amino Acid Transport in Glioblastoma Multiforme. *Cancers* **2021**, *13*, 6169. [CrossRef]
106. Yamaguchi, I.; Yoshimura, S.H.; Katoh, H. High cell density increases glioblastoma cell viability under glucose deprivation vi degradation of the cystine/glutamate transporter xCT (SLC7A11). *J. Biol. Chem.* **2020**, *295*, 6936–6945. [CrossRef]

107. Yamamoto, M.; Teramoto, K.; Katoh, H. Epidermal growth factor promotes glioblastoma cell death under glucose deprivation via upregulation of xCT (SLC7A11). *Cell. Signal.* **2021**, *78*, 109874. [CrossRef]
108. Goji, T.; Takahara, K.; Negishi, M.; Katoh, H. Cystine uptake through the cystine/glutamate antiporter xCT triggers glioblastoma cell death under glucose deprivation. *J. Biol. Chem.* **2017**, *292*, 19721–19732. [CrossRef]
109. Teramoto, K.; Katoh, H. The cystine/glutamate antiporter xCT is a key regulator of EphA2 S897 phosphorylation under glucose-limited conditions. *Cell. Signal.* **2019**, *62*, 109329. [CrossRef]
110. Forcina, G.C.; Dixon, S.J. GPX4 at the Crossroads of Lipid Homeostasis and Ferroptosis. *Proteomics* **2019**, *19*, e1800311. [CrossRef]
111. Zhan, S.; Lu, L.; Pan, S.S.; Wei, X.Q.; Miao, R.R.; Liu, X.H.; Xue, M.; Lin, X.K.; Xu, H.L. Targeting NQO1/GPX4-mediated ferroptosis by plumbagin suppresses in vitro and in vivo glioma growth. *Br. J. Cancer* **2022**, *127*, 364–376. [CrossRef]
112. Hacioglu, C.; Kar, F. Capsaicin induces redox imbalance and ferroptosis through ACSL4/GPx4 signaling pathways in U87-MG and U251 glioblastoma cells. *Metab. Brain Dis.* **2022**. [CrossRef]
113. Tan, S.; Hou, X.; Mei, L. Dihydrotanshinone I inhibits human glioma cell proliferation via the activation of ferroptosis. *Oncol. Lett.* **2020**, *20*, 122. [CrossRef]
114. Chen, T.C.; Chuang, J.Y.; Ko, C.Y.; Kao, T.J.; Yang, P.Y.; Yu, C.H.; Liu, M.S.; Hu, S.L.; Tsai, Y.T.; Chan, H.; et al. AR ubiquitination induced by the curcumin analog suppresses growth of temozolomide-resistant glioblastoma through disrupting GPX4-Mediated redox homeostasis. *Redox Biol.* **2020**, *30*, 101413. [CrossRef]
115. Song, Q.; Peng, S.; Che, F.; Zhu, X. Artesunate induces ferroptosis via modulation of p38 and ERK signaling pathway in glioblastoma cells. *J. Pharmacol. Sci.* **2022**, *148*, 300–306. [CrossRef]
116. Chen, J.; Chen, X.; Wang, F.; Gao, H.; Hu, W. Dihydroartemisinin suppresses glioma proliferation and invasion via inhibition of the ADAM17 pathway. *Neurol. Sci.* **2015**, *36*, 435–440. [CrossRef]
117. Zhang, Z.S.; Wang, J.; Shen, Y.B.; Guo, C.C.; Sai, K.E.; Chen, F.R.; Mei, X.; Han, F.U.; Chen, Z.P. Dihydroartemisinin increases temozolomide efficacy in glioma cells by inducing autophagy. *Oncol. Lett.* **2015**, *10*, 379–383. [CrossRef]
118. Yi, R.; Wang, H.; Deng, C.; Wang, X.; Yao, L.; Niu, W.; Fei, M.; Zhaba, W. Dihydroartemisinin initiates ferroptosis in glioblastoma through GPX4 inhibition. *Biosci. Rep.* **2020**, *40*, BSR20193314. [CrossRef]
119. Zhu, M.; Wu, P.; Li, Y.; Zhang, L.; Zong, Y.; Wan, M. Synergistic therapy for orthotopic gliomas via biomimetic nanosonosensitizer-mediated sonodynamic therapy and ferroptosis. *Biomater. Sci.* **2022**, *10*, 3911–3923. [CrossRef]
120. Chen, H.; Wen, J. Iron oxide nanoparticles loaded with paclitaxel inhibits glioblastoma by enhancing autophagy-dependent ferroptosis pathway. *Eur. J. Pharmacol.* **2022**, *921*, 174860. [CrossRef]
121. Liu, T.; Zhu, C.; Chen, X.; Guan, G.; Zou, C.; Shen, S.; Wu, J.; Wang, Y.; Lin, Z.; Chen, L.; et al. Ferroptosis, as the most enriched programmed cell death process in glioma, induces immunosuppression and immunotherapy resistance. *Neuro-Oncology* **2022**, *24*, 1113–1125. [CrossRef]
122. Stockwell, B.R.; Jiang, X. A Physiological Function for Ferroptosis in Tumor Suppression by the Immune System. *Cell Metab.* **2019**, *30*, 14–15. [CrossRef]
123. Da Ros, M.; De Gregorio, V.; Iorio, A.L.; Giunti, L.; Guidi, M.; de Martino, M.; Genitori, L.; Sardi, I. Glioblastoma Chemoresistance: The Double Play by Microenvironment and Blood-Brain Barrier. *Int. J. Mol. Sci.* **2018**, *19*, 2879. [CrossRef]
124. Cheng, Y.; Zhu, Y.; Xu, J.; Yang, M.; Chen, P.; Xu, W.; Zhao, J.; Geng, L.; Gong, S. PKN2 in colon cancer cells inhibits M2 phenotype polarization of tumor-associated macrophages via regulating DUSP6-Erk1/2 pathway. *Mol. Cancer* **2018**, *17*, 13. [CrossRef]
125. Choi, K.M.; Kashyap, P.C.; Dutta, N.; Stoltz, G.J.; Ordog, T.; Shea Donohue, T.; Bauer, A.J.; Linden, D.R.; Szurszewski, J.H.; Gibbons, S.J.; et al. CD206-positive M2 macrophages that express heme oxygenase-1 protect against diabetic gastroparesis in mice. *Gastroenterology* **2010**, *138*, 2399–2409. [CrossRef]
126. Lau, E.Y.; Ho, N.P.; Lee, T.K. Cancer Stem Cells and Their Microenvironment: Biology and Therapeutic Implications. *Stem Cells Int.* **2017**, *2017*, 3714190. [CrossRef]
127. Guo, X.; Zhao, Y.; Yan, H.; Yang, Y.; Shen, S.; Dai, X.; Ji, X.; Ji, F.; Gong, X.G.; Li, L.; et al. Single tumor-initiating cells evade immune clearance by recruiting type II macrophages. *Genes Dev.* **2017**, *31*, 247–259. [CrossRef]
128. Jinushi, M.; Chiba, S.; Yoshiyama, H.; Masutomi, K.; Kinoshita, I.; Dosaka-Akita, H.; Yagita, H.; Takaoka, A.; Tahara, H. Tumor-associated macrophages regulate tumorigenicity and anticancer drug responses of cancer stem/initiating cells. *Proc. Natl. Acad. Sci. USA* **2011**, *108*, 12425–12430. [CrossRef]
129. Ostrand-Rosenberg, S. Myeloid-derived suppressor cells: More mechanisms for inhibiting antitumor immunity. *Cancer Immunol. Immunother.* **2010**, *59*, 1593–1600. [CrossRef]
130. Audia, A.; Conroy, S.; Glass, R.; Bhat, K.P.L. The Impact of the Tumor Microenvironment on the Properties of Glioma Stem-Like Cells. *Front. Oncol.* **2017**, *7*, 143. [CrossRef]
131. Vignali, D.A.; Collison, L.W.; Workman, C.J. How regulatory T cells work. *Nat. Rev. Immunol.* **2008**, *8*, 523–532. [CrossRef] [PubMed]
132. Liu, L.; Li, Y.; Peng, H.; Liu, R.; Ji, W.; Shi, Z.; Shen, J.; Ma, G.; Zhang, X. Targeted exosome coating gene-chem nanocomplex as "nanoscavenger" for clearing alpha-synuclein and immune activation of Parkinson's disease. *Sci. Adv.* **2020**, *6*, eaba3967. [CrossRef]
133. Veerapathran, A.; Pidala, J.; Beato, F.; Betts, B.; Kim, J.; Turner, J.G.; Hellerstein, M.K.; Yu, X.Z.; Janssen, W.; Anasetti, C. Human regulatory T cells against minor histocompatibility antigens: Ex Vivo expansion for prevention of graft-versus-host disease. *Blood* **2013**, *122*, 2251–2261. [CrossRef]

134. Gabrusiewicz, K.; Rodriguez, B.; Wei, J.; Hashimoto, Y.; Healy, L.M.; Maiti, S.N.; Thomas, G.; Zhou, S.; Wang, Q.; Elakkad, A.; et al. Glioblastoma-infiltrated innate immune cells resemble M0 macrophage phenotype. *JCI Insight* **2016**, *1*, e85841. [CrossRef]
135. Chen, Y.; Huang, Q.; Liu, W.; Zhu, Q.; Cui, C.P.; Xu, L.; Guo, X.; Wang, P.; Liu, J.; Dong, G.; et al. Mutually exclusive acetylation and ubiquitylation of the splicing factor SRSF5 control tumor growth. *Nat. Commun.* **2018**, *9*, 2464. [CrossRef]
136. Aalipour, A.; Chuang, H.Y.; Murty, S.; D'Souza, A.L.; Park, S.M.; Gulati, G.S.; Patel, C.B.; Beinat, C.; Simonetta, F.; Martinic, I.; et al. Engineered immune cells as highly sensitive cancer diagnostics. *Nat. Biotechnol.* **2019**, *37*, 531–539. [CrossRef]
137. Chaligne, R.; Gaiti, F.; Silverbush, D.; Schiffman, J.S.; Weisman, H.R.; Kluegel, L.; Gritsch, S.; Deochand, S.D.; Gonzalez Castro, L.N.; Richman, A.R.; et al. Epigenetic encoding, heritability and plasticity of glioma transcriptional cell states. *Nat. Genet.* **2021**, *53*, 1469–1479. [CrossRef]
138. Nicholson, J.G.; Fine, H.A. Diffuse Glioma Heterogeneity and Its Therapeutic Implications. *Cancer Discov.* **2021**, *11*, 575–590. [CrossRef]
139. Lao, Y.; Wan, G.; Liu, Z.; Wang, X.; Ruan, P.; Xu, W.; Xu, D.; Xie, W.; Zhang, Y.; Xu, H.; et al. The natural compound oblongifolin C inhibits autophagic flux and enhances antitumor efficacy of nutrient deprivation. *Autophagy* **2014**, *10*, 736–749. [CrossRef]
140. Lim, M.; Xia, Y.; Bettegowda, C.; Weller, M. Current state of immunotherapy for glioblastoma. *Nat. Rev. Clin. Oncol.* **2018**, *15*, 422–442. [CrossRef]
141. Hambardzumyan, D.; Gutmann, D.H.; Kettenmann, H. The role of microglia and macrophages in glioma maintenance and progression. *Nat. Neurosci.* **2016**, *19*, 20–27. [CrossRef] [PubMed]
142. De Boeck, A.; Ahn, B.Y.; D'Mello, C.; Lun, X.; Menon, S.V.; Alshehri, M.M.; Szulzewsky, F.; Shen, Y.; Khan, L.; Dang, N.H.; et al. Glioma-derived IL-33 orchestrates an inflammatory brain tumor microenvironment that accelerates glioma progression. *Nat. Commun.* **2020**, *11*, 4997. [CrossRef] [PubMed]
143. Wang, Q.; Hu, B.; Hu, X.; Kim, H.; Squatrito, M.; Scarpace, L.; deCarvalho, A.C.; Lyu, S.; Li, P.; Li, Y.; et al. Tumor Evolution of Glioma-Intrinsic Gene Expression Subtypes Associates with Immunological Changes in the Microenvironment. *Cancer Cell* **2017**, *32*, 42–56.e6. [CrossRef]
144. Lei, G.; Zhuang, L.; Gan, B. Targeting ferroptosis as a vulnerability in cancer. *Nat. Rev. Cancer* **2022**, *22*, 381–396. [CrossRef]
145. Jiang, X.; Stockwell, B.R.; Conrad, M. Ferroptosis: Mechanisms, biology and role in disease. *Nat. Rev. Mol. Cell Biol.* **2021**, *22*, 266–282. [CrossRef]
146. Su, Y.; Zhao, B.; Zhou, L.; Zhang, Z.; Shen, Y.; Lv, H.; AlQudsy, L.H.H.; Shang, P. Ferroptosis, a novel pharmacological mechanism of anti-cancer drugs. *Cancer Lett.* **2020**, *483*, 127–136. [CrossRef]
147. Hadian, K.; Stockwell, B.R. SnapShot: Ferroptosis. *Cell* **2020**, *181*, 1188.e1. [CrossRef]
148. Fan, Y.; Bergmann, A. Multiple mechanisms modulate distinct cellular susceptibilities toward apoptosis in the developing Drosophila eye. *Dev. Cell* **2014**, *30*, 48–60. [CrossRef]
149. Barany, I.; Berenguer, E.; Solis, M.T.; Perez-Perez, Y.; Santamaria, M.E.; Crespo, J.L.; Risueno, M.C.; Diaz, I.; Testillano, P.S. Autophagy is activated and involved in cell death with participation of cathepsins during stress-induced microspore embryogenesis in barley. *J. Exp. Bot.* **2018**, *69*, 1387–1402. [CrossRef]
150. Banerjee, S.K.; Chatterjee, A.; Gupta, S.; Nagar, A. Activation and Regulation of NLRP3 by Sterile and Infectious Insults. *Front. Immunol.* **2022**, *13*, 896353. [CrossRef]
151. Yang, W.S.; SriRamaratnam, R.; Welsch, M.E.; Shimada, K.; Skouta, R.; Viswanathan, V.S.; Cheah, J.H.; Clemons, P.A.; Shamji, A.F.; Clish, C.B.; et al. Regulation of ferroptotic cancer cell death by GPX4. *Cell* **2014**, *156*, 317–331. [CrossRef] [PubMed]
152. Li, Y.; Cao, Y.; Xiao, J.; Shang, J.; Tan, Q.; Ping, F.; Huang, W.; Wu, F.; Zhang, H.; Zhang, X. Inhibitor of apoptosis-stimulating protein of p53 inhibits ferroptosis and alleviates intestinal ischemia/reperfusion-induced acute lung injury. *Cell Death Differ.* **2020**, *27*, 2635–2650. [CrossRef] [PubMed]
153. Xie, Y.; Hou, W.; Song, X.; Yu, Y.; Huang, J.; Sun, X.; Kang, R.; Tang, D. Ferroptosis: Process and function. *Cell Death Differ.* **2016**, *23*, 369–379. [CrossRef]
154. Dai, E.; Han, L.; Liu, J.; Xie, Y.; Kroemer, G.; Klionsky, D.J.; Zeh, H.J.; Kang, R.; Wang, J.; Tang, D. Autophagy-dependent ferroptosis drives tumor-associated macrophage polarization via release and uptake of oncogenic KRAS protein. *Autophagy* **2020**, *16*, 2069–2083. [CrossRef] [PubMed]
155. Chen, R.; Kang, R.; Tang, D. The mechanism of HMGB1 secretion and release. *Exp. Mol. Med.* **2022**, *54*, 91–102. [CrossRef] [PubMed]
156. Wang, H.; Xu, T.; Huang, Q.; Jin, W.; Chen, J. Immunotherapy for Malignant Glioma: Current Status and Future Directions. *Trends Pharmacol. Sci.* **2020**, *41*, 123–138. [CrossRef]
157. Shi, Y.; Lammers, T. Combining Nanomedicine and Immunotherapy. *Acc. Chem. Res.* **2019**, *52*, 1543–1554. [CrossRef]
158. Jang, H.; Kim, E.H.; Chi, S.-G.; Kim, S.H.; Yang, Y. Nanoparticles Targeting Innate Immune Cells in Tumor Microenvironment. *Int. J. Mol. Sci.* **2021**, *22*, 10009. [CrossRef]
159. Pinton, L.; Magri, S.; Masetto, E.; Vettore, M.; Schibuola, I.; Ingangi, V.; Marigo, I.; Matha, K.; Benoit, J.P.; Della Puppa, A.; et al. Targeting of immunosuppressive myeloid cells from glioblastoma patients by modulation of size and surface charge of lipid nanocapsules. *J. Nanobiotechnol.* **2020**, *18*, 31. [CrossRef]
160. Shen, Z.; Liu, T.; Li, Y.; Lau, J.; Yang, Z.; Fan, W.; Zhou, Z.; Shi, C.; Ke, C.; Bregadze, V.I.; et al. Fenton-Reaction-Acceleratable Magnetic Nanoparticles for Ferroptosis Therapy of Orthotopic Brain Tumors. *ACS Nano* **2018**, *12*, 11355–11365. [CrossRef]

61. Lee, C.; Kim, G.R.; Yoon, J.; Kim, S.E.; Yoo, J.S.; Piao, Y. In vivo delineation of glioblastoma by targeting tumor-associated macrophages with near-infrared fluorescent silica coated iron oxide nanoparticles in orthotopic xenografts for surgical guidance. *Sci. Rep.* **2018**, *8*, 11122. [CrossRef]
62. Steiniger, S.C.; Kreuter, J.; Khalansky, A.S.; Skidan, I.N.; Bobruskin, A.I.; Smirnova, Z.S.; Severin, S.E.; Uhl, R.; Kock, M.; Geiger, K.D.; et al. Chemotherapy of glioblastoma in rats using doxorubicin-loaded nanoparticles. *Int. J. Cancer* **2004**, *109*, 759–767. [CrossRef]
63. Li, Y.; Baiyang, L.; Leran, B.; Zhen, W.; Yandong, X.; Baixiang, D.; Dandan, Z.; Yufu, Z.; Jun, L.; Rutong, Y.; et al. Reduction-responsive PEtOz-SS-PCL micelle with tailored size to overcome blood-brain barrier and enhance doxorubicin antiglioma effect. *Drug Deliv.* **2017**, *24*, 1782–1790. [CrossRef]
64. Tapeinos, C.; Marino, A.; Battaglini, M.; Migliorin, S.; Brescia, R.; Scarpellini, A.; De Julian Fernandez, C.; Prato, M.; Drago, F.; Ciofani, G. Stimuli-responsive lipid-based magnetic nanovectors increase apoptosis in glioblastoma cells through synergic intracellular hyperthermia and chemotherapy. *Nanoscale* **2018**, *11*, 72–88. [CrossRef]
65. Ramalho, M.J.; Sevin, E.; Gosselet, F.; Lima, J.; Coelho, M.A.N.; Loureiro, J.A.; Pereira, M.C. Receptor-mediated PLGA nanoparticles for glioblastoma multiforme treatment. *Int. J. Pharm.* **2018**, *545*, 84–92. [CrossRef]
66. Wu, C.; Xu, Q.; Chen, X.; Liu, J. Delivery luteolin with folacin-modified nanoparticle for glioma therapy. *Int. J. Nanomed.* **2019**, *14*, 7515–7531. [CrossRef]
67. Seo, Y.E.; Suh, H.W.; Bahal, R.; Josowitz, A.; Zhang, J.; Song, E.; Cui, J.; Noorbakhsh, S.; Jackson, C.; Bu, T.; et al. Nanoparticle-mediated intratumoral inhibition of miR-21 for improved survival in glioblastoma. *Biomaterials* **2019**, *201*, 87–98. [CrossRef]
68. Guo, W.; Li, A.; Jia, Z.; Yuan, Y.; Dai, H.; Li, H. Transferrin modified PEG-PLA-resveratrol conjugates: In Vitro and In Vivo studies for glioma. *Eur. J. Pharmacol.* **2013**, *718*, 41–47. [CrossRef]
69. Lin, T.; Zhao, P.; Jiang, Y.; Tang, Y.; Jin, H.; Pan, Z.; He, H.; Yang, V.C.; Huang, Y. Blood-Brain-Barrier-Penetrating Albumin Nanoparticles for Biomimetic Drug Delivery via Albumin-Binding Protein Pathways for Antiglioma Therapy. *ACS Nano* **2016**, *10*, 9999–10012. [CrossRef]

antioxidants

MDPI

Review

Dysregulated Iron Homeostasis as Common Disease Etiology and Promising Therapeutic Target

Bruce E. Holbein [1] and Christian Lehmann [2,*]

[1] Department of Microbiology and Immunology, Dalhousie University, Halifax, NS B3H 1X5, Canada
[2] Department of Anesthesia, Pain Management and Perioperative Medicine, Dalhousie University, Halifax, NS B3H 1X5, Canada
* Correspondence: chlehmann@dal.ca

Abstract: Iron is irreplaceably required for animal and human cells as it provides the activity center for a wide variety of essential enzymes needed for energy production, nucleic acid synthesis, carbon metabolism and cellular defense. However, iron is toxic when present in excess and its uptake and storage must, therefore, be tightly regulated to avoid damage. A growing body of evidence indicates that iron dysregulation leading to excess quantities of free reactive iron is responsible for a wide range of otherwise discrete diseases. Iron excess can promote proliferative diseases such as infections and cancer by supplying iron to pathogens or cancer cells. Toxicity from reactive iron plays roles in the pathogenesis of various metabolic, neurological and inflammatory diseases. Interestingly, a common underlying aspect of these conditions is availability of excess reactive iron. This underpinning aspect provides a potential new therapeutic avenue. Existing hematologically used iron chelators to take up excess iron have shown serious limitations for use but new purpose-designed chelators in development show promise for suppressing microbial pathogen and cancer cell growth, and also for relieving iron-induced toxicity in neurological and other diseases. Hepcidin and hepcidin agonists are also showing promise for relieving iron dysregulation. Harnessing iron-driven reactive oxygen species (ROS) generation with ferroptosis has shown promise for selective destruction of cancer cells. We review biological iron requirements, iron regulation and the nature of iron dysregulation in various diseases. Current results pertaining to potential new therapies are also reviewed.

Keywords: iron homeostasis; iron dysregulation; inflammation; infection; cancer; iron chelation

Citation: Holbein, B.E.; Lehmann, C. Dysregulated Iron Homeostasis as Common Disease Etiology and Promising Therapeutic Target. *Antioxidants* **2023**, *12*, 671. https://doi.org/10.3390/antiox12030671

Academic Editor: Yan-Zhong Chang

Received: 15 February 2023
Revised: 2 March 2023
Accepted: 7 March 2023
Published: 9 March 2023

1. Introduction

When oxygen first appeared in the Earth's atmosphere, iron with its Fe^2/Fe^3 redox potential of 0.771 V represented a sweet spot for chemical catalysis of a wide range of oxidation and reduction reactions needed for biosynthesis and energy production by all living organisms, except for a few microbes. Around 2% of human genes encode for iron proteins, with more than half of these having catalytic function, with oxidoreductase class enzymes being the largest fraction [1].

1.1. Iron as Biological Catalyst

Table 1 provides examples of key iron-dependent enzymes needed by most cells for critical functions at a number of physiological levels. These enzymes are critically important for cell growth and replication and generally, iron cannot be substituted for activity in these. This underlies the irreplaceable need for iron by all higher organisms.

Table 1. Iron-dependent enzymes.

Enzyme	Function	Reference
Cytochrome oxidase EC 7.1.1.9	Energy production	Williams, 1987 [2]
Citrate aconitase EC 4.2.1.3	TCA cycle	Beinert et al., 1996 [3]
Ribonucleotide reductase EC 2.7.7.56	DNA synthesis	Jordan and Reichard, 1998 [4]
DNA polymerase EC 2.7.7.7	DNA replication and repair	Zhang et al., 2014 [5]
Oxoglutarate oxygenase EC 1.13.12.19	Lipid biosynthesis	McDonough et al., 2010 [6]
Superoxide dismutase EC 1.15.1.1	Superoxide radical detoxification	Parker and Blake, 1988 [7]
Catalase EC 1.11.1.21	Peroxide detoxification	Alfonso-Prieto et al., 2009 [8]

1.2. Excess Iron Toxicity

Seemingly paradoxical, iron is also toxic to cells because of its production of oxidizing radicals with their formation driven by redox cycling of iron. Superoxide (O_2^-) is produced constitutively by the electron transport chain in cells by slippage of an electron to O_2 with concomitant reduction of Fe^3 to Fe^2. The human body produces around 5 g O_2^-/day [9], but superoxide dismutase (an iron dependent enzyme) normally neutralizes quantities in excess of normal amounts as needed for cell regulation and signaling [10]. Superoxide is also produced by nicotinamide adenine phosphate oxidase in the phagocytic defensive cells of vertebrate animals and this is used to kill phagocytosed invading microbes [11]. Superoxide can set up a cascade cycling of Fe reduction/oxidation coupled to peroxide and hydroxyl radical ·OH production, as shown in Scheme 1, with ·OH being highly toxic through damage to DNA and membrane lipids [12].

$$Fe^{3+} + \cdot O_2^- \longrightarrow Fe^{2+} + O_2$$

$$Fe^{2+} + H_2O_2 \longrightarrow Fe^{3+} + OH^- + \cdot OH$$

Total Reaction net:

$$\cdot O_2^- + H_2O_2 \longrightarrow O_2 + OH^- + \cdot OH$$

Scheme 1. Reactive oxygen species generation.

1.3. Chemical Containment

Containment of reactive iron and its ROS products is achieved through both compartmentalization and its chemical chaperoning. Phagocytes, including macrophages and polymorphonuclear leukocytes, compartmentalize ROS production within intracellular phagosomes, which spares other intracellular organelles and the extracellular milieu from direct ROS exposure [13]. The bulk of total body iron stores represents around 55 mg/kg and is located intracellularly, safely incorporated within hemoglobin (erythrocytes), myoglobin (muscle cells) and ferritin (liver cells) [14]. Iron incorporated into heme and ferritin is not freely chemically available to participate in ROS production. Body iron stores are shuttled to sites of use in other cells by transferrin, which binds iron avidly so that it is effectively chaperoned in circulation and not available for ROS reactions. There is an additional small labile iron pool located both intracellularly and extracellularly which is readily ROS-reactive [15], and this pool is important in relation to iron dysregulation, which increases the amounts in this labile pool.

It is interesting to note that some currently used therapeutic agents, including aminoglycoside antibiotics, can bind iron not fully satisfying its chaperoning requirements [16] and can induce ROS related tissue damage [17], likely due to the mobilization of labile iron. This underscores the need for containment and chaperoning of labile ROS-reactive iron in the host as part of normal homeostatic control.

2. Regulation of Normal Iron Homeostasis

Of the 3750 mg of total iron, around 65% of this is incorporated into heme in erythrocytes, 10% within heme of myoglobin of the muscles, 14% in macrophage cells of the reticuloendothelial system (RES), 28% stored within ferritin in hepatocytes and 4% in bone marrow cells [14]. Iron taken up by enterocytes in the gut is stored transiently in small amounts in ferritin until there is demand for iron replenishment elsewhere in the body and then transferrin takes up this iron into the bloodstream, serving as a shuttle/delivery protein chaperone. The transferrin pool is normally only 30% saturated with iron and, therefore, it provides iron-holding capacity to limit amounts of labile non-transferrin-bound iron in circulation. The transferrin iron pool is highly dynamic and represents a flux of around 25 mg iron per day (Pantopoulos, 2018—see Figure 1) [18]. The body has no excretory mechanism for iron and small daily losses of 1–2 mg from skin desquamation and other sources are correspondingly compensated by uptake from the gut.

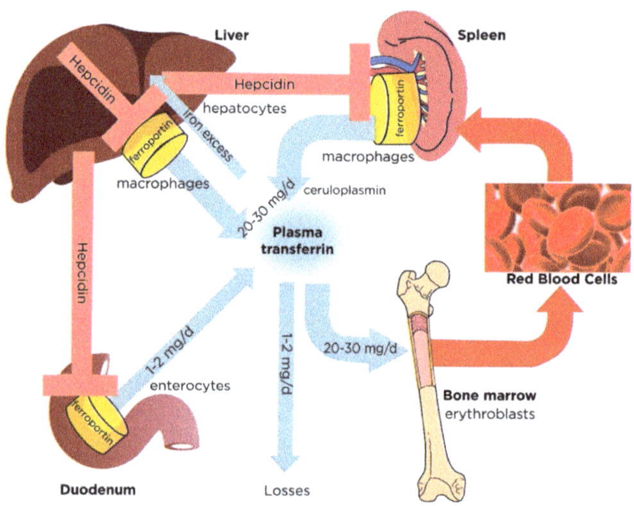

Figure 1. Dynamics of systemic iron balance. Plasma transferrin delivers iron to bone marrow erythroblasts and to other tissues. It contains a very small (∼0.1%) but highly dynamic fraction of body iron that turns over >10 times/day to meet the iron need for erythropoiesis (20–30 mg/day). The transferrin iron pool is primarily replenished with iron recycled from hepatic and splenic macrophages during erythrophagocytosis of senescent red blood cells. Duodenal enterocytes absorb dietary iron and release small amounts (1–2 mg/day) to compensate for non-specific losses. Hepatocytes store excess of body iron, which can be mobilized to plasma under iron deficiency. Iron efflux to plasma from macrophages, enterocytes or hepatocytes is negatively regulated by hepcidin, a liver-derived peptide hormone that binds to the iron exporter ferroportin and promotes its degradation (Pantopoulos, 2018—with permission) [18].

Hepcidin, a 25-amino acid peptide hormone produced primarily by liver hepatocytes, is the master regulator of iron homeostasis and it negatively regulates iron uptake by inactivating plasma membrane-bound ferroportin on enterocyte and macrophage cell membranes as needed for transfer of intracellular iron to extracellular compartments [19]. This provides tight regulation of iron homeostasis, which is critical for ensuring no excess toxic quantities of iron either intracellularly or extracellularly. Iron dysregulation often presents with elevated amounts of circulating plasma iron including transferrin-bound iron with >50% saturation along with elevated non-transferrin-bound iron (NTBI) that can reach >0.5 µM [20]. It is the labile portion of NTBI that plays a role in the pathogenesis of the various non-microbial diseases we review below, while elevated plasma iron in the form of transferrin-bound iron can also support microbial infection [21].

3. Iron Dysregulation Underlying Disease

3.1. The Nature of Iron Dysregulation

Iron dysregulation diseases often result from increased amounts of circulating labile reactive plasma iron. This can be evidenced by elevated plasma transferrin-bound iron (TBI) with transferrin saturation above its normal 30%, sometimes reaching 100% saturation [22]. With high saturation of transferrin, some of the plasma iron is not effectively chaperoned and this provides a pool of non-transferrin-bound iron (NTBI) labile reactive iron (LPI). This LPI is also more mobile and capable of entering cells, where it is toxic [23,24].

Excess plasma iron in the form of TBI and NTBI LPI influences two main categories of diseases. Elevated plasma iron, especially in the form of transferrin iron, supports cell-proliferative diseases including infection and cancer. Increased plasma iron present as non-transferrin-bound LPI is part of the pathology of other diseases where excess reactive iron triggers iron-related pathologies. The various diseases we review are summarized in Table 2.

Table 2. Diseases with iron dysregulation.

Disease Category	Disease	Example	Reference
Proliferative Cell Replication	Microbial Infection	Staphylococcal, Candidiasis	Holbein et al., 2021 [21]
	Bacterial Sepsis	*Pseudomonas aeruginosa*, *Neisseria meningitidis*	Bullen et al., 2000 [25]
	Parasitic Infection	Chagas disease, Malaria	Mach and Sutak, 2020 [26]
	Viral Infection	HIV, HBV, HCV, HCMV	Schmidt, 2020 [27]
	Cancer	Pancreatic, Liver, Lung	Torti et al., 2018 [28]
Iron-Mediated Pathology	Diabetes	Type 2 Diabetes	Harrison et al., 2023 [29]
	Cardiovascular	Cardiomyopathy	Li and Zhang, 2021 [30]
	Neurological	Alzheimer's, ALS, MS, Parkinson's	David et al., 2022 [31]
	Lung Fibrosis	Idiopathic Pulmonary Fibrosis	Ogger and Byme, 2020 [32]
	Autoimmune	Rheumatoid Arthritis, Lupus Erythematosus	Baker and Ghio, 2009 [33]
	Kidney	Fanconi Syndrome	Smith and Thévenod, 2009 [34]
	Retinal	Age-Related Macular degeneration	Loh et al., 2009 [35]
	Iron Overload	Hemochromatosis	Pantopoulos, 2018 [18]
	Cirrhosis	Non-Alcoholic Fatty Cirrhosis	Chen, 2022 [36]

3.2. Infections

All pathogenic microorganisms, with the exception of *Borrelia burgdorferi*, the agent of Lyme disease, have absolute and irreplaceable requirements for iron for their growth and replication in our bodies [37]. *B. burgdorferi* can utilize manganese in place of iron [38]. The vertebrate host normally maintains conditions of low iron bioavailability to microbes, especially in extracellular compartments such as plasma, respiratory secretions and tears where infection is often initiated, thus providing a natural nutritional immunity. Transferrin in plasma and lactoferrin in tears and other secretions maintain very low levels of freely available iron [39]. Lactoferrin concentrations can reach >3 mg/mL in tears [40]. Furthermore, when infection is first detected, the body mounts an early iron withdrawal defense response, where extracellular iron concentrations are further reduced by moving this iron to intracellular stores [19]. Successful pathogens deploy a variety of virulence mechanisms to successfully compete for iron, as reviewed elsewhere [21].

3.2.1. Pathogenic Microbes

Iron has now been found to support infection by almost all pathogenic microbes and Table 3 summarizes some important microbial infections in relation to the main host iron sources feeding the infection. Of these, some have cell-wall surface receptors to intercept host iron sources, while others deploy high-affinity siderophores to effectively strip iron from host transferrin or lactoferrin [21].

Table 3. Iron sources for pathogenic microbes.

Microbial Pathogen	Host Iron Source	Reference
Neisseria meningitidis	Transferrin	Holbein, 1981 [41]
Pseudomonas aeruginosa	Transferrin, Lactoferrin	Xiao and Kisaalita, 1997 [42]
Acinetobacter baumannii	Transferrin [1], Lactoferrin [1], Heme [2]	[1] Yamamoto et al., 1999 [43]; [2] Giardina et al., 2019 [44]
Campylobacter jejuni	Transferrin, Lactoferrin	Miller et al., 2008 [45]
Staphylococcus aureus	Transferrin [3], Heme [4]	[3] Modun et al., 1998 [46]; [4] Skaar and Schneewind, 2004 [47]
Mycobacterium tuberculosis	Transferrin	Clemens et al., 1996 [48]
Candida albicans	Transferrin	Knight et al., 2005 [49]
Aspergillus fumigatus	Transferrin	Hissen et al., 2004 [50]

Interfering with bacterial iron acquisition provides a new therapeutic avenue to fight infection, as discussed later in this review.

3.2.2. Bacterial Sepsis

Sepsis is a life-threatening condition killing around 11 million people worldwide annually [51]. It results from a severely dysregulated host response to inflammation. While sepsis can develop form viral infections such as COVID-19 [51], it most often develops through microbial infection. Bacterial sepsis can develop from serious bacterial infection with the host inflammatory response overstimulated and becoming dysregulated [52], with this often triggered by bacterial cellular components released during infection, including endotoxin [53].

During sepsis, iron metabolism is altered with increased iron uptake into cells and this has been associated with increased iron-driven oxidative injury and cell death [52]. High serum iron levels are also associated with sepsis and poor sepsis outcomes [54]. In fact, transferrin saturation reflecting serum iron availability is strongly correlated to sepsis outcomes, with increased iron availability pronounced in lethally ill patients [54,55]. This aspect, as driven by excess iron, is additional to iron feeding microbial infection, as it relates to host dysregulated inflammation. Thus, the condition of iron dysregulation provides for perfect storm conditions of increased iron availability to support rapidly proliferating microbes on the one hand and increased host damage from the associated dysregulated inflammatory response on the other hand.

Sepsis therapy presents a large unmet medical need with no currently approved therapeutics available for clinical use. However, the underlying iron dysregulation provides a new therapeutic avenue for serious infection/sepsis, as discussed later in this review.

3.2.3. Parasitic Pathogens

Multicellular parasitic organisms also require iron to invade and mount infection in vertebrate hosts [26]. Leishmania chagasi, which causes leishmaniasis, can utilize iron from transferrin, lactoferrin or heme when growing in its promastigote form, taking these up directly without deployment of iron-intercepting siderophores [56]. Trypanosoma brucei employs transferrin surface receptors and Entamoeba histolytica employs lactoferrin receptors for iron acquisition [26]. Plasmodium falciparum, which causes malaria, has also been shown to possess surface receptors for transferrin iron, which is interesting in that it resides within hemoglobin-rich erythrocytes [57]. Addition of iron has been shown to promote infection with Trypanosoma cruzi [58] and iron supplementation has been associated with increased risk of malaria [57]. Parasitic infection, in contrast to most microbial infections, is chronic in nature and has been associated with what is described as the anemia of chronic disease [59]. This also occurs in cancer, as further discussed later, and this chronic anemia is considered the result of the prolonged effort of the host to restrict iron to the invading pathogens or cancer cells. Interestingly, host iron deficiency has been found to be protective against malaria, based on epidemiological studies [57].

3.2.4. Viral Infections

Viruses have no direct requirements for iron, but it is now evident that viral replication in host cells is affected by host iron in the case of both DNA viruses, such as hepatitis B (HBV) and human cytomegalovirus (HCMV), and RNA viruses, such as hepatitis C (HCV) and human immunodeficiency virus (HIV) [27]. It remains somewhat obscure as to whether iron dysregulation predisposes the host to these viral infections, or alternatively, if the viral infection leads to host iron dysregulation. Elevated iron levels are associated with the progression of chronic HBV infection [60] and iron addition has been shown to enhance HCV viral replication in vitro [61]. Transferrin receptor-1 mRNA levels were increased due to HIV infection, leading to an increased iron uptake and higher level of cellular iron [62]. However, it has been suggested that in cirrhotic patients, HBV-related liver injury, but not the HBV infection itself, may be cause changes in serum iron markers [63].

3.3. *Cancer*

The role of iron and its dysregulation in cancer is multi-faceted, as reviewed by Torti et al. (2018) [64]. Stevens et al. (1994) reported on a cohort of 14,000 US National Health and Nutrition survey participants with a major finding that participants with higher transferrin (Tf) iron saturation levels were at higher risk of cancer than participants with lower transferrin Tf saturation levels, a finding supported by subsequent studies [65]. Cancer cells arising from various tissues such as breast or liver have increased need for iron compared to normal cells and produce increased amounts of cell-surface transferrin receptors in response to their increased and continuous iron needs [66]. Targeting transferrin receptors has been proposed for its therapeutic potential [67] as shown using antibodies directed to cancer-cell-surface transferrin receptor TfR1 [68]. This aspect has great appeal as common features to all cancers are possession of TfR1 receptors [68], their altered iron metabolism and increased needs for iron to support their rapid growth [69] and their primary dependence on host transferrin as the source of this iron. Thus, any advances in regard to beneficial iron restriction could apply to various cancers broadly.

Cancer also induces a chronic anemia, as seen in other chronic diseases, and this appears to be a host attempt to restrict cancer growth [64]. Somewhat paradoxically, cancer cells are more susceptible to ferroptosis, a programmed cell death triggered by excess reactive iron [70], suggesting host iron withdrawal defense to restrict cancer growth could also impede cancer cell killing via ferroptosis.

3.4. *Ferroptotic Cell Death*

Ferroptosis is an additional form of regulated cell death (RCD), but unlike other forms such as apoptosis and necrosis, it is caused by an iron-dependent accumulation of lipid peroxides which kill cells, although it shares other common features with the other modes of RCD [70]. ROS produced through the iron-catalyzed Fenton reaction contribute to its initiation [71] and, therefore, iron dysregulation with its excess labile iron is a predisposing factor.

Potential inhibitors of ferroptosis have been investigated including ROS-trapping antioxidants, such as α-tocopherol and ferrostatin-1, and these have shown potential for reducing ferroptotic damage [72].

Excess labile iron driving ferroptosis has also been demonstrated through its suppression by addition of the iron chelator deferoxamine [73]. However, ferric nitriloacetate (NTA) has been shown to promote ferroptosis and renal carcinoma by driving Fenton activity [71]. It is important to note that various iron chelators differ in their abilities to fully coordinate and, therefore, fully chaperone iron [21]. Deferoxamine can fully hexadentate-coordinate iron within a single deferoxamine molecule, while NTA requires two chelator molecules to fully satisfy a single iron atom. Therefore, depending on the prevailing chelator and iron concentrations, incompletely coordinated, Fenton-reactive species can be formed, a problem which appears to underlie the toxicity of some of the currently used medical chelators [21]. This problem of small molecule cell-permeable chelators is compounded in that they can more readily penetrate cells, reaching cellular iron stores and, therefore, mobilize additional labile iron supplies, this in turn exacerbating iron dysregulation.

The excess supply of labile reactive iron observed as a feature of iron dysregulation appears to underpin tissue toxicity with a number of diseases and ferroptosis appears to be at least part of the pathology of these diseases, as further discussed below.

3.5. *Inflammatory Diseases*

While inflammation typically accompanies infection and cancer, other diseases appear to be primarily inflammatory in nature and iron dysregulation and its associated ROS activity are involved in the inflammatory response.

3.5.1. Ocular

Iron dysregulation of the eye has been linked to various eye diseases affecting the cornea and retina, including corneal epithelial disease, corneal endothelial cell dysfunction, retinal pigment epithelial (RPE)-associated eye diseases, glaucoma, diabetic retinopathy (DR), retinal ischemia/reperfusion injury (RIRI), retinoblastoma, retinitis pigmentosa (RP) and age-related cataracts, and common to these, ferroptosis pathology has been implicated [74]. Ferroptotic iron-induced toxicity through ROS damage has been directly linked to corneal diseases such as cataractogenesis, inflammatory retinal diseases such as age-related macular degeneration (AMD) and optic neuropathies [35]. Transferrin mRNA levels are also upregulated in AMD, possibly in response to increased retinal iron loads [75]. This feature provides a potential avenue to new therapeutics, including the use of iron chelators to treat eye diseases that are in urgent need of new therapeutics.

3.5.2. Lung Fibrosis

Pulmonary iron content is also tightly regulated, as excess iron can catalyze ROS formation, and this has been linked to the pathogenesis of chronic inflammatory lung diseases such as idiopathic pulmonary fibrosis [32]. Iron accumulation was increased in lung sections from patients with IPF, and human lung fibroblasts show greater proliferation and cytokine and extracellular matrix responses when exposed to increased iron levels [76]. These authors provided direct evidence for iron overload affecting the progression of pulmonary fibrosis. They investigated whether changes in iron homeostasis are the cause or a consequence of pulmonary fibrosis and used mouse models of iron overload to show that iron accumulation results in impaired lung function and subsequently worse pulmonary fibrosis upon lung injury by bleomycin [76]. Lung fibrosis is a progressive irreversible disease, as fibrotic lung tissue does not repair/remodel, making anti-fibrotic agents an urgent need. Overcoming iron dysregulation to reduce iron-driven fibrosis is a new potential avenue for therapy. Iron chelation therapeutics have potential for use in slowing or stopping pulmonary fibrosis.

3.5.3. Kidney

Increased kidney proximal tubule (PT) cell cytosolic non-transferrin Tf-bound, i.e., labile iron has been shown to induce the generation of ROS in PT cells and this could contribute to the progression of proteinuric chronic kidney diseases [34]. Recent clinical studies using anti-oxidative drugs that serve to reduce labile iron suggest that chelation of iron in the kidney has beneficial effects on the course of chronic kidney disease [77]. In nephritic syndromes associated with damage to the glomerular filter, urinary Tf concentration dramatically increases and may even lead to hypo-transferrinemia, iron loss and microcytic anemia [78].

Nephritic kidney damage with urinary transferrin excretion is also part of the pathology associated with autoimmune diseases including lupus erythematosus [79] and rheumatoid arthritis [80], as further discussed further below.

3.6. Diabetes

High dietary iron and dysregulated iron metabolism are risk factors for type 2 diabetes mellitus (T2DM), affecting most of its features of decreased insulin secretion, insulin resistance and increased hepatic gluconeogenesis [29]. Dysregulated iron metabolism with increased serum levels of ferritin have been found in newly diagnosed type 2 diabetes but not in individuals with pre-diabetes [81]. Oxidative stress from ROS is now known to be a major underlying mediator of diabetic complications [82]. Iron and ferroptosis have been shown to participate in pancreatic beta cell death [83]. In model studies using T2DM, mice insulin secretion was worsened by ferroptosis-inducing compounds. However, quercetin (a natural iron chelator), ferroptosis inhibitor ferrostatin-1 and iron-chelating deferoxamine, each rescued cell viability when cells were challenged with high glucose [84]. These studies support the potential for use of iron-chelating therapeutics in the treatment of T2DM.

3.7. Cardiovascular

Ferroptosis driven by iron dysregulation has now been implicated in several cardiovascular disease conditions including cardiomyopathy, atherosclerotic disease and myocardial ischemia/perfusion injury [30,85]. Inhibition of ferroptosis by ferrostatin-1 improved cardiac function and reduced mortality in a doxorubicin-induced mouse model of cardiomyopathy, which was associated with the release of free cellular iron caused by HO-1 upregulation [86]. However, as pointed out by Li and Zhang (2021), the mechanisms of ferroptosis in heart and vasculature disease remain elusive [30].

3.8. Autoimmune

The role of iron regulation in immune-related diseases was recently reviewed by Cronin et al. (2019) [87]. Substantial evidence has now linked iron dysregulation to the pathogenesis of lupus erythematosus [88]. Importantly, the common serious complication, lupus nephritis, has been linked to renal iron accumulation [79] and ferroptosis kidney cell damage has been described as an important feature of its pathology [88].

3.9. Neurological

Many neurological conditions have been shown to coincide with altered bodily distribution of various transition series biometals, especially in the case of iron [89]. Iron dysregulation with increased labile plasma iron supply and resulting increased ROS, creating oxidative stress and damage on neurological tissues, has now been linked to Friedreich's ataxia, Alzheimer's disease (AD), Parkinson's disease, Multiple Sclerosis (MS) and Amyotrophic Lateral Sclerosis (ALS) [31]. Increased ferroptosis with corresponding destruction of neurological cells has been described in Parkinson's disease and MS [90] as well as for Alzheimer's disease [91].

For example, Parkinson's disease (PD) is characterized by progressive motor impairment attributed to progressive loss of dopaminergic neurons in the substantia nigra (SN) pars compacta. In addition to an accumulation of iron, there is also an increased production of reactive oxygen/nitrogen species (ROS/RNS) and inflammatory markers seen in this pathology [92]. In addition, abnormal iron increases are commonly detected in AD patients, although controversy continues regarding iron's association with AD plaques [91]. It now appears that a common feature of iron dysregulation underlying neurological diseases is iron-driven ferroptosis neurological cell death [93].

3.10. Iron Overload

While body iron overload can follow repeated blood transfusions, as seen in patients with Thalassemia, i.e., because the body lacks an excretory pathway for excess iron, congenital disorders of iron overload such as hemochromatosis represent iron overload disorders caused by dysregulated iron homeostasis. In hemochromatosis, disruption of the hepcidin pathway due to mutations in genes encoding auxiliary factors in iron signaling to hepcidin result in insufficient hepcidin responses to iron intake or to high body iron stores. This causes loss of hepcidin-mediated feedback inhibition in dietary iron absorption and consequently unregulated uptake of dietary iron, elevated transferrin iron saturation and appearance of labile reactive non-transferrin-bound iron [18].

3.11. Cirrhosis

Non-alcoholic fatty liver disease (NAFLD) is the most common chronic liver disease worldwide, beginning with the presence of >5% excessive lipid accumulation in the liver and typically developing into non-alcoholic steatohepatitis, fibrosis, cirrhosis and often hepatocellular carcinoma. Excess free reactive iron and its associated ROS-mediated tissue damage have been positively associated with severity of NAFLD [36].

3.12. Anemia of Chronic Infection and Inflammation

The anemia of inflammation (AI), often referred to as the anemia of chronic disease (ACD), is a secondary anemia that often develops slowly as a result of chronic inflammation during chronic infection (especially parasitic), cancer and with at least some of the other inflammatory diseases we have reviewed, such as diabetes and autoimmune diseases [59,94,95]. Overall, it appears to be a host defensive mechanism. Interestingly, a short-term hypoferremia response is often seen early in the acute stages of infection and this has also been shown to be triggered by inflammatory mediators such as ILK-6 and demonstrated to be a relatively short-term active mechanism for restricting iron supply to growing invaders [21]. The diagnostic challenge in AI/ACD is the identification of patients with concomitant true iron deficiency because they need specific evaluation for the source of blood loss and iron-targeted management strategies [Weiss et al. 2019].

4. Therapeutic Options

4.1. Iron Restriction

Higher heme iron intake and increased body iron stores were significantly associated with a greater risk of type 2 diabetes, as shown in a meta-analysis of 11 prospective studies [96]. Interestingly, blood-letting to reduce overall body iron stores has shown potential benefit in the treatment of diabetes [97].

In principle, restriction of iron uptake would also lower amounts of deleterious liable reactive iron. However, withholding (or supplying) iron has remained controversial in medicine over the years due to the delicate balance required for iron homeostasis and the consequences of triggering anemia. While supply of dietary iron would be safer administration of parenteral iron to treat anemia, especially the anemia of chronic disease has serious implications. Given the roles of dysregulated iron in infection and other diseases, this requires careful consideration and likely should be avoided if possible.

4.2. Iron Chelators

Iron chelators that bind and can sequester excess available labile reactive iron have shown potential for the therapy of several diseases associated with iron dysregulation. However, medical chelators to date have been developed to treat hematological conditions associated with high excess body iron stores (e.g., transfusional iron overload in Thalassemia patients) but were not developed as iron dysregulation therapeutics. We have provided a number of important potential attributes for ideal iron dysregulation chelator therapeutics, as summarized in Table 4.

Table 4. Iron chelators.

Chelator	**DFO Deferoxamine**	**DEF Deferasirox**	**DFP Deferiprone**	**DIBI Hydroxypyridinone**
MW (Da)	561	373	139	9000 avg. polymeric [98]
Fe(III): chelator binding complex	1:1	1:2	1:3	3:1 [98]
Fe binding constant log K	30.6 [99]	36.5 [99]	36.7	41.0 [98]
Hydrophobicity/ hydrophilicity log P	−3 [99]	4.3 [99]	−0.8 [99]	−1.87 [a]
FDA Approval	1968	2005	2011	In Development
Treatment Indication	Transfusional iron overload [100]	Transfusional iron overload [100]	Transfusional iron overload [100]	Anti-infective [101] Anti-inflammatory [102]
Serious limitations and toxicities	Promotes infections [21]	Toxicity, Renal failure [103,104]	Toxicity, Agranulocytosis [104]	None to date from oral and systemic animal testing [b]
Promotes microbial pathogen growth	Very Serious	Potentially Serious	Potentially Serious	No [21]

[a] Holbein, unpublished CRO report: JBL-DMPK-0062-0016, [b] Holbein, unpublished CRO reports: SP-JBL/NG-RIVR-142 and JBL/NG-ROR-052.

A number of clinical trials or natural products with iron-chelating properties, such as curcumin and polyphenolics, have shown promise for lowering excess iron [105] but none have been approved for clinical use.

4.3. *Hepcidin and Agonists*

Hepcidin mimetics, stimulators of its production and ferroportin inhibitors have seen early clinical stage testing as to their safety and potential efficacy [106]. The ferroportin inhibitor VIT-2763 has advanced to phase II trials and it has shown both low toxicity and good potential for lowering serum iron levels [107] Other approaches of using hepcidin mimetics including PTG-300 (rusfertide) have shown promising phase III results. As of yet, none of these have received regulatory approval for ongoing clinical use [108].

5. Future Needs and Conclusions

We identified a number of gaps in our current understanding of iron dysregulation in the pathology of diseases. A better understanding of the role of Fe dysregulation mechanisms for various diseases, especially related to cause or effect, is needed. There is a particular need for further understanding of anemia of chronic disease. Possible targeted ferroptosis induction in cancer can be considered once its mechanisms are elucidated.

A number of conclusions can be drawn at this stage. Therapeutics affecting iron dysregulation and lowering excess levels of labile reactive ion should be readily applicable to treating infection, including sepsis, and should also apply to cancer and other non-proliferative diseases from iron dysregulation. Improved chelators have immediate appeal as the broadest and preferred therapeutic approach.

Hepcidin mimetics and agonists have demonstrated potential and longer-term potential.

Author Contributions: Conceptualization, B.E.H.; writing—original draft preparation, B.E.H.; writing—review and editing, C.L. All authors have read and agreed to the published version of the manuscript.

Funding: This research received no external funding.

Acknowledgments: The authors would like to thank Cassidy Scott for the help with the manuscript.

Conflicts of Interest: The authors declare no conflict of interest.

References

1. Andreini, C.; Putignano, V.; Rosato, A.; Banci, L. The human iron-proteome. *Metallomics* **2018**, *10*, 1223–1231. [CrossRef] [PubMed]
2. Williams, R.J.P. The mechanism of cytochrome oxidase and other reaction centres for electron/proton pumping. *FEBS Lett.* **1987**, *226*, 1–7. [CrossRef] [PubMed]
3. Beinert, H.; Kennedy, M.C.; Stout, C.D. Aconitase as iron-sulfur protein, enzyme, and iron-regulatory protein. *Chem. Rev.* **1996**, *96*, 2335–2373. [CrossRef]
4. Jordan, A.; Reichard, P. Ribonucleotide reductases. *Annu. Rev. Biochem.* **1998**, *67*, 71–98. [CrossRef] [PubMed]
5. Zhang, C. Essential functions of iron-requiring proteins in DNA replication, repair and cell cycle control. *Protein Cell* **2014**, *5*, 750–760. [CrossRef]
6. McDonough, M.A.; Loenarz, C.; Chowdhury, R.; Clifton, I.J.; Schofield, C.J. Structural studies on human 2-oxoglutarate dependent oxygenases. *Curr. Opin. Struct. Biol.* **2010**, *20*, 659–672. [CrossRef]
7. Parker, M.W.; Blake, C.C.F. Iron- and manganese-containing superoxide dismutases can be distinguished by analysis of their primary structures. *FEBS Lett.* **1988**, *229*, 377–382. [CrossRef]
8. Alfonso-Prieto, M.; Biarnés, X.; Vidossich, P.; Rovira, C. The molecular mechanism of the catalase reaction. *J. Am. Chem. Soc.* **2009**, *131*, 11751–11761. [CrossRef] [PubMed]
9. Hayyan, M.; Hashim, M.A.; Alnashef, I.M. Superoxide Ion: Generation and Chemical Implications. *Chem. Rev.* **2016**, *116*, 3029–3085. [CrossRef]
10. Ray, P.D.; Huang, B.W.; Tsuji, Y. Reactive oxygen species (ROS) homeostasis and redox regulation in cellular signaling. *Cell. Signal.* **2012**, *24*, 981–990. [CrossRef]
11. Minakami, R.; Sumimoto, H. Phagocytosis-coupled activation of the superoxide-producing phagocyte oxidase, a member of the NADPH oxidase (Nox) family. *Int. J. Hematol.* **2006**, *84*, 193–198. [CrossRef] [PubMed]
12. Halliwell, B.; Adhikary, A.; Dingfelder, M.; Dizdaroglu, M. Hydroxyl radical is a significant player in oxidative DNA damage: In vivo. *Chem. Soc. Rev.* **2021**, *50*, 8355–8360. [CrossRef] [PubMed]
13. Fang, F.C. Antimicrobial reactive oxygen and nitrogen species: Concepts and controversies. *Nat. Rev. Microbiol.* **2004**, *2*, 820–832. [CrossRef] [PubMed]
14. Yiannikourides, A.; Latunde-Dada, G. A Short Review of Iron Metabolism and Pathophysiology of Iron Disorders. *Medicines* **2019**, *6*, 85. [CrossRef] [PubMed]
15. Kakhlon, O.; Cabantchik, Z.I. The labile iron pool: Characterization, measurement, and participation in cellular processes. *Free Radic. Biol. Med.* **2002**, *33*, 1037–1046. [CrossRef] [PubMed]
16. Ezraty, B.; Barras, F. The "liaisons dangereuses" between iron and antibiotics. *FEMS Microbiol. Rev.* **2016**, *40*, 418–435. [CrossRef]
17. Li, H.; Steyger, P. Synergistic ototoxicity due to noise exposure and aminoglycoside antibiotics. *Noise Health* **2009**, *11*, 26–32. [CrossRef] [PubMed]
18. Pantopoulos, K. Inherited Disorders of Iron Overload. *Front. Nutr.* **2018**, *5*, 103. [CrossRef] [PubMed]
19. Nemeth, E.; Ganz, T. Hepcidin and Iron in Health and Disease. *Annu. Rev. Med.* **2022**, *74*, 261–277. [CrossRef]
20. Vinchi, F. Non-Transferrin-Bound Iron in the Spotlight: Novel Mechanistic Insights into the Vasculotoxic and Atherosclerotic Effect of Iron. *Antioxid. Redox Signal.* **2021**, *35*, 387–414. [CrossRef]
21. Holbein, B.E.; Ang, M.T.C.; Allan, D.S.; Chen, W.; Lehmann, C. Iron-withdrawing anti-infectives for new host-directed therapies based on iron dependence, the Achilles' heel of antibiotic-resistant microbes. *Environ. Chem. Lett.* **2021**, *19*, 2789–2808. [CrossRef]
22. Akinc, A.; Chan-Daniels, A.; Sehgal, A.; Foster, D.; Bettencourt, B.R.; Hettinger, J.; Racie, T.; Aubin, J.; Kuchimanchi, S.; Epstein-Barash, H.; et al. Targeting the Hepcidin Pathway with RNAi Therapeutics for the Treatment of Anemia. *Blood* **2011**, *118*, 688. [CrossRef]
23. Cabantchik, Z.I.; Breuer, W.; Zanninelli, G.; Cianciulli, P. LPI-labile plasma iron in iron overload. *Best Pract. Res. Clin. Haematol.* **2005**, *18*, 277–287. [CrossRef] [PubMed]
24. Cabantchik, Z.I. Labile iron in cells and body fluids: Physiology, pathology, and pharmacology. *Front. Pharmacol.* **2014**, *5*, 45. [CrossRef] [PubMed]
25. Bullen, J.; Griffiths, E.; Rogers, H.; Ward, G. Sepsis: The critical role of iron. *Microbes Infect.* **2000**, *2*, 409–415. [CrossRef] [PubMed]
26. Mach, J.; Sutak, R. Iron in parasitic protists-from uptake to storage and where we can interfere. *Metallomics* **2020**, *12*, 1335–1347. [CrossRef] [PubMed]
27. Schmidt, S. The role of iron in viral infections. *Front. Biosci. (Landmark Ed.)* **2020**, *25*, 893–911. [CrossRef] [PubMed]
28. Torti, S.V.; Manz, D.; Paul, B.; Blanchette-Farra, N.; Torti, F.M. Iron and cancer. *Iron Hum. Dis.* **2018**, *38*, 97–125. [CrossRef]

9. Harrison, A.V.; Lorenzo, F.R.; McClain, D.A. Iron and the Pathophysiology of Diabetes. *Annu. Rev. Physiol.* **2023**, *85*, 339–362. [CrossRef]
0. Li, S.; Zhang, X. Iron in Cardiovascular Disease: Challenges and Potentials. *Front. Cardiovasc. Med.* **2021**, *8*, 707138. [CrossRef]
1. David, S.; Jhelum, P.; Ryan, F.; Jeong, S.Y.; Kroner, A. Dysregulation of Iron Homeostasis in the Central Nervous System and the Role of Ferroptosis in Neurodegenerative Disorders. *Antioxid. Redox Signal.* **2022**, *37*, 150–170. [CrossRef] [PubMed]
2. Ogger, P.P.; Byrne, A.J. Lung fibrosis enters the iron age. *J. Pathol.* **2020**, *252*, 1–3. [CrossRef]
3. Baker, J.F.; Ghio, A.J. Iron homoeostasis in rheumatic disease. *Rheumatology* **2009**, *48*, 1339–1344. [CrossRef] [PubMed]
4. Smith, C.P.; Thévenod, F. Iron transport and the kidney. *Biochim. Biophys. Acta-Gen. Subj.* **2009**, *1790*, 724–730. [CrossRef] [PubMed]
5. Loh, A.; Hadziahmetovic, M.; Dunaief, J.L. Iron homeostasis and eye disease. *Biochim. Biophys. Acta-Gen. Subj.* **2009**, *1790*, 637–649. [CrossRef]
6. Chen, H. Iron metabolism in non-alcoholic fatty liver disease: A promising therapeutic target. *Liver Res.* **2022**, *6*, 203–213. [CrossRef]
7. Schaible, U.E.; Kaufmann, S.H.E. Iron and microbial infection. *Nat. Rev. Microbiol.* **2004**, *2*, 946–953. [CrossRef]
8. Troxell, B.; Hassan, H.M. Transcriptional regulation by Ferric Uptake Regulator (Fur) in pathogenic bacteria. *Front. Cell. Infect. Microbiol.* **2013**, *4*, 59. [CrossRef] [PubMed]
9. Murdoch, C.C.; Skaar, E.P. Nutritional immunity: The battle for nutrient metals at the host-pathogen interface. *Nat. Rev. Microbiol.* **2022**, *20*, 657–670. [CrossRef]
0. Hanstock, H.G.; Edwards, J.P.; Walsh, N.P. Tear lactoferrin and lysozyme as clinically relevant biomarkers of mucosal immune competence. *Front. Immunol.* **2019**, *10*, 1178. [CrossRef] [PubMed]
1. Holbein, B.E. Enhancement of Neisseria meningitidis infection in mice by addition of iron bound to transferrin. *Infect. Immun.* **1981**, *34*, 120–125. [CrossRef]
2. Xiao, R.; Kisaalita, W.S. Iron acquisition from transferrin and lactoferrin by *Pseudomonas aeruginosa* pyoverdin. *Microbiology* **1997**, *143*, 2509–2515. [CrossRef]
3. Yamamoto, S.; Okujo, N.; Kataoka, H.; Narimatsu, S. Siderophore-mediated utilization of transferrin- and lactoferrin-bound iron by *Acinetobacter baumannii*. *J. Health Sci.* **1999**, *45*, 297–302. [CrossRef]
4. Giardina, B.J.; Shahzad, S.; Huang, W.; Wilks, A. Heme uptake and utilization by hypervirulent *Acinetobacter baumannii* LAC-4 is dependent on a canonical heme oxygenase (abHemO). *Arch. Biochem. Biophys.* **2019**, *672*, 108066. [CrossRef] [PubMed]
5. Miller, C.E.; Rock, J.D.; Ridley, K.A.; Williams, P.H.; Ketley, J.M. Utilization of lactoferrin-bound and transferrin-bound iron by Campylobacter jejuni. *J. Bacteriol.* **2008**, *190*, 1900–1911. [CrossRef] [PubMed]
6. Modun, B.; Evans, R.W.; Joannou, C.L.; Williams, P. Receptor-mediated recognition and uptake of iron from human transferrin by *Staphylococcus aureus* and *Staphylococcus epidermidis*. *Infect. Immun.* **1998**, *66*, 3591–3596. [CrossRef]
7. Skaar, E.P.; Schneewind, O. Iron-regulated surface determinants (Isd) of *Staphylococcus aureus*: Stealing iron from heme. *Microbes Infect.* **2004**, *6*, 390–397. [CrossRef] [PubMed]
8. Clemens, D.L.; Horwitz, M.A. The Mycobacterium tuberculosis phagosome interacts with early endosomes and is accessible to exogenously administered transferrin. *J. Exp. Med.* **1996**, *184*, 1349–1355. [CrossRef]
9. Knight, S.A.B.; Vilaire, G.; Lesuisse, E.; Dancis, A. Iron acquisition from transferrin by *Candida albicans* depends on the reductive pathway. *Infect. Immun.* **2005**, *73*, 5482–5492. [CrossRef]
0. Hissen, A.H.T.; Chow, J.M.T.; Pinto, L.J.; Moore, M.M. Survival of *Aspergillus fumigatus* in Serum Involves Removal of Iron from Transferrin: The Role of Siderophores. *Infect. Immun.* **2004**, *72*, 1402–1408. [CrossRef]
1. Olwal, C.O.; Nganyewo, N.N.; Tapela, K.; Djomkam Zune, A.L.; Owoicho, O.; Bediako, Y.; Duodu, S. Parallels in Sepsis and COVID-19 Conditions: Implications for Managing Severe COVID-19. *Front. Immunol.* **2021**, *12*, 744425. [CrossRef] [PubMed]
2. Liu, Q.; Wu, J.; Zhang, X.; Wu, X.; Zhao, Y.; Ren, J. Iron homeostasis and disorders revisited in the sepsis. *Free Radic. Biol. Med.* **2021**, *165*, 1–13. [CrossRef] [PubMed]
3. Lehmann, C.; Islam, S.; Jarosch, S.; Zhou, J.; Hoskin, D.; Greenshields, A.; Al-Banna, N.; Sharawy, N.; Sczcesniak, A.; Kelly, M.; et al. The utility of iron chelators in the management of inflammatory disorders. *Mediat. Inflamm.* **2015**, *2015*, 516740. [CrossRef]
4. Lan, P.; Pan, K.H.; Wang, S.J.; Shi, Q.C.; Yu, Y.X.; Fu, Y.; Chen, Y.; Jiang, Y.; Hua, X.T.; Zhou, J.C.; et al. High Serum Iron level is Associated with Increased Mortality in Patients with Sepsis. *Sci. Rep.* **2018**, *8*, 11072. [CrossRef]
5. Tacke, F.; Nuraldeen, R.; Koch, A.; Strathmann, K.; Hutschenreuter, G.; Trautwein, C.; Strnad, P. Iron parameters determine the prognosis of critically ill patients. *Crit. Care Med.* **2016**, *44*, 1049–1058. [CrossRef]
6. Wilson, M.E.; Vorhies, R.W.; Andersen, K.A.; Britigan, B.E. Acquisition of iron from transferrin and lactoferrin by the protozoan *Leishmania chagasi*. *Infect. Immun.* **1994**, *62*, 3262–3269. [CrossRef]
7. Clark, M.A.; Goheen, M.M.; Cerami, C. Influence of host iron status on Plasmodium falciparum infection. *Front. Pharmacol.* **2014**, *5*, 84. [CrossRef] [PubMed]
8. Lalonde, R.G.; Holbein, B.E. Role of iron in trypanosoma cruzi infection of mice. *J. Clin. Investig.* **1984**, *73*, 470–476. [CrossRef]
9. Wiciński, M.; Liczner, G.; Cadelski, K.; Kołnierzak, T.; Nowaczewska, M.; Malinowski, B. Anemia of chronic diseases: Wider diagnostics—Better treatment? *Nutrients* **2020**, *12*, 1784. [CrossRef]
0. Wei, Y.; Ye, W.; Zhao, W. Serum iron levels decreased in patients with HBV-related hepatocellular carcinoma, as a risk factor for the prognosis of HBV-related HCC. *Front. Physiol.* **2018**, *9*, 66. [CrossRef] [PubMed]

61. Kakizaki, S.; Takagi, H.; Horiguchi, N.; Toyoda, M.; Takayama, H.; Nagamine, T.; Mori, M.; Kato, N. Iron enhances hepatitis C virus replication in cultured human hepatocytes. *Liver* **2000**, *20*, 125–128. [CrossRef] [PubMed]
62. Chang, H.C.; Bayeva, M.; Taiwo, B.; Palella, F.J.; Hope, T.J.; Ardehali, H. Short communication: High cellular iron levels are associated with increased HIV infection and replication. *AIDS Res. Hum. Retrovir.* **2015**, *31*, 305–312. [CrossRef] [PubMed]
63. Mao, W.L.; Hu, Y.; Lou, Y.F.; Chen, Y.M.; Zhang, J.W. Abnormal serum iron markers in chronic hepatitis B virus infection may be because of liver injury. *Eur. J. Gastroenterol. Hepatol.* **2015**, *27*, 130–136. [CrossRef] [PubMed]
64. Torti, S.V.; Torti, F.M. Iron and cancer: More ore to be mined. *Nat. Rev. Cancer* **2013**, *13*, 342–355. [CrossRef] [PubMed]
65. Wu, T.; Sempos, C.T.; Freudenheim, J.L.; Muti, P.; Smit, E. Serum iron, copper and zinc concentrations and risk of cancer mortality in US adults. *Ann. Epidemiol.* **2004**, *14*, 195–201. [CrossRef] [PubMed]
66. Greenshields, A.L.; Power Coombs, M.R.; Fernando, W.; Holbein, B.E.; Hoskin, D.W. DIBI, a novel 3-hydroxypyridin-4-one chelator iron-binding polymer, inhibits breast cancer cell growth and functions as a chemosensitizer by promoting S-phase DNA damage. *BioMetals* **2019**, *32*, 909–921. [CrossRef]
67. Shen, Y.; Li, X.; Dong, D.; Zhang, B.; Xue, Y.; Shang, P. Transferrin receptor 1 in cancer: A new sight for cancer therapy. *Am. J. Cancer Res.* **2018**, *8*, 916–931.
68. Candelaria, P.V.; Leoh, L.S.; Penichet, M.L.; Daniels-Wells, T.R. Antibodies Targeting the Transferrin Receptor 1 (TfR1) as Direct Anti-cancer Agents. *Front. Immunol.* **2021**, *12*, 607692. [CrossRef]
69. Zhang, S.; Chang, W.; Wu, H.; Wang, Y.H.; Gong, Y.W.; Zhao, Y.L.; Liu, S.H.; Wang, H.Z.; Svatek, R.S.; Rodriguez, R.; et al. Pan-cancer analysis of iron metabolic landscape across the Cancer Genome Atlas. *J. Cell. Physiol.* **2020**, *235*, 1013–1024. [CrossRef]
70. Lei, P.; Bai, T.; Sun, Y. Mechanisms of ferroptosis and relations with regulated cell death: A review. *Front. Physiol.* **2019**, *10*, 139. [CrossRef]
71. Toyokuni, S.; Yanatori, I.; Kong, Y.; Zheng, H.; Motooka, Y.; Jiang, L. Ferroptosis at the crossroads of infection, aging and cancer. *Cancer Sci.* **2020**, *111*, 2665–2671. [CrossRef]
72. Angeli, J.P.F.; Shah, R.; Pratt, D.A.; Conrad, M. Ferroptosis Inhibition: Mechanisms and Opportunities. *Trends Pharmacol. Sci.* **2017**, *38*, 489–498. [CrossRef]
73. Yang, W.S.; Stockwell, B.R. Synthetic Lethal Screening Identifies Compounds Activating Iron-Dependent, Nonapoptotic Cell Death in Oncogenic-RAS-Harboring Cancer Cells. *Chem. Biol.* **2008**, *15*, 234–245. [CrossRef]
74. Zhang, J.; Sheng, S.; Wang, W.; Dai, J.; Zhong, Y.; Ren, J.; Jiang, K.; Li, S.; Bian, X.; Liu, L. Molecular Mechanisms of Iron Mediated Programmed Cell Death and Its Roles in Eye Diseases. *Front. Nutr.* **2022**, *9*, 811469. [CrossRef] [PubMed]
75. Chowers, I.; Wong, R.; Dentchev, T.; Farkas, R.H.; Iacovelli, J.; Gunatilaka, T.L.; Medeiros, N.E.; Presley, J.B.; Campochiaro, P.A.; Curcio, C.A.; et al. The iron carrier transferrin is upregulated in retinas from patients with age-related macular degeneration. *Investig. Ophthalmol. Vis. Sci.* **2006**, *47*, 2135–2140. [CrossRef]
76. Ali, M.K.; Kim, R.Y.; Brown, A.C.; Donovan, C.; Vanka, K.S.; Mayall, J.R.; Liu, G.; Pillar, A.L.; Jones-Freeman, B.; Xenaki, D.; et al. Critical role for iron accumulation in the pathogenesis of fibrotic lung disease. *J. Pathol.* **2020**, *251*, 49–62. [CrossRef] [PubMed]
77. Swaminathan, S.; Shah, S.V. Novel approaches targeted toward oxidative stress for the treatment of chronic kidney disease. *Curr. Opin. Nephrol. Hypertens.* **2008**, *17*, 143–148. [CrossRef] [PubMed]
78. Prinsen, B.H.C.M.T.; De Sain-van der Velden, M.G.M.; Kaysen, G.A.; Straver, H.W.H.C.; Van Rijn, H.J.M.; Stellaard, F.; Berger, R.; Rabelink, T.J. Transferrin synthesis is increased in nephrotic patients insufficiently to replace urinary losses. *J. Am. Soc. Nephrol.* **2001**, *12*, 1017–1025. [CrossRef] [PubMed]
79. Theut, L.R.; Dsouza, D.L.; Grove, R.C.; Boesen, E.I. Evidence of Renal Iron Accumulation in a Male Mouse Model of Lupus. *Front. Med.* **2020**, *7*, 516. [CrossRef]
80. Kochi, M.; Kohagura, K.; Shiohira, Y.; Iseki, K.; Ohya, Y. Chronic kidney disease, inflammation, and cardiovascular disease risk in rheumatoid arthritis. *J. Cardiol.* **2018**, *71*, 277–283. [CrossRef]
81. Venkatesan, P.; Varghese, J.; Arthi, T.S.; James, J.V.; Anura, A.; Prasad, J.; Jacob, M. Evidence of dysregulated iron homeostasis in newly diagnosed diabetics, but not in pre-diabetics. *J. Diabetes Complicat.* **2021**, *35*, 107977. [CrossRef] [PubMed]
82. Giacco, F.; Brownlee, M. Oxidative stress and diabetic complications. *Circ. Res.* **2010**, *107*, 1058–1070. [CrossRef] [PubMed]
83. Li, J.; Cao, F.; Yin, H.L.; Huang, Z.J.; Lin, Z.T.; Mao, N.; Sun, B.; Wang, G. Ferroptosis: Past, present and future. *Cell Death Dis.* **2020**, *11*, 88. [CrossRef]
84. Li, D.; Jiang, C.; Mei, G.; Zhao, Y.; Chen, L.; Liu, J.; Tang, Y.; Gao, C.; Yao, P. Quercetin alleviates ferroptosis of pancreatic β cells in type 2 diabetes. *Nutrients* **2020**, *12*, 2954. [CrossRef]
85. Fang, X.; Ardehali, H.; Min, J.; Wang, F. The molecular and metabolic landscape of iron and ferroptosis in cardiovascular disease. *Nat. Rev. Cardiol.* **2023**, *20*, 7–23. [CrossRef]
86. Fang, X.; Wang, H.; Han, D.; Xie, E.; Yang, X.; Wei, J.; Gu, S.; Gao, F.; Zhu, N.; Yin, X.; et al. Ferroptosis as a target for protection against cardiomyopathy. *Proc. Natl. Acad. Sci. USA* **2019**, *116*, 2672–2680. [CrossRef] [PubMed]
87. Cronin, S.J.F.; Woolf, C.J.; Weiss, G.; Penninger, J.M. The Role of Iron Regulation in Immunometabolism and Immune-Related Disease. *Front. Mol. Biosci.* **2019**, *6*, 116. [CrossRef]
88. Wincup, C.; Sawford, N.; Rahman, A. Pathological mechanisms of abnormal iron metabolism and mitochondrial dysfunction in systemic lupus erythematosus. *Expert Rev. Clin. Immunol.* **2021**, *17*, 957–967. [CrossRef] [PubMed]
89. Pfaender, S.; Grabrucker, A.M. Characterization of biometal profiles in neurological disorders. *Metallomics* **2014**, *6*, 960–977. [CrossRef]

90. Hadzhieva, M.; Kirches, E.; Mawrin, C. Review: Iron metabolism and the role of iron in neurodegenerative disorders. *Neuropathol. Appl. Neurobiol.* **2014**, *40*, 240–257. [CrossRef]
91. Das, N.; Raymick, J.; Sarkar, S. Role of metals in Alzheimer's disease. *Metab. Brain Dis.* **2021**, *36*, 1627–1639. [CrossRef] [PubMed]
92. Medeiros, M.S.; Schumacher-Schuh, A.; Cardoso, A.M.; Bochi, G.V.; Baldissarelli, J.; Kegler, A.; Santana, D.; Chaves, C.M.M.B.S.; Schetinger, M.R.C.; Moresco, R.N.; et al. Iron and oxidative stress in Parkinson's disease: An observational study of injury biomarkers. *PLoS ONE* **2016**, *11*, e0146129. [CrossRef] [PubMed]
93. Ren, J.X.; Sun, X.; Yan, X.L.; Guo, Z.N.; Yang, Y. Ferroptosis in Neurological Diseases. *Front. Cell. Neurosci.* **2020**, *14*, 218. [CrossRef]
94. Ismaiel, A.; Srouji, N. Al Anemia of Chronic Disease: Epidemiology and Pathophysiological Mechanisms—Literature Review. *Glob. J. Med. Ther.* **2020**, *2*, 8–16. [CrossRef]
95. Weiss, G.; Ganz, T.; Goodnough, L.T. Anemia of inflammation. *Blood* **2019**, *133*, 40–50. [CrossRef]
96. Bao, W.; Rong, Y.; Rong, S.; Liu, L. Dietary iron intake, body iron stores, and the risk of type 2 diabetes: A systematic review and meta-analysis. *BMC Med.* **2012**, *10*, 119. [CrossRef]
97. Fernández-Real, J.M.; López-Bermejo, A.; Ricart, W. Cross-talk between iron metabolism and diabetes. *Diabetes* **2002**, *51*, 2348–2354. [CrossRef]
98. Gumbau-Brisa, R.; Ang, M.T.C.; Holbein, B.E.; Bierenstiel, M. Enhanced Fe^{3+} binding through cooperativity of 3-hydroxypyridin-4-one groups within a linear co-polymer: Wrapping effect leading to superior antimicrobial activity. *BioMetals* **2020**, *33*, 339–351. [CrossRef]
99. Vlachodimitropoulou, E.; Chen, Y.L.; Garbowski, M.; Koonyosying, P.; Psaila, B.; Sola-Visner, M.; Cooper, N.; Hider, R.; Porter, J. Eltrombopag: A powerful chelator of cellular or extracellular iron(III) alone or combined with a second chelator. *Blood* **2017**, *130*, 1923–1933. [CrossRef]
100. Foley, T.L.; Simeonov, A. Targeting iron assimilation to develop new antibacterials. *Expert Opin. Drug Discov.* **2012**, *7*, 831–847. [CrossRef] [PubMed]
101. Parquet, M.d.C.; Savage, K.A.; Allan, D.S.; Davidson, R.J.; Holbein, B.E. Novel Iron-Chelator DIBI Inhibits *Staphylococcus aureus* Growth, Suppresses Experimental MRSA Infection in Mice and Enhances the Activities of Diverse Antibiotics in vitro. *Front. Microbiol.* **2018**, *9*, 1811. [CrossRef] [PubMed]
102. Fokam, D.; Dickson, K.; Kamali, K.; Holbein, B.; Colp, P.; Stueck, A.; Zhou, J.; Lehmann, C. Iron chelation in murine models of systemic inflammation induced by gram-positive and gram-negative toxins. *Antibiotics* **2020**, *9*, 283. [CrossRef] [PubMed]
103. Zeidan, A.M.; Griffiths, E.A. To chelate or not to chelate in MDS: That is the question! *Blood Rev.* **2018**, *32*, 368–377. [CrossRef] [PubMed]
104. Badeli, H.; Baghersalimi, A.; Eslami, S.; Saadat, F.; Rad, A.H.; Basavand, R.; Papkiadeh, S.R.; Darbandi, B.; Kooti, W.; Peluso, I. Early kidney damage markers after Deferasirox treatment in patients with thalassemia major: A case-control study. *Oxid. Med. Cell. Longev.* **2019**, *2019*, 5461617. [CrossRef] [PubMed]
105. Xu, T.; Zhang, X.; Liu, Y.; Wang, H.; Luo, J.; Luo, Y.; An, P. Effects of dietary polyphenol supplementation on iron status and erythropoiesis: A systematic review and meta-analysis of randomized controlled trials. *Am. J. Clin. Nutr.* **2021**, *114*, 780–793. [CrossRef] [PubMed]
106. Casu, C.; Nemeth, E.; Rivella, S. Hepcidin agonists as therapeutic tools. *Blood* **2018**, *131*, 1790–1794. [CrossRef] [PubMed]
107. Taher, A.; Kourakli-Symeonidis, A.; Tantiworawit, A.; Wong, P.; Szecsödy, P. S272: Safety and Preliminary Pharmacodynamic Effects of the Ferroportin Inhibitor Vamifeport (Vit-2763) in Patients with Non-Transfusion-Dependent Beta Thalassemia (Ntdt): Results from a Phase 2a Study. *HemaSphere* **2022**, *6*, 173–174. [CrossRef]
108. Verstovsek, S.; Kuykendall, A.T.; Hoffman, R.; Ginzburg, Y.; Pemmaraju, N.; Valone, F.; Modi, N.B.; Khanna, S.; O'Connor, P.G.; Gupta, S.K.; et al. A Phase 3 Study of the Hepcidin Mimetic Rusfertide (PTG-300) in Patients with Polycythemia Vera. *Blood* **2021**, *138*, 1504. [CrossRef]

antioxidants

MDPI

Article

Deubiquitylase OTUD3 Mediates Endoplasmic Reticulum Stress through Regulating Fortilin Stability to Restrain Dopaminergic Neurons Apoptosis

Ling Chen [1], Xuejie Huan [1], Fengju Jia [1], Zhen Zhang [1], Mingxia Bi [1], Lin Fu [1], Xixun Du [1], Xi Chen [1], Chunling Yan [1], Qian Jiao [1,*] and Hong Jiang [1,2,*]

[1] Department of Physiology, Shandong Provincial Key Laboratory of Pathogenesis and Prevention of Neurological Disorders and State Key Disciplines, Physiology, School of Basic Medicine, Qingdao University, Qingdao 266071, China

[2] College of Health and Life Science, University of Health and Rehabilitation Sciences, Qingdao 266071, China

* Correspondence: jiaoqian@qdu.edu.cn (Q.J.); hongjiang@qdu.edu.cn (H.J.); Tel.: +86-532-8595-0188 (H.J.)

Abstract: OTU domain-containing protein 3 (OTUD3) knockout mice exhibited loss of nigral dopaminergic neurons and Parkinsonian symptoms. However, the underlying mechanisms are largely unknown. In this study, we observed that the inositol-requiring enzyme 1α (IRE1α)-induced endoplasmic reticulum (ER) stress was involved in this process. We found that the ER thickness and the expression of protein disulphide isomerase (PDI) were increased, and the apoptosis level was elevated in the dopaminergic neurons of OTUD3 knockout mice. These phenomena were ameliorated by ER stress inhibitor tauroursodeoxycholic acid (TUDCA) treatment. The ratio of p-IRE1α/IRE1α, and the expression of X-box binding protein 1-spliced (XBP1s) were remarkably increased after OTUD3 knockdown, which was inhibited by IRE1α inhibitor STF-083010 treatment. Moreover, OTUD3 regulated the ubiquitination level of Fortilin through binding with the OTU domain. OTUD3 knockdown resulted in a decrease in the interaction ability of IRE1α with Fortilin and finally enhanced the activity of IRE1α. Taken together, we revealed that OTUD3 knockout-induced injury of dopaminergic neurons might be caused by activating IRE1α signaling in ER stress. These findings demonstrated that OTUD3 played a critical role in dopaminergic neuron neurodegeneration, which provided new evidence for the multiple and tissue-dependent functions of OTUD3.

Keywords: OTUD3; ER stress; IRE1α; XBP1s; Fortilin

Citation: Chen, L.; Huan, X.; Jia, F.; Zhang, Z.; Bi, M.; Fu, L.; Du, X.; Chen, X.; Yan, C.; Jiao, Q.; et al. Deubiquitylase OTUD3 Mediates Endoplasmic Reticulum Stress through Regulating Fortilin Stability to Restrain Dopaminergic Neurons Apoptosis. *Antioxidants* **2023**, *12*, 809. https://doi.org/10.3390/antiox12040809

Academic Editor: Yan-Zhong Chang

Received: 26 February 2023
Revised: 19 March 2023
Accepted: 22 March 2023
Published: 26 March 2023

1. Introduction

Parkinson's disease (PD) is a degenerative neurodegenerative disease that mainly affects middle-aged and elderly people [1,2]. The main cause of classical motor symptoms in PD is selective death of dopaminergic neurons in the substantia nigra (SN), leading to a decrease in dopamine release from SN-striatal projections [3,4]. Up to now, the pathogenesis of PD has not been fully elucidated. Our previous study has shown that OTU domain-containing protein 3 (OTUD3) prevents PD through stabilizing iron regulatory protein 2 (IRP2), and OTUD3 knockout mice showed dopaminergic neuronal death in the SN and Parkinsonian symptoms [5]. OTUD3 is a member of the OTU subfamily of the deubiquitinases (DUBs) family, which is highly correlated with tumorigenesis. In addition, there is mounting evidence that OTUD3 is involved in a variety of diseases other than tumors, such as ulcerative colitis [6], ribosome-related quality control [7], and innate antiviral immune [8]. However, the mechanisms of dopaminergic neuronal death in OTUD3 knockout mice remain unclear yet.

As one of the largest organelles in eukaryotic cells, the endoplasmic reticulum (ER) plays a vital role in protein synthesis and storage [9]. Cells will initiate ER stress when misfolded or unfolded proteins accumulate in the cytosol [10]. To alleviate the pressure of

ER stress, cells initiate the unfolded protein response (UPR). The binding immunoglobulin protein (Bip) dissociates from the three ER stress-sensing proteins: PKR-like ER protein kinase (PERK), inositol-requiring enzyme 1α (IRE1α) and activating transcription factor 6α (ATF6α) under UPR, thereby regulating the activation of downstream pathways. Bip is an ER chaperone protein and one of the key proteins that maintain the protein homeostasis of ER [11]. It corrects the misfolding and assembly and inhibits the transport of misfolded proteins or protein subunits [12–14]. It also binds to misfolded proteins and unassembled complexes, initiating ER-associated degradation (ERAD), responsible for UPR regulation [15].

Fortilin, also known as translationally controlled tumor protein (TCTP), is a multifunctional protein that contains 172 amino acids and consists of β-stranded core domain, α-helical domain, and a flexible loop. Fortilin is widely distributed in the cytoplasm, nucleus [16], mitochondria [17], and has various functions, such as anti-apoptosis, pro-survival, and pro-proliferation, as well as the removal of excess Ca^{2+} from cells [16,18–20]. Recent studies have suggested that Fortilin participates in ER stress-induced apoptosis by binding to IRE1α [21]. A study has shown mTORC1 is involved in the regulation of the level of Fortilin, the reduction of cellular Fortilin levels upon mTORC1 inhibition [22]. Overexpression of miR-27b significantly decreased Fortilin protein and gene levels in both HSC-3 and Cal-27 cell lines [23]. However, whether there are other factors regulating Fortilin still needs further research.

Our previous research demonstrated that the iron content in the SN of OTUD3 knockout mice increased by approximately two folds [5]. The abnormal iron metabolism can trigger ER stress [24]. Over the past few decades, increasing evidence has suggested that ER stress-induced cellular damage is implicated in the pathogenesis of neurodegenerative diseases [25]. In 2007, the ER stress was first discovered in dopaminergic neurons of PD patients [26]. Studies have indicated that the incidence of ER stress is closely related to the death of dopaminergic neurons in the SN [27]. Neurotoxicity of 1-methyl-4-phenylpyridinium (MPP^+) and accumulation of α-synuclein (α-syn) lead to increased expression of protein disulphide isomerase (PDI), which is a marker protein of ER stress [28]. Meanwhile, elevated levels of p-PERK, C/EBP homologous protein (CHOP), and p-IRE1α can be detected in PD cell models induced by 6-hydroxydopamine (6-OHDA), MPP^+, and rotenone [27,29,30]. p-PERK promotes the phosphorylation of eukaryotic translation initiation factor 2 alpha (eIF2α), which significantly reduces protein synthesis in the ER. p-eIF2α selectively enhances the translation of activating transcription factor 4 (ATF4) to mitigate ER stress and restore protein synthesis [31]. Under acute ER stress, the hyperactivation of PERK can increase the expression of pro-apoptotic factor CHOP by upregulating ATF4 [32–35]. p-IRE1α can catalyze the unconventional splicing of X-box binding protein 1 (XBP1) mRNA to form XBP1-spliced (XBP1s), which stimulates the expansion of ER and the synthesis of secreted proteins [36–38]. This evidence suggests that ER stress is a key factor contributing to dopaminergic neuron death. However, the relationship between OTUD3 and ER stress has not been reported. Whether OTUD3 affected ER stress through novel targets, thereby affecting the survival of dopaminergic neurons, remains unclear.

In this study, we investigated the role of OTUD3 in dopaminergic neuron death and further explore its underlying mechanisms. We first reported that ER stress is present in SN dopaminergic neurons in *OTUD3*$^{-/-}$ mice. Furthermore, we explored OTUD3-induced ER stress through IRE1α pathway, which was caused by the deubiquitination of Fortilin.

2. Materials and Methods

2.1. Plasmids and Viruses

Full-length OTUD3 WT was cloned into the pCMV-Myc vectors as indicated. Lentiviruses carrying shRNA targeting human OTUD3 (Target sequence: GGACAATAACAGAAGCGAA) were from OBIO (Shanghai, China). The plasmids Myc-Fortilin, Flag-Fortilin, Flag-OTU, Flag-UBA, Flag-UBA + Tail and Flag-Tail were purchased from OBIO Technology.

2.2. Cell Culture and Transfection

SH-SY5Y cells were purchased from the National Infrastructure of Cell Line Resource (Shanghai, China) and cultured in MEM/F12 medium supplemented with 15% fetal bovine serum (Gibco, Billings, MT, USA) and 1% penicillin-streptomycin at 37 °C with 5% CO_2. SH-SY5Y cell line is the subline of SK-N-SH cell line after three clones, which possesses moderate dopamine β hydroxylase activity. This cell line is used for PD research because of its human origin, catecholaminergic neuronal properties, and ease of maintenance [39]. Human embryonic kidney cell HEK293T was purchased from the National Collection of Authenticated Cell Cultures (Shanghai, China) and cultured in DMEM-High Glucose supplemented with 10% fetal bovine serum (Gibco, Billings, MT, USA) and 1% penicillin-streptomycin at 37 °C with 5% CO_2. Cell transfection was performed using the Lipofectamine 2000 (Invitrogen, Waltham, MA, USA) reagent according to the manufacturer's protocol.

2.3. Cell Viability Assessment

Cell viability was measured using Cell Counting Kit-8 (CCK-8, K1018, APExBIO, Houston, TX, USA). Cells were seeded in 96-well plates, and at the end of treatment, add 10 μL of CCK-8 solution to each well of the plate. Incubate the plate for 2 h. Measure the absorbance at 450 nm using a multimode plate reader (PerkinElmer, VICTOR Nivo, Waltham, MA, USA).

2.4. Trypan Blue Assay

SH-SY5Y cells that were treated with lentivirus were washed with Hank's Balanced Salt Solution (HBSS, 14025-092, Gibco, Billings, MT, USA), then the cells were incubated with a 0.4% trypan blue (T8154, Sigma Aldrich, St. Louis, MO, USA) mixture and stained for 3–5 min in an incubator at 37 °C. Finally, all cells were rewashed with HBSS and observed under a microscope. Dead cells were stained blue.

2.5. Propidium Iodide (PI) Assay

SH-SY5Y cells which treated with lentivirus were seeded at a density of 1×10^5 cells/well in 48-well plates. On the following day, PI (556547, BD Biosciences, Franklin Lakes, NJ, USA) solution was added into each well for 10 min in an incubator at 37 °C and then washed three times with HBSS solution. Thereafter, the cells were visualized by an inverted fluorescence microscope.

2.6. Measurement of Intracellular MDA Levels

The MDA concentration measurement was based on the protocol of the MDA assay kit (Beyotime, Shanghai, China). Briefly, cells were lysed with RIPA lysis buffer and quantified protein, then 200 μL of MDA working solution was added to 100 μL of protein homogenate, then heated the mixtures at 100 °C for 15 min. After cooling to room temperature, centrifuge at 1000 g for 10 min, pipette 200 μL of supernatant into a 96-well plate, and measure the absorbance at 532 nm with a multimode plate reader (PerkinElmer, VICTOR Nivo, Waltham, MA, USA). MDA content in cells is the concentration of MDA per unit mass.

2.7. Flow Cytometric Measurement of Apoptosis

The cells were washed with PBS and stained with 7-Amino-Actinomycin and PE Annexin according to the manufacturer's protocol (Bioscience, Bellingham, WA, USA). Early apoptotic cells (PE Annexin V positive, 7-Amino-Actinomycin negative), end-stage apoptosis, and death (PE Annexin V positive, 7-Amino-Actinomycin positive) were then determined by flow V cytometry.

2.8. Animals and In Vivo Treatment

The OTUD3 transgenic mice model was generated by the Model Animal Research Center of Nanjing University. 6–8 months old OTUD3 transgenic mice were applied in the study. Animals were maintained in controlled temperature and humidity rooms on a 12 h light/dark cycle with free access to food and water. The *OTUD3$^{-/-}$* mice were injected daily with TUDCA (150 mg/kg/day) (Sigma Aldrich, St. Louis, MO, USA) by intraperitoneal injection once a day for 2 weeks. Controls received injections with an equivalent volume of vehicle (0.9% NaCl). Animal experiments were carried out according to the guidelines of the National Institutes of Health Guidelines for the Care and Use of Laboratory Animals. All protocols were approved by the Animal Ethics Committee of Qingdao University (QDU-AEC-2023026).

2.9. RT-PCR and Quantitative PCR

Total cell RNA was prepared using Trizol reagent (Invitrogen) following the manufacturer's instructions. 1 μg RNA was used for cDNA synthesis in a 20 μL reaction with the reverse transcription kit (Vazyme, Nanjing, China). PCRs were performed in 20 μL reaction volumes with SYBR Green PCR master mix (Vazyme, Nanjing, China) and 0.2 μM specific primers. The primer sequences used for all qPCRs are described below. Mouse Bip: 5′-TGATGCCCAGCGACAAGC-3′ and 5′-CACCCAGGTCAAACACAAG GAT-3′; Human Bip: 5′-ACCGCTGAGGCTTATTTGGG-3′ and 5′-GCTGCCGTAGGCTC GTTGA-3′. GAPDH: 5′-GCACCGTCAAGGCTGAGAAC-3′ and 5′-TGGTGAAGACGCCAGTGGA-3′.

2.10. Transmission Electron Microscopy Measurement of Changes in Endoplasmic Reticulum

Transmission electron microscopy was performed using a Tecnai 10 microscope (FEI, Hillsboro, OR, USA) at the Electron Microscopy Core Facility, Qingdao University. The ER thickness was quantified by measuring the distance between two membranes of the ER lumen in the Image J photographs. Briefly, values for the distance between the two membranes of ER lumen from randomly chosen five areas of each TEM photograph were averaged.

2.11. Western Blotting and Immunoprecipitation

Samples from cells and animals were lysed in RIPA lysis buffer containing a protease inhibitor and phosphatase inhibitor cocktail. The protein concentration was detected by BCA kits (Thermo Fisher Scientific, Waltham, MA, USA). Samples were separated by SDS-poly acrylamide gelelectrophoresis and transferred to a PVDF membrane. After blocked with 100 g/L non-fat milk for 2 h at room temperature. The samples were then subjected to immunoblotting with the indicated antibodies. For immunoprecipitation assays, cells were lysed with HEPES lysis buffer (20 mM HEPES, pH 7.2, 50 mM NaCl, 1 mM NaF, 0.5% Triton X-100) supplemented with a protease-inhibitor cocktail (Roche, Basel, Switzerland). Immunoprecipitations were performed using the indicated primary antibody and protein A/G agarose beads (Santa Cruz Biotechnology, Santa Cruz, CA, USA) at 4 °C. The immunocomplexes were then washed with HEPES lysis buffer five times. Both lysates and immunoprecipitates were examined using the indicated primary antibodies followed by detection with the related secondary antibody. The following primary antibodies were used at the indicated dilutions/concentrations: Mouse anti-ATF6α (1:1000, sc-166659, Santa Cruz Biotechnology, Santa Cruz, CA, USA); Rabbit anti-phospho-eIF2α (Ser51) (1:000, #3398, Cell Signaling Technologies, Danvers, MA, USA); Rabbit anti-eIF2α (1:1000, #5324, Cell Signaling Technologies, Danvers, MA, USA); Rabbit anti-phospho-PERK (Thr 982) (1:000, abs137056, Absin, Shanghai, China); Rabbit anti-PERK (1:000, 3192, Cell Signaling Technologies, Danvers, MA, USA); Rabbit anti-phospho-IRE1α (1:000, ab48187, Abcam, UK), Rabbit anti-IRE1α (1:000, #3294, Cell Signaling Technologies, Danvers, MA, USA); Rabbit anti-XBP1s (1:000, #40435, Cell Signaling Technologies, Danvers, MA, USA); Mouse anti-CHOP (1:000, #2895, Cell Signaling Technologies, Danvers, MA, USA); Rabbit anti-Fortilin monoclonal antibody (1:000, ab133568, Abcam, UK); Rabbit anti-ATF4 (1:500, #11815, Cell Signaling Technologies, Danvers, MA, USA); Mouse anti-ubiquitin (1:1000, #3936, Cell Signaling Technologies); Rabbit anti-Bip (1:000, #3177, Cell Signaling Technologies, Danvers, MA, USA). Mouse-anti-OTUD3 (1:000, MABS1819, Merck

Millipore, Burlington, MA, USA); Rabbit-anti-OTUD3 (1:1000, HPA028544, Sigma Aldrich, Gillingham, UK). Anti-DDDDK-tag mAb (1:1000, #M185-3L, MBL); Anti-Myc-tag mAb (1:1000, #M192-3, MBL, Tsukuba, Japan). The secondary antibodies were used for Western blotting analysis: anti-rabbit IgG-HRP and anti-mouse IgG-HRP (Santa Cruz Biotechnology, Santa Cruz, CA, USA).

2.12. *In Vitro Ubiquitin Conjugation Assay*

Cells were treated with 20 μM of the proteasome inhibitor MG132 (Calbiochem) for 8 h before sample collection. The cells were washed with PBS, pelleted, and lysed in HEPES buffer plus 1% DTT and 1% protease inhibitor. The lysates were centrifuged to obtain cytosolic proteins and incubated with anti-Fortilin antibodies for 5 h and with protein A/G agarose beads for a further 8 h at 4 °C. The beads were then washed five times with HEPES buffer. The proteins were released from the beads by boiling them in SDS-PAGE sample buffer and analyzed by immunoblotting with an anti-ubiquitin monoclonal antibody.

2.13. *Immunofluorescence Staining*

6–8-month-old $OTUD3^{-/-}$ mice with 50 mg/kg sodium pentobarbital anesthetized were perfused intracardially with 0.9% NaCl followed by 4% (w/v) paraformaldehyde solution (PFA). Brains were removed and post-fixed in PFA overnight at 4 °C, then gradually transferred to 20% (w/v), and 30% (w/v) sucrose until sectioning. Sections (20 μm) were cut on a freezing microtome (CM1905, Leica, Wetzlar, Germany). After being washed three times in PBST (0.3% Tween-20 in 0.1 M PBS), sections were blocked by 10% goat serum and then incubated overnight with primary antibody of mouse anti-TH (1:500, MAB318, Merck Millipore, Burlington, MA, USA) and rabbit anti-PDI (1:200, #3501, Cell Signaling Technologies, Danvers, MA, USA) at 4 °C overnight. Furthermore, the sections were incubated in the second antibody, Alexa Fluor 555 donkey anti-rabbit IgG (dilution used = 1:500, A31572, Invitrogen, Waltham, MA, USA), Alexa Fluor 488 donkey anti-mouse IgG (1:500, A21202, Invitrogen, Waltham, MA, USA), for 1 h at room temperature, and then added DAPI for 5 min.

2.14. *Regents*

TUDCA (100 μM) (Sigma Aldrich, St. Louis, MO, USA) was dissolved in ddH_2O. STF-083010 (40μM) (Sigma Aldrich, St. Louis, MO, USA) was dissolved in DMSO.

2.15. *Statistical Analysis*

Data sets with only two independent groups were analyzed for statistical significance using an unpaired, two-tailed *t*-test. One-way analysis of variance (ANOVA) followed by the Student Newman-Keuls test was used for comparing the difference in more than two groups. Data are presented as mean ± SEM and analyzed by SPSS and GraphPad Prism 8.0 (GraphPad Software Inc., San Diego, CA, USA). A probability of $p < 0.05$ was considered statistically significant.

3. Results

3.1. *OTUD3 Deletion Induced Neuronal Apoptosis*

To observe the effect of OTUD3 on cell survival, Lv-shRNA-OTUD3 was constructed to knockdown the expression of OTUD3 in SH-SY5Y cells (Figure 1A,B). Cell viability was decreased in the OTUD3 knockdown group (Figure 1C). Meanwhile, the ratio of propidium iodide (PI) and trypan blue positive cells was significantly increased in the OTUD3 knockdown group compared with the control (Figure 1D–G). The MDA release was significantly increased in the OTUD3 knockdown group, which indicated the lipid oxidation level was up-regulated (Figure 1H). We further explored the mode of cell death induced by OTUD3 knockdown. Flow cytometry results showed that the Annexin V positive cell rate was evidently increased in OTUD3 knockdown cells (Figure 1I,J). In OTUD3 knockdown cells, the ratio of cleaved-caspase 3/caspase 3 was significantly increased, as well as in the

SN of *OTUD3*$^{-/-}$ mice (Figure 1K–N). These data indicated that knockdown of OTUD3 partially induced dopaminergic neurons deaths through apoptosis.

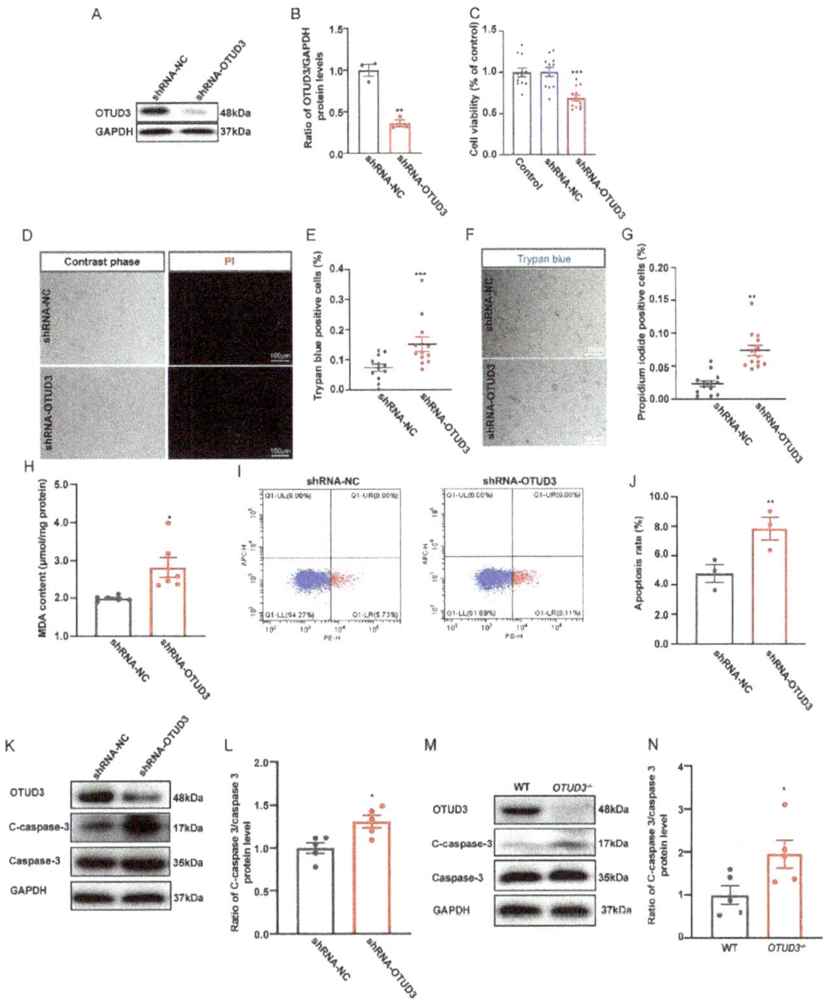

Figure 1. OTUD3 knockdown induces neuronal apoptosis. (**A**,**B**) SH-SY5Y cells with decreased OTUD3 expression were generated by lentivirally introducing shRNA-OTUD3 into the cells and characterized by western blotting, n = 3. (**C**) Cell viability was determined by CCK-8, n = 12. (**D**,**E**) PI staining and statistical analysis, n = 12. (**F**,**G**) Trypan blue staining and statistical analysis, n = 12. (**H**) The MDA content were determined by lipid peroxidation MDA assay kit, n = 6. (**I**) Cell apoptosis analyzed by flow cytometer with PE Annexin V/7-ADD double staining. (**J**) Statistical analysis the apoptosis rate, n = 3. (**K–N**) Western blotting and statistical analysis of the ratio of cleaved-caspase3/caspase3 in the SN of *OTUD3*$^{-/-}$ mice and OTUD3 knockdown cells, n = 5. Data were mean ± SEM, t-test, * $p < 0.05$, ** $p < 0.01$, *** $p < 0.001$.

3.2. OTUD3 Knockdown Induced ER Stress

To clarify whether OTUD3 knockdown-induced apoptosis was associated with ER stress, we observed the ER morphology and the expression of PDI (marker protein for ER stress) both in vivo and in vitro. As the main morphological manifestation of ER stress, ER expansion was significantly enlarged in OTUD3 knockdown cells. Likewise, we observed the same phenomenon in dopaminergic neurons of *OTUD3*$^{-/-}$ mice (Figure 2A–C). Additionally, we treated OTUD3 knockdown cells with ER stress inhibitor tauroursodeoxycholic acid (TUDCA) for 24 h, the proportion of Annexin V positive cells of that was decreased (Figure 2D,E). After TUDCA treatment for two weeks, the ER lumen was markedly decreased in the dopaminergic neuron of *OTUD3*$^{-/-}$ mice (Figure 2F,G), and the expression of PDI in dopaminergic neurons was also decreased (Figure 2H). There was an upward trend for the expression of TH protein in the SN of *OTUD3*$^{-/-}$ mice after TUDCA treatment (Figure 2I,J), and the ratio of cleaved-caspase 3/caspase 3 showed a downward trend after TUDCA treatment (Figure 2I–K). These results suggested that OTUD3 knockdown would induce ER stress in dopaminergic neurons both in vivo and in vitro.

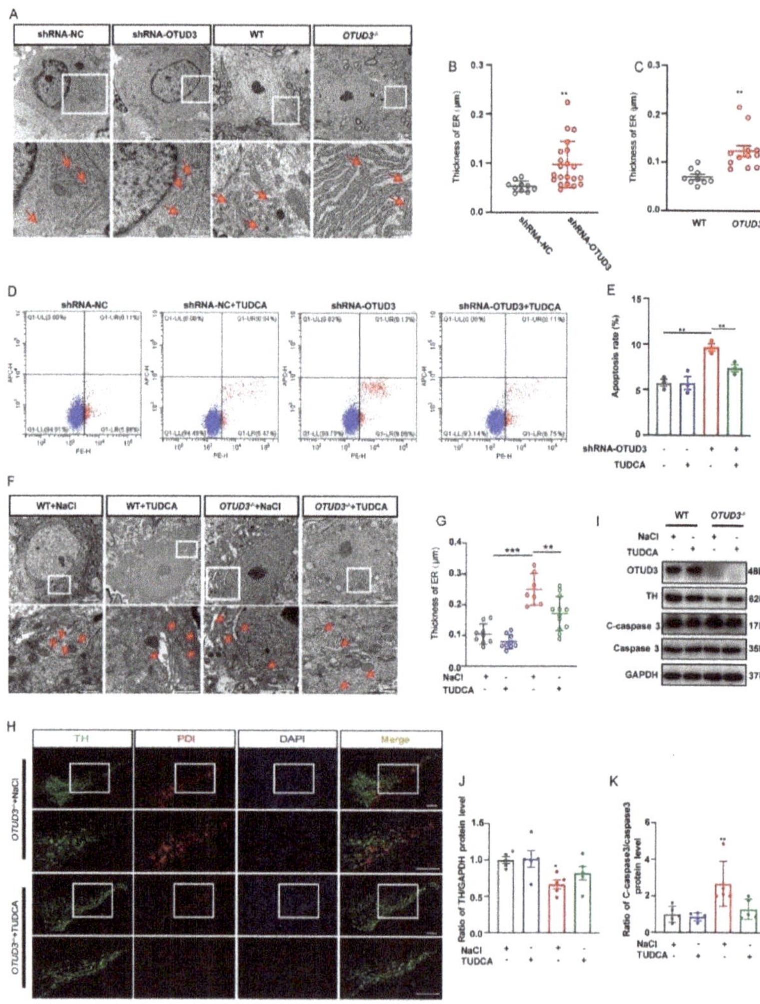

Figure 2. Effects of knockdown OTUD3 on ER stress. (**A–C**) Transmission electron microscope was applied to assessment of ER shape and statistical analysis of ER thickness, $n = 9$; red arrows represent the ER. (**D**,**E**) Cell apoptosis and statistical analysis of apoptosis rate by flow cytometer with PE-Annexin V/7-ADD double staining, $n = 3$. (**F**,**G**) ER shape and statistical analysis of ER thickness in *OTUD3*$^{-/-}$ mice after TUDCA treatment, $n = 9$. (**H**) Protein disulfide isomerase immunofluorescence staining of SN in *OTUD3*$^{-/-}$ mice. Scale bar = 100μm. (**I–K**) Western blotting and statistical analysis of the expression of TH and the ratio of cleaved-caspase3/caspase3 proteins in *OTUD3*$^{-/-}$ mice after TUDCA treatment, $n = 5$. Data were mean ± SEM, *t*-test, * $p < 0.05$, ** $p < 0.01$, *** $p < 0.001$.

3.3. The Expression of Bip Was Not Changed In Vivo and In Vitro

Previous studies have shown that Bip is the target protein of OTUD3 in lung cancer cells [40]. In order to make it clear whether Bip expression changed in vivo and in vitro, we detected protein and mRNA expression levels of Bip both in OTUD3 knockdown cells and the SN of *OTUD3*$^{-/-}$ mice. However, there were no changes in the expression levels of Bip (Figure 3A–C,F–H). Furthermore, the protein levels of Bip were also unchanged in both OTUD3 overexpression cells and the SN of OTUD3 transgenic (*OTUD3*^{TG}) mice (Figure 3D,E,I,J). These data suggested that OTUD3 knockdown-induced ER stress did not affect the protein expression of Bip in the SN of mice, which might be related to the fact that the function of OTUD3 is tissue-dependent.

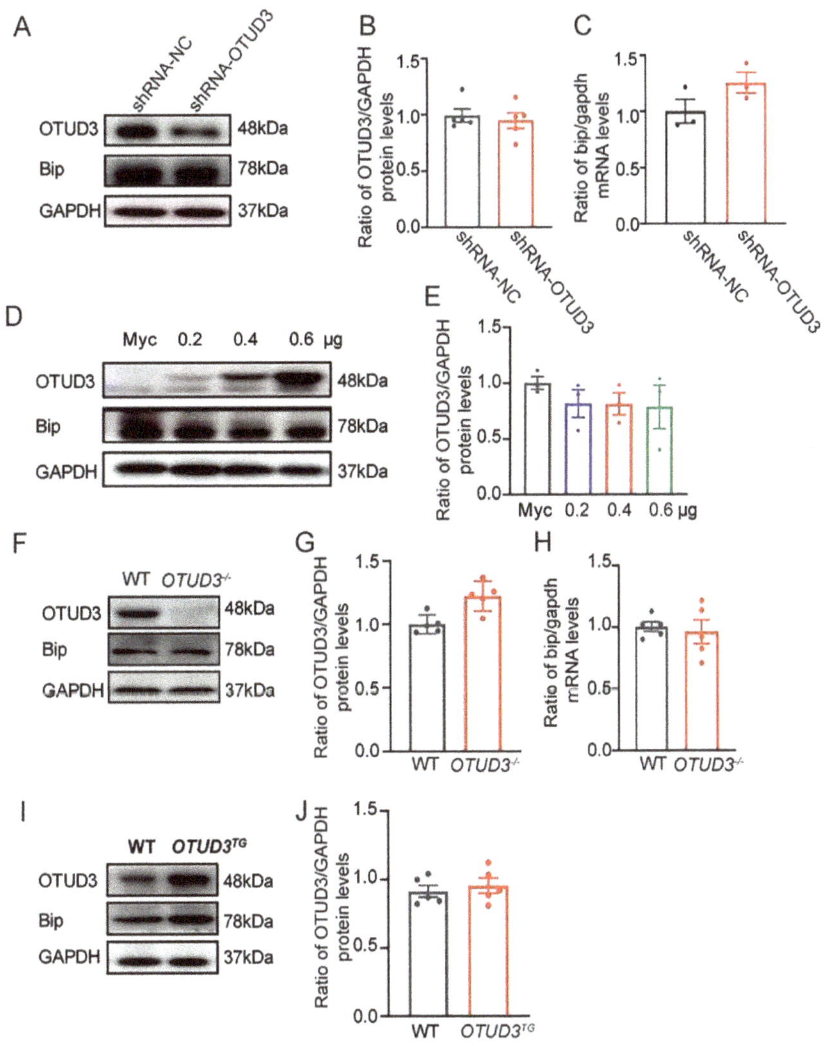

Figure 3. OTUD3 knockdown-induced ER stress was independent of Bip expression change. (**A**,**B**,**D**,**E**) Western blotting and statistical analysis of the expression of Bip in SH-SY5Y cells, $n = 3$. (**C**,**H**) The expression of bip mRNA in OTUD3 knockdown cells and the SN of *OTUD3*$^{-/-}$ mice, $n = 3, 5$. (**F**,**G**,**I**,**J**) Western blotting and statistical analysis of the expression of Bip in the SN of mice, $n = 5$. Data were mean ± SEM.

3.4. OTUD3 Knockdown Induced ER Stress via Activating IRE1α Pathway

Three classical ER stress transmembrane sensors (IRE1α, PERK and ATF6α) might be activated when ER stress occurs. To explore which ER stress pathway was activated by OTUD3 knockdown, we detected the protein expression of the three pathways separately. We found no significant changes in the expression of ATF6α-N in both OTUD3 knockdown cells and the SN of *OTUD3*$^{-/-}$ mice (Supplementary Figure S1A–H). Although the ratio of p-PERK/PERK and the ratio of p-eIf2α/eIf2α were significantly increased in OTUD3 knockdown cells and the SN of *OTUD3*$^{-/-}$ mice, however, the expression of PERK downstream target protein ATF4 and CHOP protein was failed to increase (Supplementary Figure S1I–P). These results suggested that the pathway of ATF6α and PERK were not activated in OTUD3 knockdown cells and the SN of *OTUD3*$^{-/-}$ mice.

Finally, we observed a significant increase in the ratio of p-IRE1α/IRE1α and the expression levels of XBP1s both in OTUD3 knockdown cells and the SN of *OTUD3*$^{-/-}$ mice (Figure 4A–F), which were significantly reduced after TUDCA treatment (Figure 4G–L). STF-083010 blocked the increase of p-IRE1α and XBP1's protein expression after 24 h treatment in OTUD3 knockdown cells. There was no difference between the control and the OTUD3 knockdown-STF-083010 group for the ratio of p-IRE1α/IRE1α and the expression of XBP1s (Figure 4M–O). STF-083010 treatment also inhibited the cell apoptosis of OTUD3 knockdown cells but partially reversed apoptosis after STF-083010 treatment (Figure 4P,Q). Moreover, the thickness of the ER lumen in OTUD3 knockdown cells was reduced after STF-083010 treatment, and there was no difference between OTUD3 knockdown-STF-083010 and control (Figure 4R,S). These results showed that OTUD3 knockdown-induced ER stress activated the IRE1α pathway.

3.5. OTUD3 Was Involved in ER Stress by Regulating the Ubiquitination Level of IRE1α Binding Protein Fortilin

To explore the underlying mechanism by which OTUD3 knockdown activated the IRE1α pathway, we performed co-immunoprecipitation (Co-IP) to detect protein interactions. Our findings indicated that OTUD3 did not directly interact with IRE1α, as endogenous OTUD3 was not immunoprecipitated by IRE1α antibody in shRNA-NC and OTUD3 knockdown cells (Figure 5A). We know that the activity of IRE1α is regulated by two binding proteins, Bip and Fortilin [21]. The co-precipitated signal showed that the binding ability of IRE1α to Bip was decreased in OTUD3 knockdown cells; however, the expression of Bip in whole cell lysate was not changed (Figure 5A). Meanwhile, we observed that the co-precipitated signal of IRE1α and Fortilin was decreased in OTUD3 knockdown cells (Figure 5A,B). The endogenous Fortilin could co-precipitate with OTUD3 (Figure 5B,C). To further explore the relationship between OTUD3 and Fortilin, we validated the expression of Fortilin both in OTUD3 knockdown and overexpression conditions. In OTUD3 knockdown cells and the SN of *OTUD3*$^{-/-}$ mice, the expression of Fortilin was significantly decreased (Figure 5D–G), whereas it was increased in OTUD3 overexpression cells and *OTUD3*^{TG} mice (Figure 5H–K). Moreover, co-localization studies indicated that Fortilin and OTUD3 were present in the cytoplasm of HEK293T and SH-SY5Y cells (Figure 5L).

Ubiquitylation assays showed that OTUD3 knockdown significantly enhanced the ubiquitylation level of Fortilin in SH-SY5Y cells and HEK293T cells (Figure 5M,N). Subsequently, an exogenous Co-IP assay showed that the OTU domain of OTUD3 is sufficient to interact with Fortilin (Figure 5P). Overexpression of WT OTUD3, rather than the C76A mutant, reversed the decreased Fortilin level induced by OTUD3 knockdown (Figure 5Q). Treatment with the proteasome inhibitor MG132 also reversed the decrease of Fortilin (Figure 5R). Overall, our findings illuminated that OTUD3 did not directly bind to IRE1α, but regulated the ubiquitination level of Fortilin through its OTU domain.

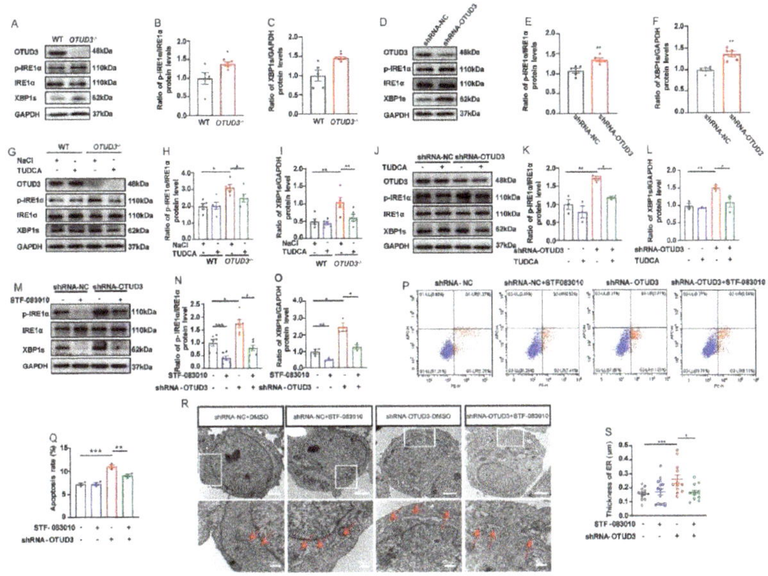

Figure 4. OTUD3 knockdown-induced ER stress by actives IRE1α pathway. (**A–C**) Western blotting and statistical analysis of the expression of p-IRE1α, IRE1α and XBP1s protein in $OTUD3^{-/-}$ mice, $n = 5$. (**D–F**) Western blotting and statistical analysis of the expression of p-IRE1α, IRE1α and XBP1s protein in OTUD3 knockdown cells, $n = 5$. (**G–L**) Western blotting and statistical analysis of the expression of p-IRE1α/IRE1α, XBP1s in $OTUD3^{-/-}$ mice ($n = 5$) and OTUD3 knockdown cells ($n = 3$) after TUDCA treatment. (**M–O**) Western blotting and statistical analysis of the expression of p-IRE1α/IRE1α and XBP1s in OTUD3 knockdown cells after STF-083010 treatment ($n = 3$). (**P**) Cell apoptosis analyzed by flow cytometer with PE Annexin V/7-ADD double staining. (**Q**) Statistical analysis the apoptosis rate after STF-083010 treatment, $n = 3$. (**R**,**S**) Transmission electron microscope was applied to assessment of ER shape and statistical analysis of ER lumen after STF-083010 treatment, $n = 13, 15$; red arrows represent the ER. Data were mean ± SEM, t−test, * $p < 0.05$, ** $p < 0.01$, *** $p < 0.001$.

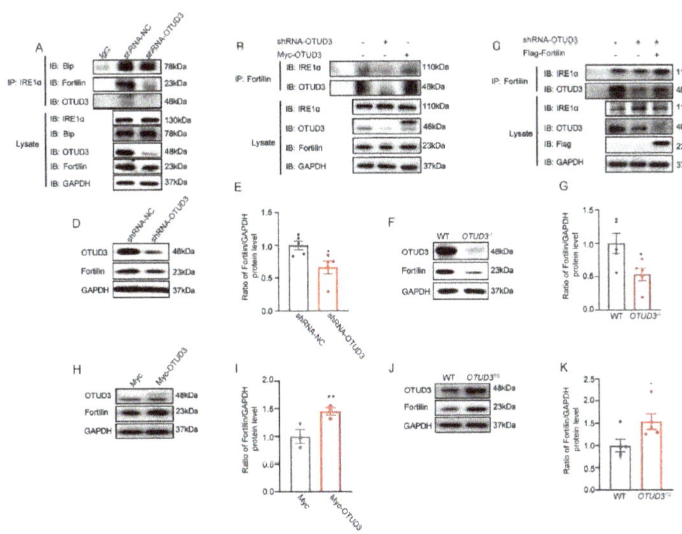

Figure 5. *Cont.*

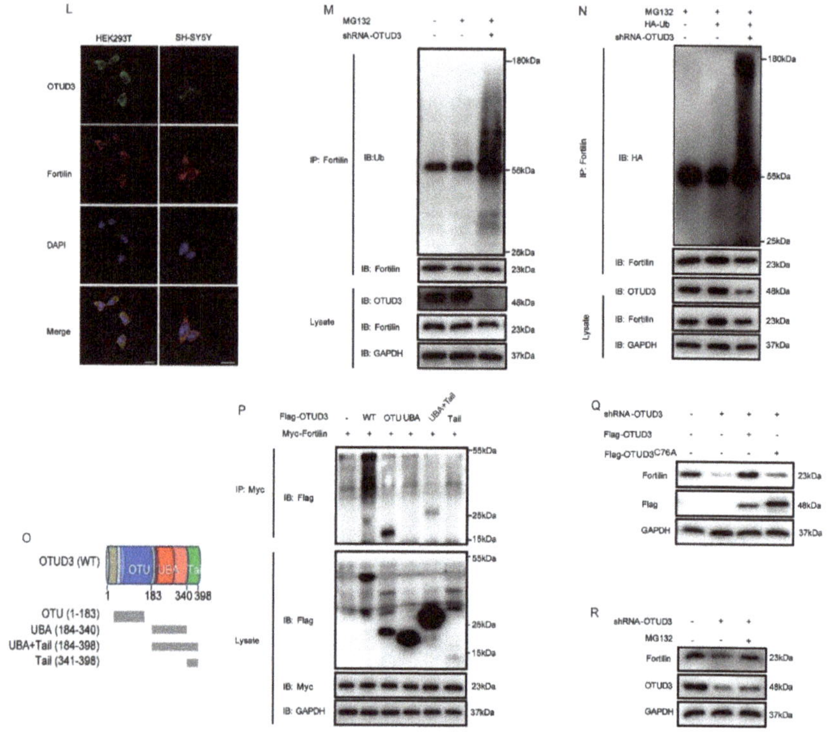

Figure 5. OTUD3 regulates the ubiquitination level of Fortilin protein. (**A**) The interaction between IRE1α and Bip, Fortilin and OTUD3 were examined immunoprecipitated with anti-IRE1α antibody. The whole-cell lysate was subjected to immunoblot with anti-IRE1α, anti-Bip, anti-OTUD3 and anti-Fortilin antibody. (**B**,**C**) The interaction between Fortilin and IRE1α and OTUD3 were examined immunoprecipitated with anti-Fortilin antibody. The whole-cell lysate was subjected to immunoblot with anti-IRE1α, anti-OTUD3 and anti-Fortilin antibody. (**D**–**K**) Western blotting and statistical analysis of the expression of Fortilin in OTUD3 knockdown cells, *OTUD3*$^{-/-}$ mice, SH-SY5Y cells transfected with Myc-OTUD3 and OTUD3 transgenic mice, $n = 5$. (**L**) OTUD3 and Fortilin colocalized in the cytoplasm of HEK293T and SH-SY5Y cells. (**M**,**N**) The ubiquitylation levels of Fortilin was determined by in vitro ubiquitin conjugation assay. $n = 3$. (**O**,**P**) Co-IP assays were performed to map the domain of Fortilin required for interaction with OTUD3 ($n = 3$). (**Q**) Western blotting was examined the expression of Fortilin in OTUD3 knockdown cells after transfected with Flag-OTUD3^{WT} and Flag-OTUD3^{C76A}. $n = 3$. (**R**) Western blotting was examined the expression of Fortilin in OTUD3 knockdown cells after treatment with MG132. $n = 3$. Data were mean $\pm$ SEM. *t*-test, * $p < 0.05$, ** $p < 0.01$.

3.6. Fortilin Alleviated OTUD3 Knockdown Induced ER Stress

To investigate the role of Fortilin in the regulation of IRE1α activity by OTUD3, we examined the expression of the associated proteins after Fortilin overexpression. We observed that higher levels of the ratio of p-PERK/PERK and the ratio of p-elf2α/elf2α in OTUD3 knockdown cells were not affected by Fortilin overexpression. Notably, the ratio of p-IRE1α/IRE1α, cleaved-caspase3/caspase3, and the expression of XBP1s was increased in OTUD3 knockdown cells, and these effects were abated by Fortilin overexpression; there was no difference between the OTUD3 knockdown + Flag-Fortilin and control (Figure 6A–I). Additionally, we observed that the thickness of the ER lumen in OTUD3 knockdown cells was reduced after Fortilin overexpression (Figure 6J,K). The results indicated that Fortilin can relieve ER stress by inhibiting IRE1α pathway.

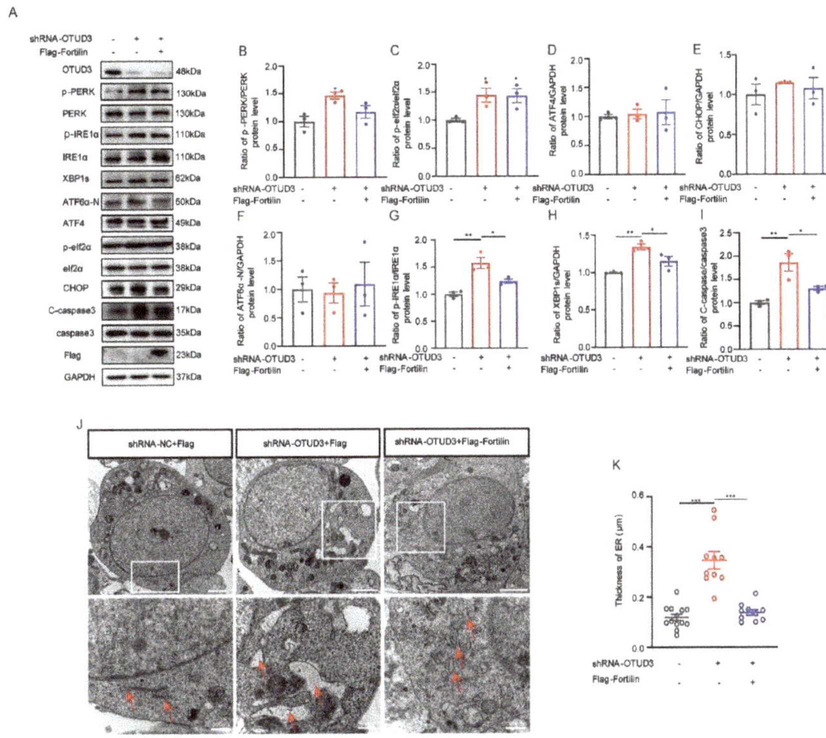

Figure 6. Fortilin alleviate ER stress induced by OTUD3 knockdown. (**A**) Western blotting to detect the ER stress-associated proteins expression. (**B–E**) Statistical analysis of the expression of PERK pathway-associated proteins after Fortilin overexpression in OTUD3 knockdown cells. $n = 3$. (**F**) Statistical analysis of the expression ATF6α-N proteins after Fortilin overexpression in OTUD3 knockdown cells. $n = 3$. (**G–I**) Statistical analysis of the ratio of p-IRE1α/IRE1α, cleaved-caspase 3/caspase 3 and the expression of XBP1s after Fortilin overexpression in OTUD3 knockdown cells. (**J**,**K**) Transmission electron microscope was applied to assessment of ER shape and statistical analysis of ER lumen after Fortilin overexpression, $n = 10, 12$; red arrows represent the ER. Data were mean ± SEM, *t*-test, * $p < 0.05$, ** $p < 0.01$, *** $p < 0.001$.

4. Discussion

The current work revealed that OTUD3 was involved in the regulation of ER stress by regulating IRE1α activity. OTUD3 knockdown increased the ratio of p-IRE1α/IRE1α, and the expression of XBP1s, which could be inhibited by IRE1α inhibitor STF-083010. Further investigation revealed that Fortilin was a target protein of OTUD3. Downregulation of OTUD3 reduced the binding ability of Fortilin to IRE1α, thereby activating the IRE1α pathway and inducing ER stress and neuronal apoptosis.

OTUD3 has been identified as a tumor suppressor that is highly associated with tumorigenesis [41,42]. Our previous study found that OTUD3 knockout mice display motor deficits and nigrostriatal dopaminergic neurodegeneration, resembling the pathology of PD [5]. In the present study, we observed a significant increase in cell apoptosis after OTUD3 knockdown, consistent with our previous reports [5]. More and more evidence has shown that ER stress-induced apoptosis is an important cell death pathway for dopaminergic neurons [29,30]. We also observed abnormal expansion of ER morphology by transmission electron microscopy, both in $OTUD3^{-/-}$ mice and OTUD3 knockdown cells, consistent with the previously reported ER stress-induced morphological changes [43]. At the same time, we also found that PDI, another marker protein of ER stress, was colocalized with dopaminergic neurons in $OTUD3^{-/-}$ mice [44]. These results suggested that deleting OTUD3 would induce ER stress in dopaminergic neurons.

As a deubiquitylase, there are many target proteins of OTUD3, such as phosphatase and tensin homolog deleted on chromosome 10 (PTEN) [42], Bip [40], p53 [41], and actinin-4 (ACTN4) [45]. ER chaperone Bip, as one target protein of OTUD3, is a major regulator of ER stress [11]. Bip can maintain the permeability barrier of the ER during protein translocation and target misfolded proteins for retrograde translocation so that they can be degraded by the proteasome, sensing conditions of stress in the ER to activate the UPR [46]. Under physiological conditions, IRE1α, PERK, and ATF6α remain inactive by binding to Bip. Under ER stress, Bip dissociates from the three sensors, activating downstream pathways and determining cell fate [47]. In the present study, we observed that the binding ability of Bip and IRE1α was decreased after OTUD3 knockdown. However, there was no significant change in the expression of Bip in either OTUD3 knockdown or overexpression, indicating that OTUD3 did not affect the level of Bip in dopaminergic neurons. OTUD3 can stabilize Bip and promote lung tumorigenesis [40], whereas it suppresses breast tumorigenesis through stabilizing the PTEN protein [42]. This suggests that the function of OTUD3 for target proteins is distinctive in different tissues and tissue-dependent contexts.

Under ER stress, cells activate the IRE1α, PERK, and ATF6α pathways to restore ER homeostasis or induce cell death [48–50]. In the present study, our results demonstrated that OTUD3 knockdown-induced ER stress and cell apoptosis through the activation of the IRE1α pathway but not the PERK or ATF6α pathways. OTUD3 knockdown increased the activation of IRE1α and the expression of XBP1s. Reversal of OTUD3 protein expression or treatment with IRE1α inhibitor STF-083010 could ameliorate IRE1α signal activation and inhibit apoptosis. The activation of IRE1α signal is closely related to apoptosis [51]. The activation of PERK and ATF6α was also related to cell apoptosis, we did not observe changes of ATF6-N protein. Although we observed an evident increase in the activation of PERK and eIf2α, the expression of downstream target proteins ATF4 and CHOP did not change. Therefore, we considered that OTUD3 knockdown could induced ER stress and cell apoptosis by activating the IRE1α pathway.

IRE1α is known to remain inactive by binding to Bip. Recent studies have shown that the IRE1α also binds to Fortilin, and when Fortilin dissociates from IRE1α, it can activate the IRE1α and lead to apoptosis [21]. As an anti-apoptotic factor, Fortilin can play anti-apoptotic function by regulating the activity of various proteins [52,53]. To verify whether the function of OTUD3 on IRE1α activity was mediated by directly regulating its ability to bind to IRE1α, Co-IP was used to detect the proteins interactions. The results showed that there was no interaction between IRE1α and OTUD3. We observed that OTUD3 knockdown could decrease the binding ability between Fortilin/Bip and IRE1α, and overexpression of OTUD3 could increase the binding ability of Fortilin and IRE1α. Additionally, we found that the expression and ubiquitination level of Fortilin protein were regulated by OTUD3, while the protein level of Bip was not influenced by OTUD3. We further found that the Fortilin was interacted with the OTU domain. These findings suggested that OTUD3 can regulate the expression of Fortilin rather than Bip, and adjust the binding ability of Fortilin to IRE1α, thus regulating IRE1α activity. In previous studies, Fortilin overexpression could inhibit the cell death and ER stress induced by thapsigargin [21]. As expected, we observed that the ER stress and cell apoptosis were alleviated by Fortilin overexpression. Our data indicated that the ER stress caused by OTUD3 knockdown might be related the Fortilin expression. Our previous research showed that the iron content is increased in the SN of $OTUD3^{-/-}$ mice [5], and abnormal iron metabolism can induce ER stress [24]. These evidences suggest that OTUD3 knockout may induce ER stress by reducing the expression of IRP2 protein and up-regulating the contents of iron in neurons. Meanwhile, OTUD3 may also be involved in regulating ER stress by regulating the expression of Fortilin protein. However, there were some limitations in our present study: the ubiquitination site of Fortilin specifically regulated by OTUD3 remains to be further explored, and the changes of OTUD3 in the brain of PD patients were not clarified.

5. Conclusions

The present study highlights a novel role of OTUD3 in regulating ER stress. Our findings demonstrate that OTUD3 plays a crucial role in protecting cells against ER stress-induced apoptosis by regulating the level of ubiquitination of Fortilin and inhibiting the activation of IRE1α signaling. Therefore, targeting the inhibition of IRE1α signal activation could represent a promising therapeutic for PD induced by OTUD3 elimination.

Supplementary Materials: The following supporting information can be downloaded at: https://www.mdpi.com/article/10.3390/antiox12040809/s1, Figure S1: OTUD3 knockdown-induced ER stress was not depend on ATF6α and PERK pathway.

Author Contributions: H.J., Q.J. and L.C. conceived and designed research. L.C. performed major experiments and analyzed, F.J. helped to perform ubiquitylation assay. Z.Z. helped to perform animal experiments. M.B., X.H., X.D., X.C., L.F. and C.Y. assisted to analyze data. L.C. wrote the manuscript. H.J. and Q.J. revised the manuscript. All authors have read and agreed to the published version of the manuscript.

Funding: This work was supported by the National Natural Science Foundation of China (32171131, 82071429), Shandong Province Natural Science Foundation (2021ZDSYS11, ZR2019ZD31, ZR2020MC072, ZR2020QH125), Taishan Scholars Construction Project.

Institutional Review Board Statement: The animal study protocol was approved by the Ethics Committee of Qingdao University (protocol code QDU-AEC- 2023026, February 2023).

Informed Consent Statement: Not applicable.

Data Availability Statement: All data generated or analyzed during this study are included.

Acknowledgments: We thank all the participants enrolled in the current study and Lingqiang Zhang (State Key Laboratory of Proteomics, National Center of Protein Sciences (Beijing, China), Beijing Institute of Lifeomics, Beijing, China) for kindly gifting the *OTUD3*$^{-/-}$ and *OTUD3*^{TG} mice and the plasmids of Myc-OTUD3 and Flag-OTUD3.

Conflicts of Interest: The authors declare no conflict of interest.

References

1. Tolosa, E.; Garrido, A. Challenges in the diagnosis of Parkinson's disease. *Lancet Neurol.* **2021**, *20*, 385–397. [CrossRef] [PubMed]
2. Bloem, B.R.; Okun, M.S. Parkinson's disease. *Lancet* **2021**, *397*, 2284–2303. [CrossRef] [PubMed]
3. Mahoney-Sánchez, L.; Bouchaoui, H. Ferroptosis and its potential role in the physiopathology of Parkinson's Disease. *Prog. Neurobiol.* **2021**, *196*, 101890. [CrossRef]
4. Bi, M.; Du, X. Deficient immunoproteasome assembly drives gain of α-synuclein pathology in Parkinson's disease. *Redox Biol.* **2021**, *47*, 102167. [CrossRef] [PubMed]
5. Jia, F.; Li, H. Deubiquitylase OTUD3 prevents Parkinson's disease through stabilizing iron regulatory protein 2. *Cell Death Dis.* **2022**, *13*, 418. [CrossRef] [PubMed]
6. Yang, S.K.; Hong, M. Genome-wide association study of ulcerative colitis in Koreans suggests extensive overlapping of genetic susceptibility with Caucasians. *Inflamm. Bowel Dis.* **2013**, *19*, 954–966. [CrossRef]
7. Garshott, D.M.; Sundaramoorthy, E. Distinct regulatory ribosomal ubiquitylation events are reversible and hierarchically organized. *Elife* **2020**, *9*, e54023. [CrossRef]
8. Zhang, Z.; Fang, X. Acetylation-Dependent Deubiquitinase OTUD3 Controls MAVS Activation in Innate Antiviral Immunity. *Mol. Cell* **2020**, *79*, 304–319.e7. [CrossRef]
9. Fagone, P.; Jackowski, S. Membrane phospholipid synthesis and endoplasmic reticulum function. *J. Lipid Res.* **2009**, *50*, S311–S316. [CrossRef]
10. van Anken, E.; Braakman, I. Versatility of the endoplasmic reticulum protein folding factory. *Crit. Rev. Biochem. Mol. Biol.* **2005**, *40*, 191–228. [CrossRef]
11. Pobre, K.F.R.; Poet, G.J. The endoplasmic reticulum (ER) chaperone BiP is a master regulator of ER functions: Getting by with a little help from ERdj friends. *J. Biol. Chem.* **2019**, *294*, 2098–2108. [CrossRef]
12. Hendershot, L.M.; Valentine, V.A. Localization of the gene encoding human BiP/GRP78, the endoplasmic reticulum cognate of the HSP70 family, to chromosome 9q34. *Genomics* **1994**, *20*, 281–284. [CrossRef] [PubMed]
13. Haas, I.G. BiP—A heat shock protein involved in immunoglobulin chain assembly. *Curr. Top. Microbiol. Immunol.* **1991**, *167*, 71–82. [PubMed]
14. Gething, M.J.; Sambrook, J. Protein folding in the cell. *Nature* **1992**, *355*, 33–45. [CrossRef] [PubMed]

15. Little, E.; Ramakrishnan, M. The glucose-regulated proteins (GRP78 and GRP94): Functions, gene regulation, and applications. *Crit. Rev. Eukaryot. Gene Expr.* **1994**, *4*, 1–18. [CrossRef]
16. Li, F.; Zhang, D. Characterization of fortilin, a novel antiapoptotic protein. *J. Biol. Chem.* **2001**, *276*, 47542–47549. [CrossRef]
17. Susini, L.; Besse, S. TCTP protects from apoptotic cell death by antagonizing bax function. *Cell Death Differ.* **2008**, *15*, 1211–1220. [CrossRef]
18. Cans, C.; Passer, B.J. Translationally controlled tumor protein acts as a guanine nucleotide dissociation inhibitor on the translation elongation factor eEF1A. *Proc. Natl. Acad. Sci. USA* **2003**, *100*, 13892–13897. [CrossRef]
19. Gu, X.; Yao, L. TCTP promotes glioma cell proliferation in vitro and in vivo via enhanced β-catenin/TCF-4 transcription. *Neuro-Oncology* **2014**, *16*, 217–227. [CrossRef]
20. Hao, S.; Qin, Y. Serum translationally controlled tumor protein is involved in rat liver regeneration after hepatectomy. *Hepatol. Res.* **2016**, *46*, 1392–1401. [CrossRef]
21. Pinkaew, D.; Chattopadhyay, A. Fortilin binds IRE1α and prevents ER stress from signaling apoptotic cell death. *Nat. Commun.* **2017**, *8*, 18. [CrossRef]
22. Jeong, M.; Jeong, M.H. TCTP protein degradation by targeting mTORC1 and signaling through S6K, Akt, and Plk1 sensitizes lung cancer cells to DNA-damaging drugs. *Sci. Rep.* **2021**, *11*, 20812. [CrossRef] [PubMed]
23. Lo, W.Y.; Wang, H.J. miR-27b-regulated TCTP as a novel plasma biomarker for oral cancer: From quantitative proteomics to post-transcriptional study. *J. Proteomics* **2012**, *77*, 154–166. [CrossRef] [PubMed]
24. Yoshitake, Y.; Shinozaki, D. Autophagy triggered by iron-mediated ER stress is an important stress response to the early phase of Pi starvation in plants. *Plant J.* **2022**, *110*, 1370–1381. [CrossRef]
25. da Silva, D.C.; Valentão, P. Endoplasmic reticulum stress signaling in cancer and neurodegenerative disorders: Tools and strategies to understand its complexity. *Pharmacol. Res.* **2020**, *155*, 104702. [CrossRef]
26. Hoozemans, J.J.; van Haastert, E.S. Activation of the unfolded protein response in Parkinson's disease. *Biochem. Biophys. Res. Commun.* **2007**, *354*, 707–711. [CrossRef] [PubMed]
27. Ryu, E.J.; Harding, H.P. Endoplasmic reticulum stress and the unfolded protein response in cellular models of Parkinson's disease. *J. Neurosci.* **2002**, *22*, 10690–10698. [CrossRef] [PubMed]
28. Lehtonen, Š.; Jaronen, M. Inhibition of Excessive Oxidative Protein Folding Is Protective in MPP(+) Toxicity-Induced Parkinson's Disease Models. *Antioxid. Redox Signal.* **2016**, *25*, 485–497. [CrossRef]
29. Holtz, W.A.; O'Malley, K.L. Parkinsonian mimetics induce aspects of unfolded protein response in death of dopaminergic neurons. *J. Biol. Chem.* **2003**, *278*, 19367–19377. [CrossRef]
30. Hu, L.W.; Yen, J.H. Luteolin modulates 6-hydroxydopamine-induced transcriptional changes of stress response pathways in PC12 cells. *PLoS ONE* **2014**, *9*, e97880. [CrossRef]
31. Devi, L.; Ohno, M. PERK mediates eIF2α phosphorylation responsible for BACE1 elevation, CREB dysfunction and neurodegeneration in a mouse model of Alzheimer's disease. *Neurobiol. Aging* **2014**, *35*, 2272–2281. [CrossRef] [PubMed]
32. Pytel, D.; Majsterek, I. Tumor progression and the different faces of the PERK kinase. *Oncogene* **2016**, *35*, 1207–1215. [CrossRef] [PubMed]
33. Wang, Y.; Alam, G.N. The unfolded protein response induces the angiogenic switch in human tumor cells through the PERK/ATF4 pathway. *Cancer Res.* **2012**, *72*, 5396–5406. [CrossRef]
34. Pillai, S. Birth pangs: The stressful origins of lymphocytes. *J. Clin. Investig.* **2005**, *115*, 224–227. [CrossRef]
35. Endres, K.; Reinhardt, S. ER-stress in Alzheimer's disease: Turning the scale? *Am. J. Neurodegener. Dis.* **2013**, *2*, 247–265. [PubMed]
36. Ron, D.; Walter, P. Signal integration in the endoplasmic reticulum unfolded protein response. *Nat. Rev. Mol. Cell Biol.* **2007**, *8*, 519–529. [CrossRef]
37. Shaffer, A.L.; Shapiro-Shelef, M. XBP1, downstream of Blimp-1, expands the secretory apparatus and other organelles, and increases protein synthesis in plasma cell differentiation. *Immunity* **2004**, *21*, 81–93. [CrossRef] [PubMed]
38. Lee, A.H.; Chu, G.C. XBP-1 is required for biogenesis of cellular secretory machinery of exocrine glands. *Embo J.* **2005**, *24*, 4368–4380. [CrossRef] [PubMed]
39. Xicoy, H.; Wieringa, B. The SH-SY5Y cell line in Parkinson's disease research: A systematic review. *Mol. Neurodegener.* **2017**, *12*, 10. [CrossRef]
40. Du, T.; Li, H. The deubiquitylase OTUD3 stabilizes GRP78 and promotes lung tumorigenesis. *Nat. Commun.* **2019**, *10*, 2914. [CrossRef]
41. Pu, Q.; Lv, Y.R. Tumor suppressor OTUD3 induces growth inhibition and apoptosis by directly deubiquitinating and stabilizing p53 in invasive breast carcinoma cells. *BMC Cancer* **2020**, *20*, 583. [CrossRef] [PubMed]
42. Yuan, L.; Lv, Y. Deubiquitylase OTUD3 regulates PTEN stability and suppresses tumorigenesis. *Nat. Cell Biol.* **2015**, *17*, 1169–1181. [CrossRef] [PubMed]
43. Wu, H.; Wei, L. Integration of Hippo signalling and the unfolded protein response to restrain liver overgrowth and tumorigenesis. *Nat. Commun.* **2015**, *6*, 6239. [CrossRef]
44. Hetzer, S.M.; Guilhaume-Correa, F. Traumatic Optic Neuropathy Is Associated with Visual Impairment, Neurodegeneration, and Endoplasmic Reticulum Stress in Adolescent Mice. *Cells* **2021**, *10*, 996. [CrossRef]
45. Xie, P.; Chen, Y. The deubiquitinase OTUD3 stabilizes ACTN4 to drive growth and metastasis of hepatocellular carcinoma. *Aging (Albany NY)* **2021**, *13*, 19317–19338. [CrossRef]

46. Hetz, C.; Zhang, K. Mechanisms, regulation and functions of the unfolded protein response. *Nat. Rev. Mol. Cell Biol.* **2020**, *21*, 421–438. [CrossRef] [PubMed]
47. Hetz, C.; Papa, F.R. The Unfolded Protein Response and Cell Fate Control. *Mol. Cell* **2018**, *69*, 169–181. [CrossRef]
48. Hetz, C.; Martinon, F. The unfolded protein response: Integrating stress signals through the stress sensor IRE1α. *Physiol. Rev.* **2011**, *91*, 1219–1243. [CrossRef]
49. Liu, X.; Chen, Y. Microglia-derived IL-1β promoted neuronal apoptosis through ER stress-mediated signaling pathway PERK/eIF2α/ATF4/CHOP upon arsenic exposure. *J. Hazard. Mater.* **2021**, *417*, 125997. [CrossRef] [PubMed]
50. Wang, X.; Zhuang, Y. Endoplasmic reticulum stress aggravates copper-induced apoptosis via the PERK/ATF4/CHOP signaling pathway in duck renal tubular epithelial cells. *Environ. Pollut.* **2021**, *272*, 115981. [CrossRef]
51. Merksamer, P.I.; Papa, F.R. The UPR and cell fate at a glance. *J. Cell Sci.* **2010**, *123 Pt 7*, 1003–1006. [CrossRef] [PubMed]
52. Chen, Y.; Fujita, T. Physical and functional antagonism between tumor suppressor protein p53 and fortilin, an anti-apoptotic protein. *J. Biol. Chem.* **2011**, *286*, 32575–32585. [CrossRef] [PubMed]
53. Zhang, D.; Li, F. Physical and functional interaction between myeloid cell leukemia 1 protein (MCL1) and Fortilin. The potential role of MCL1 as a fortilin chaperone. *J. Biol. Chem.* **2002**, *277*, 37430–37438. [CrossRef] [PubMed]

antioxidants

MDPI

Article

Inhibiting NLRP3 Inflammasome Activation by CY-09 Helps to Restore Cerebral Glucose Metabolism in 3×Tg-AD Mice

Shuangxue Han [1,2], Zhijun He [1,6], Xia Hu [3], Xiaoqian Li [1], Kaixin Zheng [1], Yingying Huang [1], Peng Xiao [3], Qingguo Xie [3,7,*], Jiazuan Ni [1,4,5] and Qiong Liu [1,4,5,*]

1 Shenzhen Key Laboratory of Marine Biotechnology and Ecology, College of Life Sciences and Oceanography, Shenzhen University, Shenzhen 518055, China
2 College of Physics and Optoelectronic Engineering, Shenzhen University, Shenzhen 518060, China
3 College of Life Science and Technology, Huazhong University of Science and Technology, Wuhan 430074, China
4 Shenzhen Bay Laboratory, Shenzhen 518055, China
5 Shenzhen-Hong Kong Institute of Brain Science-Shenzhen Fundamental Research Institutions, Shenzhen 518055, China
6 National R&D Center for Se-Rich Agricultural Products Processing, Hubei Engineering Research Center for Deep Processing of Green Se-Rich Agricultural Products, School of Modern Industry for Selenium Science and Engineering, Wuhan Polytechnic University, Wuhan 430023, China
7 Department of Medical Physics and Biomedical Engineering, Istituto Neurologico Mediterraneo, Neuromed IRCCS, 86077 Pozzilli, Italy
* Correspondence: qgxie@hust.edu.cn (Q.X.); liuqiong@szu.edu.cn (Q.L.)

Abstract: The reduction of the cerebral glucose metabolism is closely related to the activation of the NOD-like receptor protein 3 (NLRP3) inflammasome in Alzheimer's disease (AD); however, its underlying mechanism remains unclear. In this paper, ^{18}F-flurodeoxyglucose positron emission tomography was used to trace cerebral glucose metabolism in vivo, along with Western blotting and immunofluorescence assays to examine the expression and distribution of associated proteins. Glucose and insulin tolerance tests were carried out to detect insulin resistance, and the Morris water maze was used to test the spatial learning and memory ability of the mice. The results show increased NLRP3 inflammasome activation, elevated insulin resistance, and decreased glucose metabolism in 3×Tg-AD mice. Inhibiting NLRP3 inflammasome activation using CY-09, a specific inhibitor for NLRP3, may restore cerebral glucose metabolism by increasing the expression and distribution of glucose transporters and enzymes and attenuating insulin resistance in AD mice. Moreover, CY-09 helps to improve AD pathology and relieve cognitive impairment in these mice. Although CY-09 has no significant effect on ferroptosis, it can effectively reduce fatty acid synthesis and lipid peroxidation. These findings provide new evidence for NLRP3 inflammasome as a therapeutic target for AD, suggesting that CY-09 may be a potential drug for the treatment of this disease.

Keywords: Alzheimer's disease; glucose metabolism; NLRP3 inflammasome; insulin resistance; oxidative stress; CY-09

Citation: Han, S.; He, Z.; Hu, X.; Li, X.; Zheng, K.; Huang, Y.; Xiao, P.; Xie, Q.; Ni, J.; Liu, Q. Inhibiting NLRP3 Inflammasome Activation by CY-09 Helps to Restore Cerebral Glucose Metabolism in 3×Tg-AD Mice. *Antioxidants* **2023**, *12*, 722. https://doi.org/10.3390/antiox12030722

Academic Editor: Yan-Zhong Chang

Received: 31 January 2023
Revised: 6 March 2023
Accepted: 8 March 2023
Published: 15 March 2023

1. Introduction

Alzheimer's disease (AD) is a common neurodegenerative disease in elder people. Deposits of amyloid-β (Aβ) and hyperphosphorylated tau protein are its main characteristics [1,2]. Due to the failure of anti-amyloid and anti-tau aggregation drugs, neuroinflammation has been considered as a new therapeutic target for AD treatment.

NLRP3 inflammasome, composed of nucleotide-binding oligomerization domain (NOD)-like receptor protein 3 (NLRP3) and apoptosis-associated speck-like proteins of CARD (caspase recruitment domain) (ASC) and pro-caspase-1, plays an important role in neuroinflammation. Activation of NLRP3 inflammasome causes the increase in caspase-1 and the release of interleukin-1β (IL-1β) [3,4]. Increased activation of NLRP3 inflammasome is closely related with the reduction of cerebral glucose metabolism. Studies show that the translocation of hexokinase (HK), a key enzyme in glucose metabolism, can activate NLRP3 inflammasome, and inhibition of NLRP3 inflammasome can restore the expression and distribution of HK in AD model cells [5,6]. In AD mice, NLRP3 binds to the mitochondria and is then activated by mitochondrial reactive oxygen species (mtROS) [7,8]. HK binding to mitochondria reduces mtROS transport into the cytoplasm and further reduces the activation of NLRP3 inflammasome [9,10]. A positive correlation occurs between glucose metabolism and neuroinflammation in early AD, but disappears as the pathological course progresses [11]. In addition, increased neuroinflammation leads to a shift in energy metabolism from oxidative phosphorylation (OxPhos) to aerobic glycolysis in microglia. Activated microglia compete with neurons for glucose, which limits the energy availability of neurons [12]. Our previous studies have shown that inhibiting the activation of NLRP3 inflammasome can increase the expression and distribution of HK in vitro [5], but determining whether it could restore cerebral glucose metabolism in this process requires further exploration in vivo.

Glucose metabolism provides over 70% of energy for the brain. Glucose is transported into cells via glucose transporters (GLUTs), is phosphorylated by HK, and then catalyzed by a series of enzymes to produce energy. GLUT1 and GLUT3 are expressed in the brain and are responsible for the entry of glucose from the blood to the intracellular environment [13–15]. In AD, decreased expression levels of GLUT1 and GLUT3 lead to reduced glucose transport. GLUT4 is also expressed in the brain. Unlike GLUT1 and GLUT3, it is sensitive to insulin, and when insulin levels rise, it is transferred to cell membranes for the transport of glucose, [16,17]. The transfer of GLUT4 is regulated by the insulin-PI3K-AKT pathway [18]. Several studies have shown that insulin resistance is a key event leading to AD pathology [17,19,20]. Insulin receptors (IR) recognize insulin and then self-phosphorylate to recruit insulin receptor substrate (IRS) and activate the IRS-AKT-AS160 pathway [21,22]. AS160 is a guanosine triphosphate enzyme activating protein, the phosphorylation of which promotes translocation of GLUT4 [18,23]. In AD, lower insulin levels in cerebrospinal fluid result in decreased expression of p-IR, p-AKT, and p-AS160, a s well as increased IR [24]. A positron emission tomography (PET) study confirmed that the expression of IR was increased in the lower glucose metabolism region.

Oxidative stress also contributes to the pathology of AD. The generation of an excessive amount of ROS results in oxidative stress and the activation of NLRP3 inflammasome. Moreover, oxidative stress causes iron metabolism disorders and leads to ferroptosis [25,26]. Ferroptosis manifests as increased cellular Fe^{2+}, decreased glutathione peroxidase 4 (GPX4), and increased lipid peroxidation [27]. Elevated Fe^{2+} results from increased transferrin transport and ferritinophagy [28]. Transferrin (TF), transferrin receptor (TFR), and ferroportin (FPN) are responsible for the transport of Fe, and nuclear receptor coactivator 4 (NCOA4) participates in ferritinophagy. Long-chain acyl-CoA synthetase 4 (ACSL4) is a key protein linking ferroptosis and lipid peroxidation. Increased ACSL4 is consistent with increased ferroptosis and lipid peroxidation in AD [29–31]. Decreased GPX4 and Solute Carrier Family 7, Member 11 (SLC7A11) are also related to lipid peroxidation.

In the study, we injected CY-09, a specific inhibitor of NLRP3, into non-transgenic (NTg) and triple transgenic AD (3×Tg-AD) mice daily, with a dose of 2.5 mg/kg for six weeks, according to the reference [32]. Then, we explored the effect of NLRP3 inflammasome inactivation on glucose transport, insulin resistance, and glucose metabolic enzymes in vivo. Simultaneously, we investigated whether NLRP3 inflammasome inactivation by CY-09 could reduce AD classical pathology and oxidative stress and improve cognitive deficits. Overall, this study aimed to determine the effect of NLRP3 inflammasome activation on glucose metabolism and to investigate the potentiality of CY-09 as a therapeutic drug for AD treatment.

2. Materials and Methods

2.1. Reagents and Antibodies

CY-09 (Cat#S5774) was purchased from Selleck Chemicals LLC, Houston, TX, USA. Polyoxyl 15 hydroxystearate (Cat#HY-136349) and DMSO (Cat#HY-Y0320) were purchased from MedChemExpress LLC, Deer Park Dr, Suite Q, Monmouth Junction, NJ, USA. Saline (Cat#R22172) was purchased from Shanghai yuanye Bio-Technology Co., Ltd., Shanghai, China. ^{18}F-FDG was provided by Union Hospital, Tongji Medical College, Huazhong University of Science and Technology, Wuhan, China. Antibodies against NLRP3 (Cat#AG-20B-0014-C100) and caspase-1 (P20) (Cat#AG-20B-0042-C100) were purchased from Adipogen Corporation, San Diego, CA, USA. Antibodies against IL-1β (Cat#16806-1-AP), GLUT1 (Cat#21829-1-AP), GLUT4 (Cat#66846-1-Ig), GLUT3 (Cat#20403-1-AP), HK2 (Cat#22029-1-AP), voltage-dependent anion-selective channel protein 1 (VDAC1) (Cat#10866-1-AP), and SLC7A11 (Cat#26864-1-AP) were purchased from Proteintech Group, Inc., Chicago, IL, USA. Antibodies against IRS (Cat#3407), p-IRS-Ser1101 (Cat#2385), AKT (Cat#9272), p-AKT-Ser473 (Cat#4060), GSK3β (Cat#12456), p-GSK3β-Ser9 (Cat#5558), FAS (Cat#3180), ACC (Cat#3676), p-ACC-Ser79 (Cat#11818), and LRP1 (Cat#64099) were purchased from Cell Signaling Technology, Inc., Boston, MA, USA. Antibodies against IR (Cat#sc-57342), p-IR-Tyr1150 (Cat#sc-81500), and ACSL4 (Cat#sc-271800) were purchased from Santa Cruz Biotechnology, Inc., Dallas, TX, USA. The antibodies against AS160 (Cat#ab189890), p-AS160-T642 (Cat#ab131214), HK1 (Cat#ab150423), PDHE1α (Cat#ab168379), COX IV (Cat#ab16056), APP (Cat#ab32136), BACE1 (Cat#ab108394), PSD95 (Cat#ab18258), synaptophysin (Cat#ab32127), tau5 (Cat#ab80579), p-tau-Ser404 (Cat#ab92676), MDA (Cat#ab27642), GPX4 (Cat#ab125066), and HMGCS1 (Cat#ab155787) were purchased from Abcam plc., Cambridge, UK. Antibodies against 6E10 (Cat#803002), sAPPα (Cat#813501), and sAPPβ (Cat#813401) were purchased from BioLegend, Inc., San Diego, CA, USA. The antibody against Aβ1-42 (Cat#AB5078P) was purchased from Millipore Corporation, Boston, MA, USA. Antibodies against HT7 (Cat#MN1000), TF (Cat#PA5-27306), TFR (Cat#13-6800), FPN (Cat#PA5-22993), NOCA4 (Cat#PA5-96398), and SREBP2 (Cat#PA1-338) were purchased from Thermo Fisher Scientific, Waltham, MA, USA. The antibody against β-actin (Cat#AB0033) was purchased from Abways Technology, Inc., Shanghai, China. The antibody against HMGCR (Cat#T56640S) was purchased from Abmart Shanghai Co., Ltd., Shanghai, China. Goat Anti-Rabbit IgG H&L (HRP) secondary antibody (Cat#ab6721) and Goat Anti-Mouse IgG H&L (HRP) secondary antibody (Cat#ab6789) were purchased from Abcam plc., Cambridge, UK. Alexa Fluor® 488 AffiniPure Goat Anti-Rabbit IgG (H+L) secondary antibody (Cat#111-545-003) and Alexa Fluor® 594 AffiniPure Goat Anti-Mouse IgG (H+L) secondary antibody (Cat#111-585-003) were purchased from Jackson ImmunoResearch Inc., West Grove, PA, USA. DAPI Staining Solution (Cat#C1002) was purchased from Beyotime Biotech Inc., Shanghai, China.

2.2. Animals and Treatment

The impact of sex on AD pathology has been reported in many references. Senior females are more likely to develop AD due to their lower estrogen levels. Ovarian hormone loss causes a bioenergetic deficit and a shift in metabolic fuel availability in AD model mice [33,34]. Thus, we select 9-month-old female C57BL/6J mice (NTg mice) and 3×Tg-AD mice for this study. NTg mice were purchased from Guangdong Medical Laboratory Animal Center and 3×Tg-AD mice which harbor the mutated human genes amyloid precursor protein (APP) (SWE), PS1 (M146V), and Tau (P301L) were purchased from the Jackson Laboratory. Mice were housed in conditions under 22 °C in a 12:12 h light/dark cycle with food and water ad libitum. All animal experimental protocols were reviewed and approved by the Institutional Animal Care and Use Committee of Tongji Medical College of Huazhong University of Science and Technology (IACUC Number: 2390; Approved Date: 27 February 2018).

To observe the impact of NLRP3 inflammasome activation on cerebral glucose metabolism in AD, we used CY-09 to inhibit NLRP3 inflammasome activation. Here, 24 female mice were randomly divided into four groups: NTg mice, CY-09-treated NTg mice (NTg + CY-09 mice), 3×Tg-AD mice, and CY-09-treated 3×Tg-AD mice (3×Tg-AD + CY-09 mice). CY-09 was dissolved in a vehicle containing 10% DMSO, 10% Polyoxyl 15 hydroxystearate, and 80% saline. It was then injected into the mice daily, with a dose of 2.5 mg/kg for six weeks, according to the reference [32].

2.3. ^{18}F-FDG PET

^{18}F-FDG PET experiments were applied to examine the glucose metabolism in the mouse brain [35]. All the mice were fasted for 12–16 h and then injected by vein with 7.4 MBq ^{18}F-FDG before the PET scan. For the static PET scan, the mice were scanned for 10 min after 60 min of free metabolism. However, in the dynamic scan, the mice were scanned for 60 min immediately after injection. PET data were acquired with Trans-PET Discoverist 180 (Raycan Technology Co., Ltd., Suzhou, China) and reconstructed with a 3D OSEM algorithm. The time segmentary schemes for dynamic data reconstruction were as follows: 5 s × 6, 10 s × 3, 30 s × 4, 60 s × 2, 120 s × 5, 300 s × 3, 600 s × 1, 300 s × 1, 900 s × 1. To calculate the cerebral metabolic rate of glucose (CMRglu), blood glucose was measured from the second drop of tail blood using a Roche blood glucose meter (Roche Pharma (Schweiz) Ltd., Basel, Switzerland). Standard uptake values (SUVs) of the whole brain, cortex, and hippocampus, as well as the CMRglu of the whole brain, were quantified by Amide 1.0.4 software (Crump Institute for Molecular Imaging, UCLA School of Medicine, CA, USA), Crimas 2.9 (Turku PET center, Turku, Finland), MATLAB 2019b (MathWorks Inc., Natick, MA, USA), and SPM12 (The Wellcome Centre for Human Neuroimaging, UCL Queen Square Institute of Neurology, London, UK).

2.4. Morris Water Maze Test

The Morris water maze test is one of the behavioral experiments used to detect the spatial learning and memory ability of mice. In the study, we mainly refer to the previous protocol, with slight modifications [36,37]. None of the mice were trained on the test before experimentation. However, an additional process was carried out to familiarize the mice with the environment before the formal experiment started. The Morris water maze test includes place navigation tests and spatial probe tests. In the place navigation test, the mice were first placed on the platform for 1 min to familiarize themselves with the environment. Subsequently, the mice were placed into the water from the four quadrants. The time (escape latency) from entering the water to finding the platform was recorded within 60 s. The place navigation test lasted for five days. In the spatial probe test, the platform was removed and the mice were placed into the diagonal quadrant. Then, the time spent by the mice in the target quadrant in 24 h and 72 h was recorded. The Morris water maze WMT-200A (Chengdu Techman Software Co., Ltd., Chengdu, China) and the Animal Behavior Analysis System BAS-100 (Chengdu Techman Software Co., Ltd., Chengdu, China) were used for the recording of data.

2.5. Glucose Tolerance Test (GTT) and Insulin Tolerance Test (ITT)

GTT and ITT were performed to respectively detect the ability to regulate blood glucose and insulin sensitivity [32,38]. In GTT, the mice were fasted for 12 h and then injected with glucose in the dose of 2 g glucose per Kg mice (glucose was purchased from Beyotime Biotech Inc., Shanghai, China, Cat#ST1228). In ITT, mice were fasted for 4 h and then injected with insulin in the dose of 0.75 U insulin per Kg mice (insulin was purchased from Beyotime Biotech Inc., Shanghai, China, Cat#P3376). All experiments were required to detect and record the blood glucose in time points of 0, 30, 60, 90, 120, and 150 min after the injection. Blood glucose was measured from the second drop of tail blood using a Roche blood glucose meter (Roche Pharma (Schweiz) Ltd., Basel, Switzerland).

2.6. Brain Tissues Extraction and Preservation

After a week of recovery, the mice were euthanized after isoflurane anesthesia to extract the brain and blood. The hemibrain of each mouse was used for immunofluorescence experiments and stored with 4% paraformaldehyde. The total proteins of the rest of the hippocampus were extracted using the Phosphorylated Protein Extraction Kit (Cat#KGP950, Jiangsu Keygen Biotech Corp., Ltd., Nanjing, China) for Western blot experiments. The of the rest cortex was used for LC-MS/MS analysis, ROS measurement, and mitochondrion isolation. All tissues were stored at −80 °C before the experiments. Blood was tested for insulin directly after the extraction.

2.7. ELISA Assay and Reactive Oxygen Species Measurement

All experiments were conducted following the instructions of each kit. Fasting blood insulin was detected using a Highly Sensitive Mouse Insulin Immunoassay Kit (Cat#HMS200, EZassay Ltd., Shenzhen, China). The secretion of IL-1β was detected using a Mouse IL-1β ELISA Kit (Cat#EMC001b, Neobioscience Technology Company, Shenzhen, China). ROS levels were measured using the Reactive Oxygen Species Assay Kit (Cat#E004-1-1, Nanjing Jiancheng Bioengineering Institute, Nanjing, China). These data were detected by a SpectraMax® L Microplate Reader (Molecular Devices, LLC, San Jose, CA, USA).

2.8. LC-MS/MS

LC-MS/MS was performed by Triple TOF 6500 (AB Sciex LLC, Framingham, MA, USA) to determine whether CY-09 crosses the BBB and enters the brains of the mice. In this experiment, metabolites were extracted from brain tissues using a 300 μL methanol acetonitrile mixture (methanol: acetonitrile = 2:1), as previously described [39]. After vortexing for 1 min, sonicating for 10 min at 0 °C, and centrifuging for 15 min at 13,000× *g*, the samples were held at −20 °C until further detection. The conditions for the chromatographic separation of metabolites were as follows—flow rate: 0.3 mL/min; injection volume: 6 μL; mobile phase: phase A, water (containing 0.1% formic acid); phase B, acetonitrile (containing 0.1% formic acid). The ion source parameters of mass spectrometry were as follows—curtain gas: 35 psi; ion spray voltage: −4500 V; source temperature: 550 °C; ion source gas1: 55psi; ion source gas2: 55psi. MultiQuant 3.02 (AB Sciex LLC, Framingham, MA, USA) and ProteoWizard 1.3.5.0 (ProteoWizard, Palo Alto, CA, USA) were used to process the data.

2.9. Mitochondrion Isolation and Hexokinase Activity

The brain mitochondria were isolated following the product instructions for the mitochondrial isolation kit (Cat#C3606, Beyotime Biotech Inc, Shanghai, China). Isolated mitochondria and cytoplasm were collected and stored at −80 °C. A NanoDrop 2000c (Thermo Fisher Scientific, Waltham, MA, USA) was used to detect the mitochondria concentration. The hexokinase activity was measured by Micro Hexokinase Assay Kit (Cat#BC0745, Beijing Solarbio Science & Technology Co., Ltd., Beijing, China). All experiments were carried out according to the product instructions.

2.10. Western Blot Analysis

The total proteins of the brain tissues were extracted using the Phosphorylated Protein Extraction Kit (Cat#KGP950, Jiangsu Keygen Biotech Corp., Ltd., Nanjing, China). The PierceTM BCA Protein Assay Kit (Cat#23225, Thermo Fisher Scientific, Waltham, MA, USA) was used for the detection of protein concentrations. Experimental protocols were the same as previously described [5]. In short, 10% and 12% sodium dodecyl sulfate polyacrylamide gel electrophoresis (SDS-PAGE) gels were used in the experiments. Proteins larger than 100 kD were separated using a 10% SDS-PAGE gel, while proteins smaller than 100 kD were separated using a 12% SDS-PAGE gel. Equal amounts of total protein lysates (20 μg per well) of each sample were loaded for the electrophoresis. Then, the proteins were transferred to a 0.45 μm polyvinylidene difluoride (PVDF) membrane (Cat#ISEQ00010, Millipore Corporation, Boston, MA, USA). After the electrotransfer, the membranes were blocked in 5% non-fat milk for 2 h and washed four times (10 min each time) with TBST buffer (150 mM NaCl, 10 mM Tris, 0.1% Tween-20, pH 7.4). Afterwards, the membranes were incubated with primary antibodies overnight at 4 °C. The next day, the membranes were incubated with secondary antibodies at room temperature for 2 h after the washing of the primary antibody with TBST buffer. After the washing of the secondary antibodies, immunoreactive bands were visualized by the Tanon 5200 Series Image Analysis System (Tanon Science & Technology Co., Ltd., Shanghai, China). Grayscale analysis was used to quantify the protein expression. Grayscale values of the target protein were normalized to the grayscale values of β-actin or VDAC1. ImageJ 1.53C (NIH, Bethesda, MD, USA) was used to quantify the grayscale values of each immunoreactive band.

2.11. Immunofluorescence Assay

An immunofluorescence assay was performed for the brain slices following the previously established protocol [40]. Briefly, the brain slices were dewaxed and the xylene removed with dimethylbenzene and different concentrations of alcohol (100%, 95%, 85%, 70%, 50%, 30%, and 0%), respectively. Next, the brain slices were incubated with primary antibodies overnight after antigen retrieval, membrane rupture, and blocking. The next day, brain slices were washed using PBST for 30 min and then incubated with a second antibody for 2 h in the dark. After the washing of the second antibody, the brain slices were placed on slides and observed using an Olympus BX53 microscope (Olympus Corporation, Tokyo, Japan).

2.12. Statistical Analysis

In this study, semi-quantitative analysis and absolute quantification analysis were used to quantify the SUVs and the CMR_{glu}, respectively [41]. The equation for *SUV* is as follows:

$$SUV = \frac{C_T}{D_{Inj}} \cdot W_s \quad (1)$$

C_T, D_{Inj}, and W_S represent the radioactivity in the region of interest, the injected dose of radioactivity, and the weight of the mouse, respectively.

The simplified equation for CMR_{glu} is as follows:

$$CMR_{glu} = \frac{C_g}{LC} \cdot K \quad (2)$$

Cg is the concentration of blood glucose. LC is the lumped constant, and it reflects the difference between the metabolism of ^{18}F-FDG and glucose. K is the uptake rate of ^{18}F-FDG.

All data were presented as mean ± SD and analyzed with GraphPad Prism 9 (GraphPad Software Inc., La Jolla, CA, USA). The normality of distribution of the results was checked by Shapiro–Wilk test. A normal distribution of the data was indicated by the test; thus, a one-way ANOVA, followed by Bonferroni's multiple comparisons test, were performed in the study. Every possible comparison was explored, and significant differences ($p < 0.05$) between each group were shown in the figures.

3. Results

3.1. CY-09 Could Cross the Blood-Brain Barrier In Vivo

CY-09 inhibits the assembly and activation of the NLRP3 inflammasome by its combination with the ATP-binding motif of the NLRP3 NACHT domain to inhibit the ATPase activity. To investigate whether it is possible for CY-09 to cross the blood–brain barrier (BBB) in mice, initially, we adopted LC-MS/MS to detect the content of CY-09 in the brain of each group, i.e., the NTg, NTg + CY-09, 3×Tg-AD, and 3×Tg-AD + CY-09 mice. The results were shown in Figures 1a and S1; CY-09 was found in the brain tissues of NTg + CY-09 and 3×Tg-AD + CY-09 mice, which indicated that CY-09 could cross the BBB in vivo.

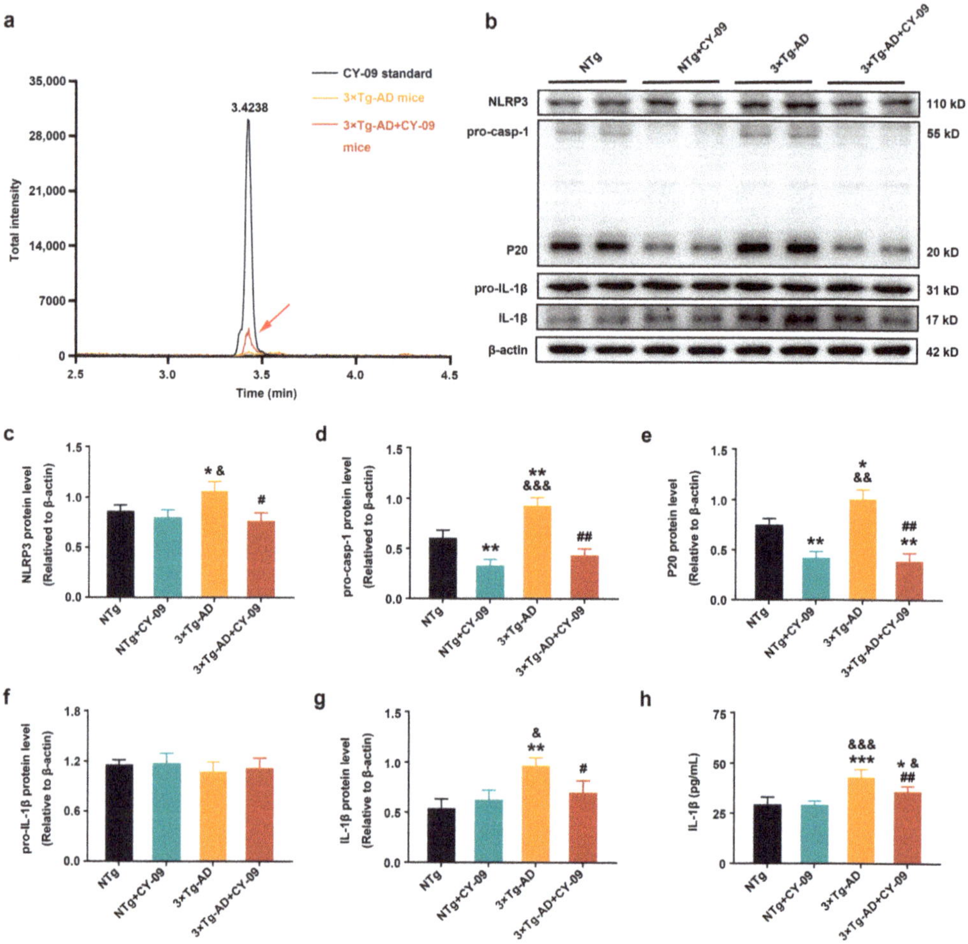

Figure 1. CY-09 attenuated activation of NOD-like receptor protein 3 (NLRP3) inflammasome in triple transgenic AD (3×Tg-AD) mice. (**a**) Content of CY-09 in 3×Tg-AD and 3×Tg-AD + CY-09 mice brains; (**b**–**g**) Western blot analysis of NLRP3, pro-caspase-1, caspase-1 (P20), pro-interleukin-1β (pro-IL-1β), and IL-1β in non-transgenic (NTg), NTg + CY-09, 3×Tg-AD, and 3×Tg-AD + CY-09 mice; (**h**) detection of IL-1β by ELISA in the four groups of mice. (n = 6, mean ± SD, one-way ANOVA and Bonferroni post hoc test; * $p < 0.05$, ** $p < 0.01$, *** $p < 0.001$ vs. NTg mice, & $p < 0.05$, && $p < 0.01$, &&& $p < 0.001$ vs. NTg + CY-09 mice, # $p < 0.05$, ## $p < 0.01$ vs. 3×Tg-AD mice).

3.2. *CY-09 Inhibited NLRP3 Inflammasome Activation in 3×Tg-AD Mice*

Then, Western blot and ELISA were then used to detect the expressions of NLRP3 inflammasome-related proteins to confirm the effect of CY-09 on NLRP3 inflammasome activation. As demonstrated in Figure 1b–h, compared with NTg mice, the expressions of NLRP3, pro-caspase-1, caspase-1(P20), and IL-1β were significantly increased in 3×Tg-AD mice, with the *p*-values lower than 0.05 and 0.01 respectively. Simultaneously, a notable reduction in the expressions of these proteins was found in CY-09 treated 3×Tg-AD mice with the *p*-values lower than 0.05 and 0.01, respectively. Moreover, the expressions of pro-caspase-1 and caspase-1 were remarkably decreased in NTg + CY-09 mice, with the *p*-values all lower than 0.01. Except for a significant difference between NTg mice, NTg + CY-09 mice and 3×Tg-AD + CY-09 mice, the results of IL-1β secretion measured by ELISA were consistent with those of Western blotting. No differences were found in the expression of pro-IL-1β among the four groups of mice. Together, these data suggested that CY-09 has an inhibiting effect on NLRP3 inflammasome activation in the brain of triple transgenic AD mice.

3.3. CY-09 Increased Cerebral Glucose Metabolism in 3×Tg-AD Mice

Next, we evaluated the effect of NLRP3 inflammasome activation on cerebral glucose metabolism using static and dynamic PET. ^{18}F-FDG is the most commonly used PET tracer for glucose metabolism. As an analog of glucose, ^{18}F-FDG is transported into cells by GLUTs from blood after the i.v. injection and is then phosphorylated by HK. Due to the differences in structure, 6-P-^{18}F-FDG cannot be catalyzed by glucose-6-phosphate isomerase and must remain in the cytoplasm. The amount and distribution of 6-P-^{18}F-FDG represent the glucose metabolism levels in different brain regions.

Static PET results are shown in Figure 2a–e; compared with NTg mice, the standard uptake values (SUVs) of the whole brain, the cortex, and the hippocampus were greatly decreased in 3×Tg-AD mice, with the *p*-value lower than 0.05. After CY-09 treatment, the SUVs were significantly higher in 3×Tg-AD mice than those in non-treated AD mice, with a *p*-value lower than 0.05. There were no differences in the weight of the mice in the four groups. Consistent with the static PET results, the cerebral SUVs of NTg and CY-09 treated 3×Tg-AD mice were also higher than those of the 3×Tg-AD mice, even though the SUVs of the four groups of mice increased in dynamic PET over time (Figure 2f–h). Moreover, the results of the cerebral metabolic rate of glucose (CMRglu) showed a notable reduction in the 3×Tg-AD mice and a remarkable increase in the CY-09 treated 3×Tg-AD mice, with the *p*-values all lower than 0.05. Overall, the PET data demonstrated that inhibiting NLRP3 inflammasome activation helps to restore cerebral glucose metabolism in the 3×Tg-AD mice.

3.4. CY-09 Increased Glucose Transport in 3×Tg-AD Mice

GLUTs are responsible for glucose transport, which is the basis of glucose metabolism. GLUT1, GLUT3, and GLUT4 can be expressed in the brain. Here, we used Western blotting to analyze the expression of GLUT1, GLUT3, and GLUT4. As shown in Figure 3a–d, the expression of GLUT1, GLUT3, and GLUT4 were lower in 3×Tg-AD mice than in NTg mice, with the *p*-value lower than 0.01, 0.05, and 0.01 respectively. However, the expressions of GLUT1 and GLUT4 were significantly increased in CY-09 treated 3×Tg-AD mice than in non-treated AD mice, with the *p*-value lower than 0.05. The expression of GLUT3 was increased with a *p*-value of 0.077. Further, immunostaining of GLUT4 demonstrated the decreased distribution in 3×Tg-AD mice and the increased distribution in CY-09 treated 3×Tg-AD mice (Figures 3e and S2). Hence, these data exhibited that inhibiting NLRP3 inflammasome activation can increase the expression and distribution of GLUTs in the 3×Tg-AD mice.

3.5. CY-09 Attenuated Insulin Resistance in 3×Tg-AD Mice

GLUT4 is regulated by the insulin signaling pathway to participate in glucose metabolism. Insulin resistance manifests itself as insensitivity to insulin and a higher level of insulin and glucose in the blood, eventually leading to impaired insulin signaling pathways and glucose metabolism. It had been confirmed to exist in AD.

To evaluate the effect of NLRP3 inflammasome activation on insulin resistance in AD, we first detected blood glucose and blood insulin levels. As shown in Figure 4, higher fasting and basal blood glucose and fasting insulin level were found in 3×Tg-AD mice compared with NTg mice, with the *p*-values all lower than 0.05 and 0.01, respectively. Results of the GTT and ITT showed an increased glucose tolerance and decreased insulin tolerance in 3×Tg-AD mice. However, after CY-09 treatment, blood glucose and insulin levels were decreased and better insulin sensitivity was found in CY-09-treated 3×Tg-AD mice. Thus, the results indicated that inhibiting NLRP3 inflammasome activation can attenuate insulin resistance in the 3×Tg-AD mice.

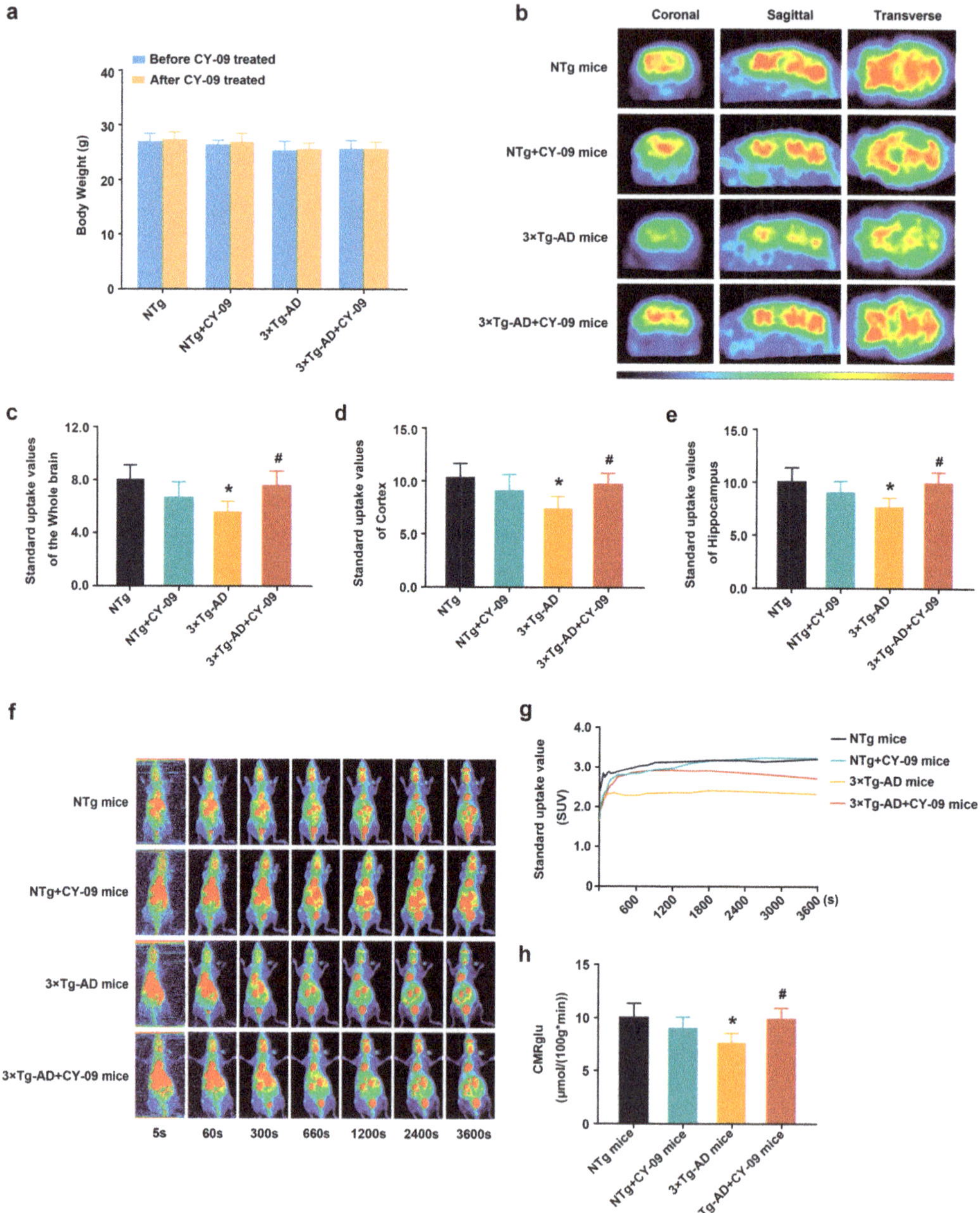

Figure 2. ^{18}F-FDG positron emission tomography (PET) images of NTg, NTg + CY-09, 3×Tg-AD, and 3×Tg-AD + CY-09 mice. (**a**) Body weight of mice, before and after CY-09 treatment; (**b**) static PET images of the four groups of mice; (**c**–**e**) standard uptake values in the whole brain, cortex, and hippocampus in the four groups of mice; (**f**) dynamic PET images of the four groups of mice; (**g**) cerebral time–activity curves (TAC) and (**h**) cerebral metabolic rate of glucose (CMRglu) in the four groups of mice. (n = 6, mean ± SD, one-way ANOVA and Bonferroni post hoc test; * $p < 0.05$ vs. NTg mice, # $p < 0.05$ vs. 3×Tg-AD mice).

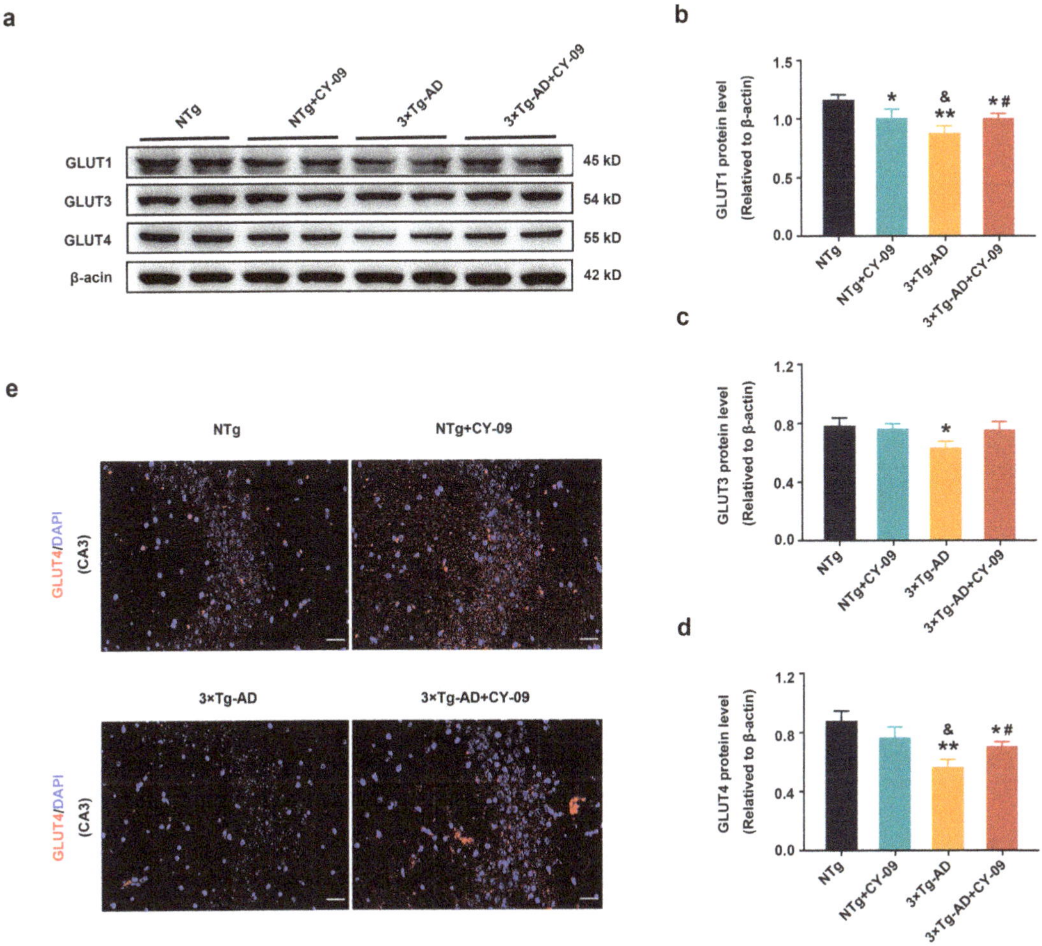

Figure 3. Increased expression of glucose transporters in CY-09-treated 3×Tg AD mice. (**a**) Representative Western blots and quantification of (**b**) glucose transporters 1 (GLUT1), (**c**) GLUT3, and (**d**) GLUT4 expression in NTg, NTg + CY-09, 3×Tg-AD and 3×Tg-AD + CY-09 mice (n = 6, mean ± SD, one-way ANOVA and Bonferroni post hoc test; * $p < 0.05$, ** $p < 0.01$ vs. NTg mice, [&] $p < 0.05$ vs. NTg + CY-09 mice, # $p < 0.05$ vs. 3×Tg-AD mice); (**e**) distribution of GLUT4 in the CA3 region of the four groups of mice (scale bar: 50 μm).

Then, to further explore the underlying mechanism by which NLRP3 inflammasome activation affects insulin resistance, we detected the expression and distribution of the IR-AKT-AS160 insulin signaling pathway-related proteins in NTg, NTg + CY-09, 3×Tg-AD, and 3×Tg-AD + CY-09 mice. As shown in Figure 5a,b, compared with NTg mice, expression of p-IR-Tyr1150 (phosphorylation protein of IR at Tyr1150) relative to IR was significantly decreased in 3×Tg-AD mice, with a p-value lower than 0.01, but the expression was increased after the CY-09 treatment, with a p-value lower than 0.05. The distribution of p-IR-Tyr1150 was found to be lower in the 3×Tg-AD mice and higher in the 3×Tg-AD + CY-09 mice. By contrast, the expression and distribution of IR were not different among the four groups of mice (Figures 5i and S3). These findings suggested that inhibiting NLRP3 inflammasome activation can enhance the self-phosphorylation of IR in the 3×Tg-AD mice.

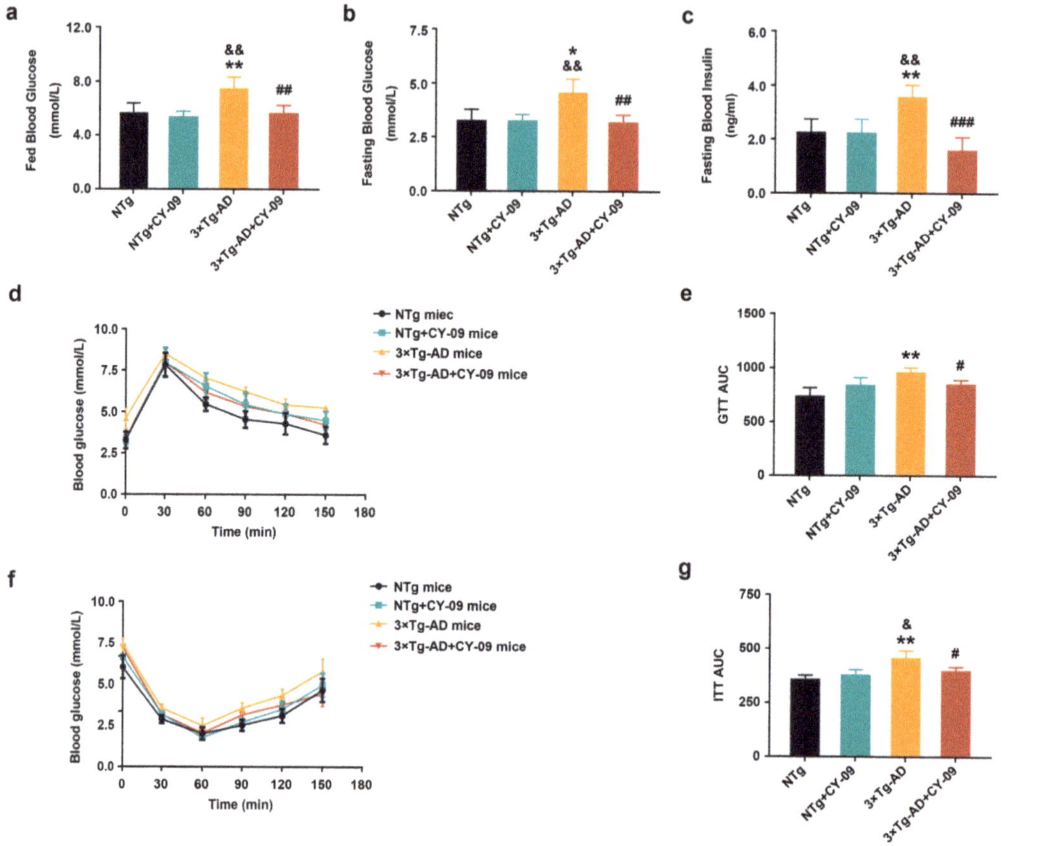

Figure 4. CY-09 relieved the insulin resistance in 3×Tg-AD mice. (**a**–**c**) Fasting blood glucose, fed blood glucose, and fasting blood insulin were measured in NTg, NTg + CY-09, 3×Tg-AD and 3×Tg-AD + CY-09 mice; (**d**,**e**) glucose tolerance tests (GTT) and (**f**,**g**) insulin tolerance tests (ITT) were conducted to determine the insulin sensitivity of the four groups of mice. (n = 6, mean ± SD, one-way ANOVA and Bonferroni post hoc test; * $p < 0.05$, ** $p < 0.01$ vs. NTg mice, & $p < 0.05$, && $p < 0.01$ vs. NTg + CY-09 mice, # $p < 0.05$, ## $p < 0.01$, ### $p < 0.001$ vs. 3×Tg-AD mice).

Self-phosphorylation of IR recruits IRS and starts the IRS-AKT-AS160 insulin signaling pathway. In parallel, we detected the expression of IRS, AKT, and AS160 and their phosphorylated levels in the four groups of mice. The results reported in Figure 5c–h show a notable reduction of p-AKT-Ser473 and p-AS160-T642, while significantly increased p-IRS-Ser1101 was found in the 3×Tg-AD mice, with the p-value lower than 0.01, 0.01 and 0.05, respectively. Treatment with CY-09 helped to reverse the expression of these proteins in 3×Tg-AD + CY-09 mice. There was also a significant reduction in IRS between the NTg + CY-09 mice and 3×Tg-AD + CY-09 mice, with a p-value less than 0.01. No differences were found in the expressions of AKT and AS160 among the four groups of mice. All the results showed that inhibiting NLRP3 inflammasome activation can restore the IR-IRS-AKT-AS160 insulin signaling pathway to alleviate insulin resistance in the 3×Tg-AD mice.

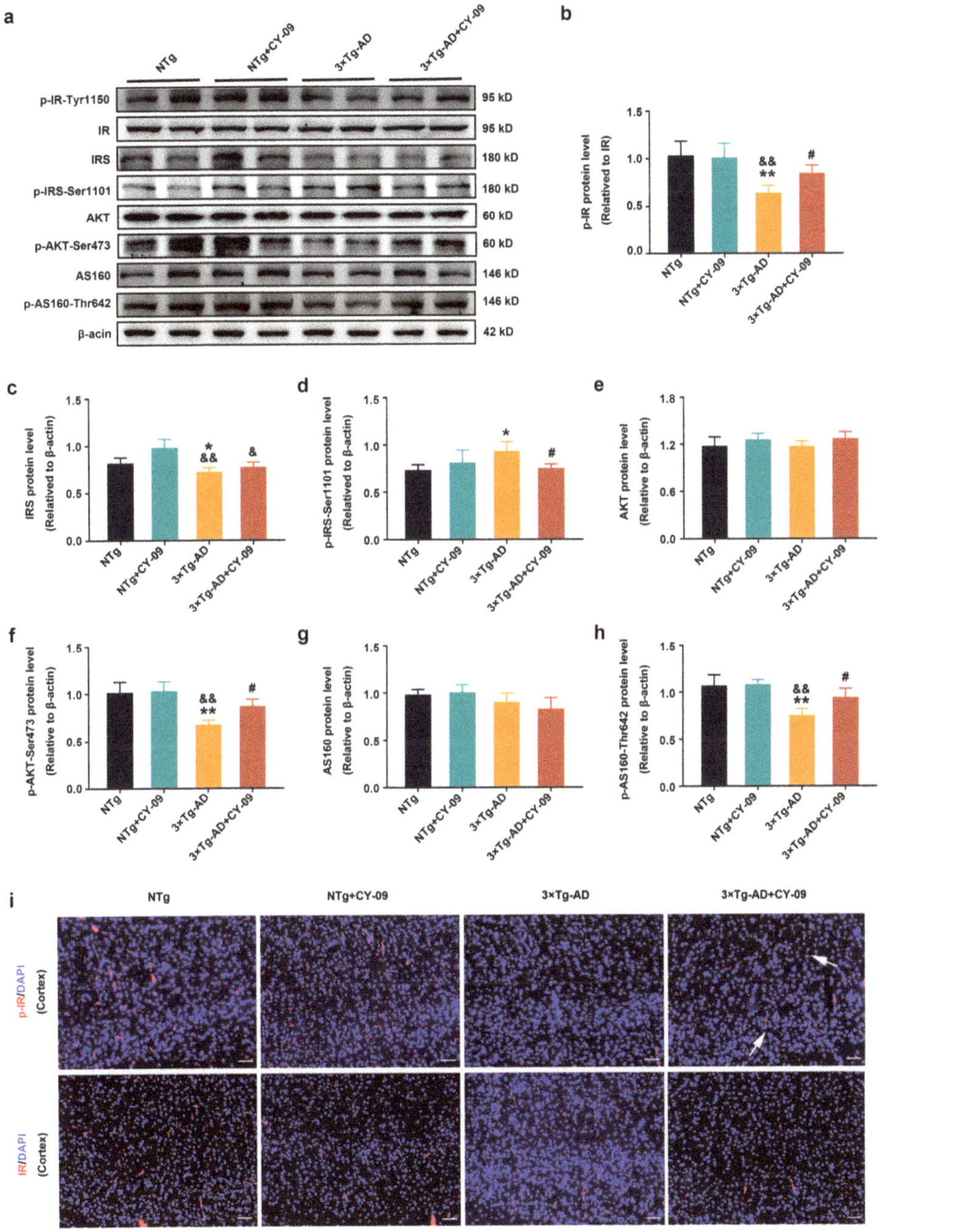

Figure 5. CY-09 restored the insulin signaling pathway in 3×Tg-AD mice. (**a–h**) Western blot analysis of insulin receptors (IR), p-IR-Tyr1150, insulin receptor substrate (IRS), p-IRS-Ser1101, AKT, p-AKT-Ser473, AS160 and p-AS160-Thr642 in NTg, NTg + CY-09, 3×Tg-AD, and 3×Tg-AD + CY-09 mice. (n = 6, mean ± SD, one-way ANOVA and Bonferroni post hoc test; * $p < 0.05$, ** $p < 0.01$ vs. NTg mice, & $p < 0.05$, && $p < 0.01$ vs. NTg + CY-09 mice, # $p < 0.05$ vs. 3×Tg-AD mice). (**i**) Distribution of IR and p-IR-Tyr1150 in the brains of the four groups of mice (scale bar: 100 μm; arrows indicate p-IR in 3×Tg-AD + CY-09 mice).

3.6. CY-09 Increased the Expression and Distribution of Metabolic Enzymes in 3×Tg-AD Mice

Increased glucose transport and improved insulin resistance were found in CY-09-treated 3×Tg-AD mice; we also detected the expressions and distribution of HK, which is the first enzyme that phosphorylates glucose when associated with VDAC1 in the mitochondria. We previously reported that the expression of cHK1 (HK1 in the cytoplasm) was significantly increased in 3×Tg-AD mice, while the expression of mHK1 (HK1 in the mitochondria) was remarkably decreased. However, increased HK1 expression and HK activity were found in CY-09-treated N2a-sw cells (a model cell of AD). These results prompted us to examine the expression and activity of HK in CY-09-treated 3×Tg-AD mice. Here, we isolated the mitochondria and cytoplasm from NTg, NTg + CY-09, 3×Tg-AD, and 3×Tg-AD + CY-09 mice to detect the expression and distribution of HK1.

As shown in Figure 6, consistent with our previous results, increased cHK1 and significantly decreased mHK1 and mHK2 were found in the 3×Tg-AD mice, with the p-value lower than 0.01 and 0.05. However, the expression of mHK1 was increased, while cHK1 was decreased in the CY-09-treated 3×Tg-AD mice, and the differences were significant compared to the untreated 3×Tg-AD mice, with the p-values lower than 0.01 and 0.05. cHK2 was also found to be notably decreased in the CY-09-treated 3×Tg-AD mice, with a p-value lower than 0.05. Although HK activity was remarkably decreased in 3×Tg-AD + CY-09 mice than in NTg + CY-09 mice, it was notably increased in the CY-09-treated 3×Tg-AD mice compared to the non-treated 3×Tg-AD mice, with a p-value lower than 0.05 (Figure 6b). Besides, we also detected the expressions of pyruvate dehydrogenase α 1 (PDHE1α) and cytochrome c oxidase subunit IV (COX4). They were all significantly reduced in 3×Tg-AD mice, with p-values lower than 0.05. Meanwhile, the expression of PDHE1α was remarkably increased in CY-09-treated 3×Tg-AD mice. In short, the data demonstrated an increase in the expression and distribution of HK by inhibiting NLRP3 inflammasome activation in the 3×Tg-AD mice.

3.7. CY-09 Relieved Cognitive Impairment and Pathological Injury in 3×Tg-AD Mice

Previous studies showed that NLRP3 inflammasome activation contributes to the pathology of AD. The results of this study demonstrated that inhibition of the NLRP3 inflammasome by CY-09 significantly recovered glucose metabolism. Next, we focused on the effects of CY-09 on cognitive impairment and classically pathological biomarkers in AD and confirmed whether CY-09 is a potential therapeutic drug for AD.

The Morris water maze was used to evaluate the learning and memory abilities of the four groups of mice. As shown in Figure 7, on the fifth day, the escape latency of 3×Tg-AD mice was remarkably longer than that of the NTg mice, but notably reduced after CY-09 treatment, with the p-values all lower than 0.05. Moreover, the time in the target quadrants of the 3×Tg-AD mice was less than that of the NTg mice, while increasing after CY-09 treatment in 24 h and 72 h space exploration experiments, with a p-value lower than 0.05 and a p-value of 0.1282, respectively. No differences were found in swimming speed between the four groups of mice. The results exhibited that inhibition of the NLRP3 inflammasome by CY-09 helped to relieve the cognitive impairment of the 3×Tg-AD mice.

Afterward, we detected the expression and distribution of pathological proteins in AD model mice. As shown in Figure 8, compared with the NTg mice, expressions of APP, beta-site app cleaving enzyme 1 (BACE1), sAPPβ, and Aβ1-42 were significantly increased in 3×Tg-AD mice, with the p-value lower than 0.01, 0.05, 0.05, and 0.05. Expressions of sAPPα, post-synaptic density protein 95 (PSD95) and synaptophysin were greatly decreased in 3×Tg-AD mice, with the p-values all lower than 0.05. However, in CY-09 treated 3×Tg-AD mice, the expressions of these proteins were reversed, with increased expression of sAPPα, PSD95, and synaptophysin; decreased levels of APP, BACE1, sAPPβ and Aβ1-42 were found in comparison to those in the untreated 3×Tg-AD mice, with the p-value lower than 0.05 and 0.01, respectively. Immunostaining of 6E10 confirmed the decrease in the expression and distribution of Aβ in CY-09 treated 3×Tg-AD mice (Figures 8i and S4).

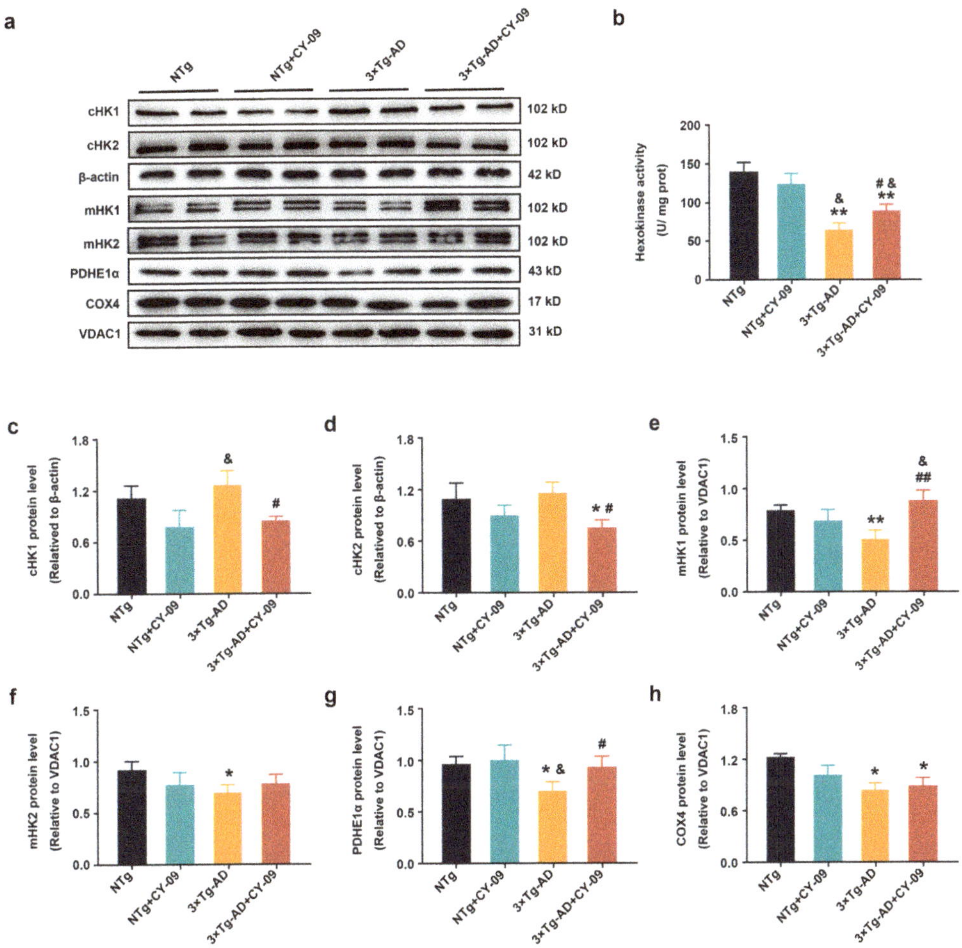

Figure 6. CY-09 reversed the expression and distribution of metabolic enzymes in 3×Tg-AD mice. (**a**,**c**–**h**) Western blot analysis of cHK1 (HK1 in the cytoplasm), cHK2 (HK2 in the cytoplasm), mHK1 (HK1 in mitochondria), mHK2 (HK2 in mitochondria), pyruvate dehydrogenase α 1 (PDHE1α) and cytochrome c oxidase subunit IV (COX4) in NTg, NTg + CY-09, 3×Tg-AD, and 3×Tg-AD + CY-09 mice. (**b**) Detection of hexokinase activity in the brain tissue of the four groups of mice. (n = 6, mean ± SD, one-way ANOVA and Bonferroni post hoc test; * $p < 0.05$, ** $p < 0.01$ vs. NTg mice, & $p < 0.05$ vs. NTg + CY-09 mice, # $p < 0.05$, ## $p < 0.01$ vs. 3×Tg-AD mice).

Furthermore, as demonstrated in Figures 9 and S5, we observed increased expression and distribution of pS404-tau in 3×Tg-AD mice, but treatment with CY-09 greatly reduced the expression and distribution, with the p-values lower than 0.05 and 0.01. There were no differences in the expressions of Tau5 among the four groups of mice. However, remarkably increased total human tau was found in the 3×Tg-AD mice and the 3×Tg-AD + CY-09 mice, with a p-value lower than 0.001, while no reduction was found in the CY-09-treated 3×Tg-AD mice. By contrast, expression of p-GSK3β-Ser9 relative to GSK3β exhibited a decrease in 3×Tg-AD mice ($p = 0.0834$), but it significantly increased after CY-09 treatment, with a p-value lower than 0.05. Together, these results demonstrated that CY-09 can reverse the expression and distribution of pathological proteins and alleviate cognitive impairment in the 3×Tg-AD mice.

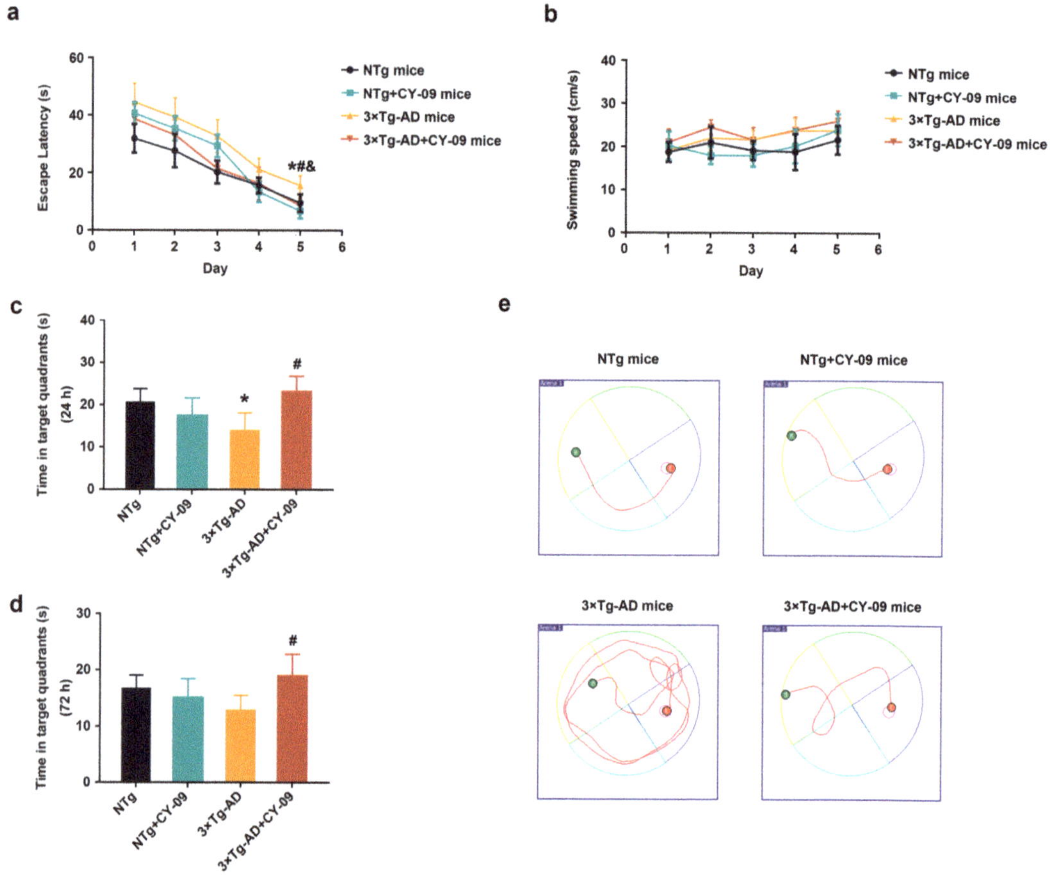

Figure 7. CY-09 improved the learning and memory ability of 3×Tg-AD mice. (**a**) Escape latency and (**b**) swimming speed during training of the four groups of mice. (n = 6, mean ± SD, one-way ANOVA and Bonferroni post hoc test; * $p < 0.05$: 3×Tg-AD mice vs. NTg mice, & $p < 0.05$: 3×Tg-AD mice vs. NTg + CY-09 mice, # $p < 0.05$: 3×Tg-AD + CY-09 mice vs. 3×Tg-AD mice). (**c**,**d**) Time in target quadrants in 24 h and 72 h space exploration experiments. (n = 6, mean ± SD, one-way ANOVA and Bonferroni post hoc test; * $p < 0.05$ vs. NTg mice, # $p < 0.05$ vs. 3×Tg-AD mice). (**e**) Schematic diagram of the swimming paths of the four groups of mice.

3.8. CY-09 Decreased Oxidative Stress in 3×Tg-AD Mice

Finally, we explored the effects of CY-09 on oxidative stress and ferroptosis. First we detected ROS and malondialdehyde (MDA) in the four groups of mice. As shown in Figure 10, significantly increased ROS levels and MDA were found in 3×Tg-AD mice, with the *p*-value lower than 0.001 and 0.01. After the CY-09 treatment, ROS and MDA were notably decreased, with the *p*-value lower than 0.001 and 0.05. These data reflected that CY-09 can reduce oxidative stress in 3×Tg-AD mice.

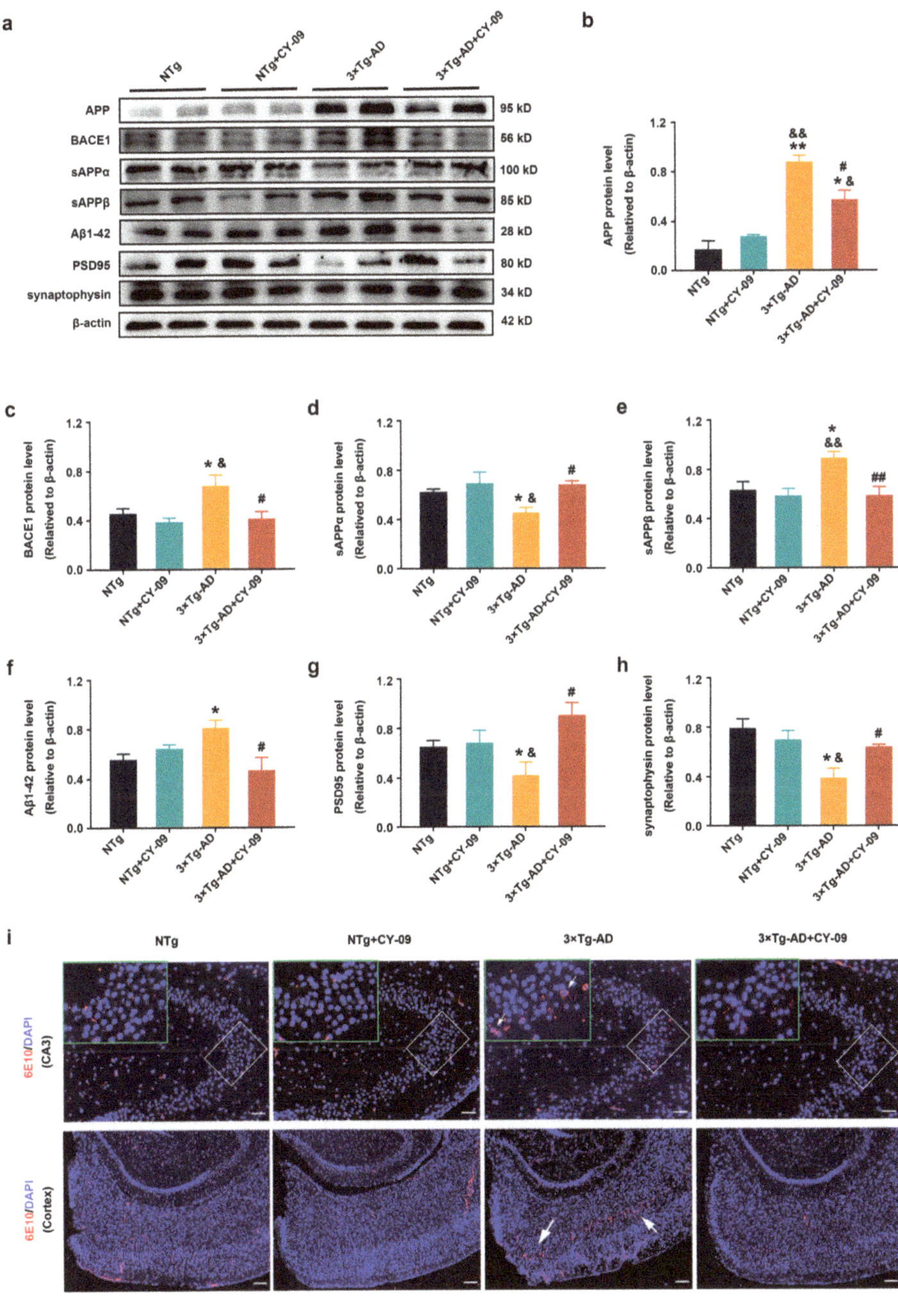

Figure 8. CY-09 reversed the expression and distribution of Aβ related proteins in 3×Tg-AD mice. (**a**–**h**) Western blot analysis of amyloid precursor protein (APP), beta-site app cleaving enzyme 1 (BACE1), sAPPα, sAPPβ, Aβ1-42, post-synaptic density protein 95 (PSD95), and synaptophysin in NTg, NTg + CY-09, 3×Tg-AD, and 3×Tg-AD + CY-09 mice. ($n = 6$, mean ± SD, one-way ANOVA and Bonferroni post hoc test; * $p < 0.05$, ** $p < 0.01$ vs. NTg mice, & $p < 0.05$, && $p < 0.01$ vs. NTg + CY-09 mice, # $p < 0.05$, ## $p < 0.01$ vs. 3×Tg-AD mice). (**i**) Distribution of Aβ in the CA3 region and cortex of the four groups of mice (scale bar: 100 μm and 250 μm; arrows indicate Aβ in 3×Tg-AD mice).

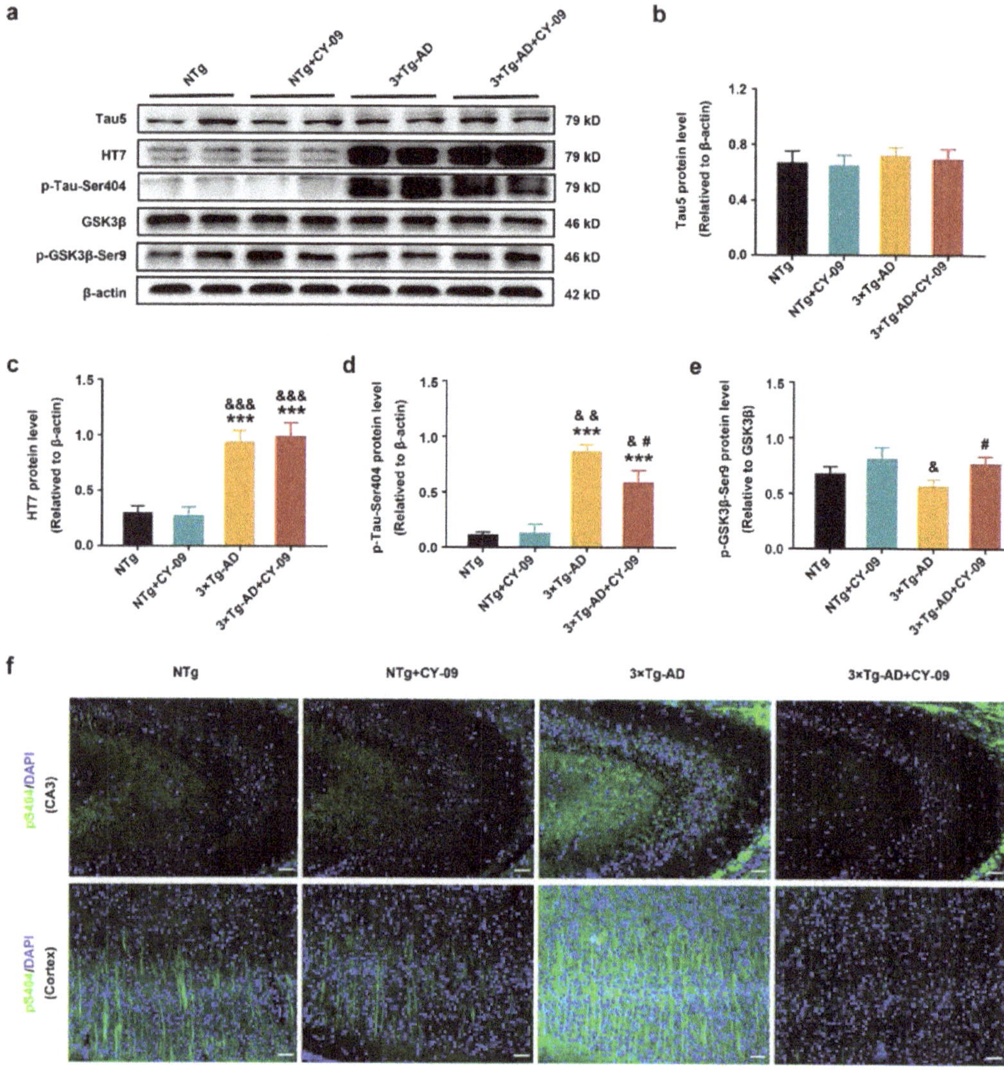

Figure 9. CY-09 reversed the expression and distribution of tau-related proteins in 3×Tg-AD mice. (**a**–**e**) Western blot analysis of Tau5, HT7, p-Tau-Ser404, GSK3β, and p-GSK3β-Ser9 in NTg, NTg+CY-09, 3×Tg-AD, and 3×Tg-AD + CY-09 mice. (n = 6, mean ± SD, one-way ANOVA and Bonferroni post hoc test; *** $p < 0.001$ vs. NTg mice, & $p < 0.05$, && $p < 0.01$, &&& $p < 0.001$ vs. NTg + CY-09 mice, # $p < 0.05$ vs. 3×Tg-AD mice). (**f**) Distribution of p-Tau-Ser404 in the CA3 region and cortex of the four groups of mice (scale bar: 100 μm).

Then, we tested the ferroptosis-related proteins, including TF, TFR, FPN, NCOA4, ACSL4, GPX4, and SLC7A11. TF and TFR were responsible for the transport of Fe^{2+} to the cell, while FPN transported Fe^{2+} outside the cell. NCOA4 participated in the ferritinophagy. GPX4, SLC7A11, and ACSL4 were critical enzymes in lipid peroxidation. As reported in Figure 11, compared with NTg mice, expressions of TFR and ACSL4 were increased, while expressions of FPN, NCOA4, GPX4, and SLC7A11 were decreased in the 3×Tg-AD mice, with the p-values all lower than 0.05. Furthermore, when compared with the NTg + CY-09 mice, the expressions of NCOA4, GPX4, and SLC7A11 were significantly decreased in the 3×Tg-AD mice, with the *p*-values all lower than 0.05. However, except for ACSL4, there were no differences in these proteins between the 3×Tg-AD mice and the 3×Tg-AD + CY-09 mice. The expression of ASCL4 was reduced in the 3×Tg-AD + CY-09 mice, with a *p*-value lower than 0.05. No differences in TF expression were found between the four groups of mice except for NTg mice and 3×Tg-AD + CY-09 mice. The results implied that the inactivation of the NLRP3 inflammasome by CY-09 cannot reverse the ferroptosis in the 3×Tg-AD mice.

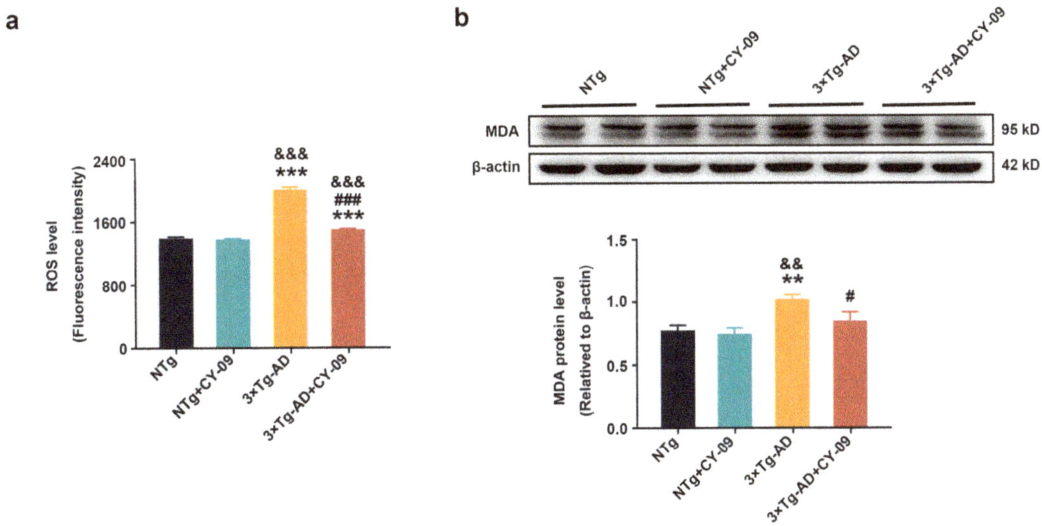

Figure 10. CY-09 decreased reactive oxygen species (ROS) levels and malondialdehyde (MDA) expression in the four groups of mice. (**a**) ROS levels in NTg, NTg + CY-09, 3×Tg-AD, and 3×Tg-AD + CY-09 mice; (**b**) expression of MDA in the four groups of mice. (n = 6, mean ± SD, one-way ANOVA and Bonferroni post hoc test; ** $p < 0.01$, *** $p < 0.001$ vs. NTg mice, && $p < 0.01$, &&& $p < 0.001$ vs. NTg + CY-09 mice, # $p < 0.05$, ### $p < 0.001$ vs. 3×Tg-AD mice).

Reduced expressions of MDA and ACSL4 in the 3×Tg-AD + CY-09 mice prompted us to detect the fatty metabolism changes in the four groups of mice. Data presented in Figure 12, reveal that in comparison to NTg mice, the expressions of acetyl coA carboxylase (ACC), 3-hydroxy-3-methylglutaryl-coenzyme A synthase 1 (HMGCS1), and 3-hydroxy-3-methylglutaryl-coenzyme A reductase (HMGCR) were greatly increased in the 3×Tg-AD mice, with the *p*-values all lower than 0.05. A significant decrease was found in the expression of p-ACC in the 3×Tg-AD mice, with a *p*-value lower than 0.05. After CY-09 treatment, expressions of ACC, p-ACC, HMGCS1, and HMGCR were reversed. Fatty acid synthase (FAS) was also increased in the 3×Tg-AD mice, but with no significant difference in protein levels. Similarly, there were no differences in low-density lipoprotein receptor related protein 1 (LRP1) and sterol regulatory element binding protein 2 (SREBP2) between the four groups of mice, except for LRP1 in CY-09-treated NTg and 3×Tg-AD mice. Increased ACC and FAS and decreased p-ACC indicated increased fatty acid synthesis, consistent with the increased MDA and ACSL4. HMGCS1 and HMGCR are responsible for the synthesis of cholesterol. LRP1 and SREBP2 regulate lipid metabolism homeostasis and cholesterol levels, respectively. The above results suggested that increased fatty acid synthesis may be the main cause of increased lipid peroxidation. CY-09 could reduce the synthesis of fatty acid and lipid peroxidation by inhibiting NLRP3 inflammasome activation in the 3×Tg-AD mice.

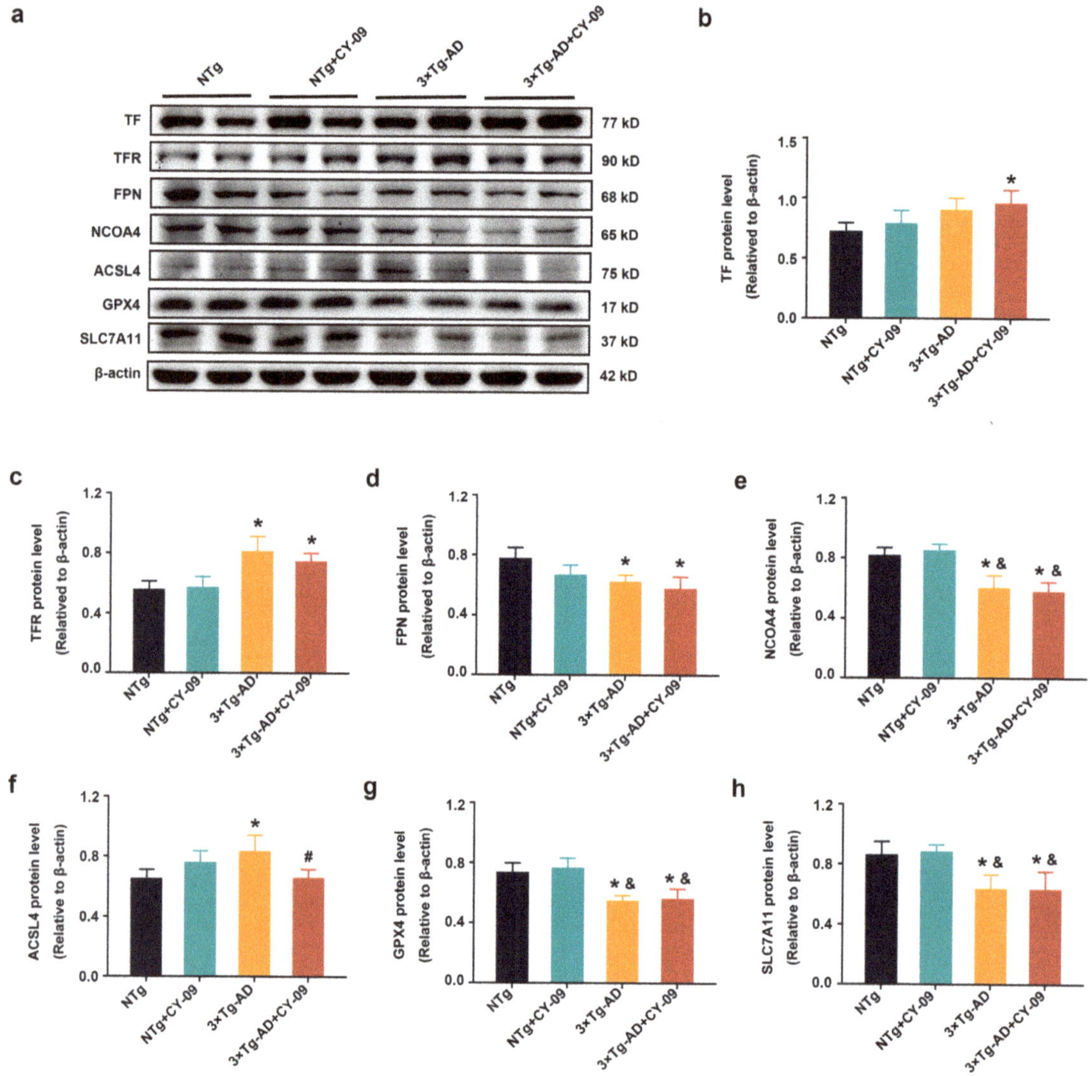

Figure 11. CY-09 could not reverse the ferroptosis in 3×Tg-AD mice. (**a–h**) Western blot analysis of transferrin (TF), transferrin receptor (TFR), ferroportin (FPN), nuclear receptor coactivator 4 (NCOA4), long-chain acyl-CoA synthetase 4 (ACSL4), glutathione peroxidase 4 (GPX4), Solute Carrier Family 7 and Member 11 (SLC7A11) in the four groups of mice. (n = 6, mean ± SD, one-way ANOVA and Bonferroni post hoc test; * $p < 0.05$ vs. NTg mice, $^{\&}$ $p < 0.05$ vs. NTg + CY-09 mice, # $p < 0.05$ vs. 3×Tg-AD mice).

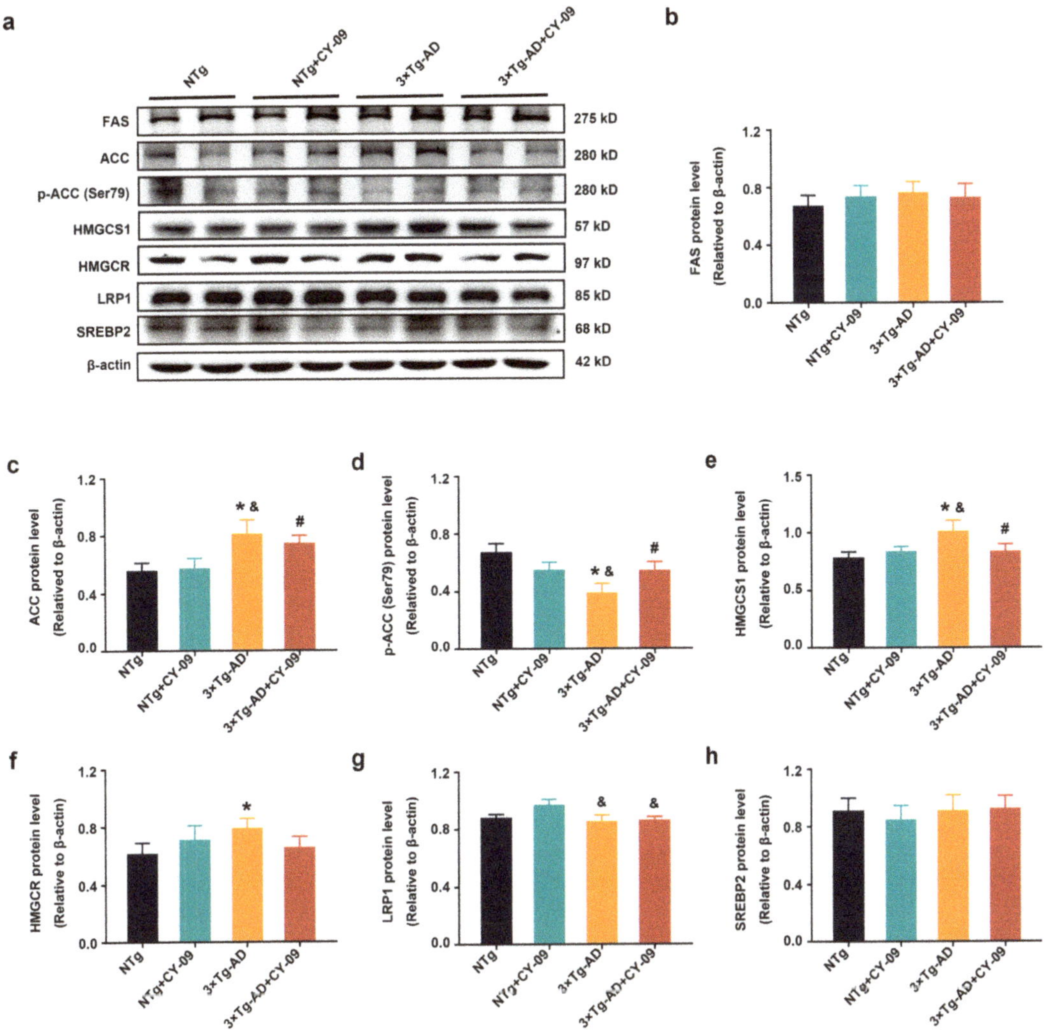

Figure 12. CY-09 decreased the synthesis of fatty acids in 3×Tg-AD mice. (**a–h**) Western blot analysis of fatty acid synthase (FAS), acetyl CoA carboxylase (ACC), p-ACC, 3-hydroxy-3-methylglutaryl-Coenzyme A synthase 1 (HMGCS1), 3-hydroxy-3-methylglutaryl-Coenzyme A reductase (HMGCR), low-density lipoprotein receptor related protein 1 (LRP1), and sterol regulatory element binding protein2 (SREBP2) in the four groups of mice. (n = 6, mean ± SD, one-way ANOVA and Bonferroni post hoc test; * $p < 0.05$ vs. NTg mice, & $p < 0.05$ vs. NTg + CY-09 mice, # $p < 0.05$ vs. 3×Tg-AD mice).

4. Discussion

Neuroinflammation has been recognized as a key event in inducing the onset and progression of AD, and it is closely linked to Aβ accumulation, tau hyperphosphorylation, and decreased cerebral glucose metabolism [3]. NLRP3 inflammasome is the most important inflammasome involved in neuroinflammation. In this study, CY-09 was used to inhibit the activation of NLRP3 inflammasome, leading to the restoration of glucose metabolism, as demonstrated by the increases in the expression and distribution of glucose transporters and related enzymes and the attenuation of insulin resistance. Moreover, inhibition of NLRP3 inflammasome activation by CY-09 could also reduce oxidative stress and improve cognitive ability in AD model mice.

In recent years, several studies have shown that NLRP3 inflammasome activation contributes to the progress of AD pathologies, specifically Aβ and tau proteins [42,43]. Inhibition of NLRP3 inflammasome activation helps to reduce Aβ deposition and improve cognitive impairment [43]. Most inhibitors of NLRP3 inflammasome activation do not specifically target NLRP3, except for CY-09, which combines directly with the ATP-binding motif of the NLRP3 NACHT domain [32]. Thus, CY-09 was selected in this study to inhibit NLRP3 inflammasome activation. As crossing the BBB and binding to NLRP3 in the brain is the prerequisite for inhibition, we used LC-MS/MS to characterize the molecular structure of CY-09 and detected its level in the brains of the mice. These results proved that CY-09 crossed the BBB to the brain to exert its biological function. Then, we further detected the expression and distribution of Aβ and tau proteins in CY-09-treated and non-treated 3×Tg-AD mice. Consistent with previous studies [43], our results showed reversed expression levels of pathological proteins in CY-09 treated 3×Tg-AD mice. In addition, CY-09 could also relieve cognitive deficits in the AD mice.

Reduced cerebral glucose metabolism has been reported in the AD brain due to the decrease in glucose transport, insulin resistance, and reduced metabolic enzymes [44–46]. We used ^{18}F-FDG PET to analyze the alteration of glucose metabolism in CY-09-treated and non-treated 3×Tg-AD mice. Our results showed increased glucose uptake and metabolism rates in CY-09-treated 3×Tg-AD mice. GLUT1, GLUT3 and GLUT4 are responsible for glucose uptake and transport in the brain. Decreased expressions of GLUT1 and GLUT3 in AD indicated reduced glucose uptake and transport [47–50]. In this work, GLUT1 and GLUT3 expression levels were found to be increased in CY-09-treated 3×Tg-AD mice, while GLUT4 also exhibited higher expression and distribution. These results implied that inhibiting NLRP3 inflammasome activation by CY-09 may improve insulin resistance. Insulin is a critical hormone that regulates blood glucose [20,51]. It is transported into the brain and binds with insulin receptors, which are expressed in the cell membranes to activate the insulin signaling pathway [21,52,53]. In AD, insulin levels increase in the blood and decrease in the brain, the insulin signal pathway is suppressed, and GLUT4 cannot localize to the cell membrane, leading to reduced glucose uptake. Intranasal insulin injection helps to recover the insulin signaling pathway, stimulate the transfer of GLUT4 to the cell membranes, and thus increase the uptake of glucose in AD [54]. CY-09 has been reported to inhibit NLRP3 inflammasome activation in order to improve insulin resistance in obesity and non-alcoholic fatty liver disease. In this study, elevated insulin levels and recovered insulin signal pathways were detected in the CY-09-treated AD mice. Increased insulin sensitivity was also measured by GTT and ITT in the CY-09-treated AD mice. We also detected the recovered insulin signal pathway in CY-09-treated 3×Tg-AD mice. These data suggested that the inhibition of NLRP3 inflammasome activation by CY-09 helps to alleviate insulin resistance in AD.

Insulin resistance and glucose metabolism dysfunction are two hallmarks of diabetes mellitus. As they were also found in AD, most researchers considered AD as type 3 diabetes mellitus. However, it should be noted that in diabetes, decreased insulin secretion or insulin resistance leads to increased blood glucose [20,47]. Abnormal insulin level in the blood is the main cause of diabetes. While in AD, the higher insulin levels in the blood and lower levels in the brain indicated a damaged insulin signaling pathway. Therefore, maintaining the stability of insulin levels and blood glucose is the key point for diabetes, but increase the insulin levels and glucose metabolism in the brain is more important in AD.

Glucose was transported into the cells by GLUTs and phosphorylated by mitochondria-bound HK to initiate glucose metabolism. Decreased expression and abnormal distribution of HK were found in AD. Glycolysis and oxidative phosphorylation (OxPhos) are two main pathways of glucose metabolism [55]. Glycolysis generates little ATP, while OxPhos generates a large amount of ATP to meet the energy needs of the neurons [56]. The dividing point between the two pathways is the metabolic selection of pyruvate. Pyruvate is catalyzed by lactate dehydrogenase to generate lactate in glycolysis, while it is catalyzed by PDHE to generate Acetyl-CoA in OxPhos. An increasing number of studies have demonstrated that the metabolic pattern of neurons changes from OxPhos to aerobic glycolysis in the AD brain [25,56,57]. Reduced PDHE expression may be one reason for this shift. Here, we detected the increased expression levels of HK1 and PDHE1α in CY-09 treated 3×Tg-AD mice, which are beneficial for maintaining glucose metabolism and ATP production in the brain.

Oxidative stress is another important pathological characteristic that is closely related to NLRP3 inflammasome and glucose metabolism in AD. ROS was released from the oxidative respiratory chain and can activate the NLRP3 inflammasome [58,59]. Here, we found that the inactivation of the NLRP3 inflammasome by CY-09 decreased ROS levels in CY-09-treated 3×Tg-AD mice. Oxidative stress causes iron metabolism disorders and leads to ferroptosis, which is a new area in the research of AD pathogenesis [25,26]. Ferroptosis manifests as increased cellular Fe^{2+}, decreased GPX4, and increased lipid peroxidation [27]. Elevated Fe^{2+} results from increased transferrin transport and ferritinophagy [28]. Consistent with the studies, we detected increased TF, TFR, and FPN and decreased GPX4 and SLC7A11 in the 3×Tg-AD mice. Unfortunately, there was no significant difference in these proteins between the AD and CY-09-treated AD mice. These results indicate that the inactivation of NLRP3 inflammasome does not reduce ferroptosis. Moreover, a contradiction between decreased ferritinophagy and increased ferroptosis was found in AD. Generally, decreased NCOA4 represents decreased ferritinophagy [28,60]. Thus, the expression and function of this protein remain to be clarified.

As a marker of ferroptosis, ACSL4 also participates in lipid peroxidation. Some studies showed that lipid peroxidation is related to Aβ deposition and the hyperphosphorylation of tau [29–31]. In this paper, the levels of ACSL4 and MDA were increased in 3×Tg-AD mice and decreased after CY-09 treatment. The results prompted us to explore the effect of CY-09 on lipid metabolism, mainly the synthesis and metabolism of fatty acids and cholesterol. FAS and ACC are rate-limiting enzymes of fatty acid synthesis. In AD mice, the expression of FAS and ACC increased, whereas the phosphorylation level of ACC deceased, thus indicating an increase in fatty acids synthesis. HMGCS1 and HMGCR are important for cholesterol synthesis [61]. Increased HMGCS1 and HMGCR were also found in AD mice. Surprisingly, expressions of ACC, HMGCS1, and HMGCR were decreased in CY-09-treated AD mice. This suggests that decreased MDA and lipid peroxidation may be related to the decreased synthesis of fatty acids and cholesterol in CY-09-treated AD mice. LRP1 plays an important role in lipid metabolism, glucose metabolism, insulin signaling, and the elimination of Aβ in AD [26,62]. Increased LRP1 helps to improve cognitive ability in AD [63]. SREBP2 is a negative regulator of LRP1 [64]. However, in our results, no difference in LRP1 and SREBP2 was found between AD mice and CY-09-treated AD mice. Here, a limitation should be noted. In this study, we used only one strain of AD model mice to study the effect of NLRP3 inflammasome activation on glucose metabolism and the role of CY-09. Using two or more strains of mice would help to validate the experimental results. Therefore, the conclusion drawn in this paper is restricted to only the 3×Tg-AD mice, and it should be verified using additional AD models in the near future.

5. Conclusions

Summarily, inhibiting NLRP3 inflammasome activation by CY-09 helps to restore cerebral glucose metabolism, improve memory and learning ability, and reduce fatty acid synthesis and lipid peroxidation in the 3×Tg-AD mice. Thus, CY-09 has the potential to be developed for the treatment of AD. Further studies are required regarding the shift between glycolysis and OxPhos pathways in AD in order to better understand the mechanism of neuroinflammation and glucose metabolism in the development of AD pathology.

Supplementary Materials: The following are available online at https://www.mdpi.com/article 10.3390/antiox12030722/s1, Figure S1: Content of CY-09 in NTg and NTg + CY-09 mice brain Figure S2: Fluorescence intensity of GLUT4 in NTg, NTg + CY-09, 3×Tg-AD and 3×Tg-AD + CY-09 mice (n = 6, mean ± SD, one-way ANOVA and Bonferroni post hoc test; * $p < 0.05$, ** $p < 0.01$ vs. NTg mice, & $p < 0.05$, &&& $p < 0.001$ vs. NTg + CY-09 mice, ## $p < 0.01$ vs. 3×Tg-AD mice), Figure S3: Fluorescence intensity of (**a**) IR and (**b**) p-IR in NTg, NTg + CY-09, 3×Tg-AD and 3×Tg-AD + CY-09 mice (n = 6, mean ± SD, one-way ANOVA and Bonferroni post hoc test; ** $p < 0.01$ vs. NTg mice && $p < 0.01$ vs. NTg + CY-09 mice, # $p < 0.05$ vs. 3×Tg-AD mice), Figure S4: Fluorescence intensity of Aβ (6E10) in (**a**) CA3 region and (**b**) cortex of NTg, NTg + CY-09, 3×Tg-AD and 3×Tg-AD + CY-09 mice (n = 6, mean ± SD, one-way ANOVA and Bonferroni post hoc test; * $p < 0.05$ vs. NTg mice & $p < 0.05$ vs. NTg + CY-09 mice, # $p < 0.05$, ## $p < 0.01$ vs. 3×Tg-AD mice), Figure S5: Fluorescence intensity of p-tau-S404 in (**a**) CA3 region and (**b**) cortex of NTg, NTg + CY-09, 3×Tg-AD and 3×Tg-AD + CY-09 mice (n = 6, mean ± SD, one-way ANOVA and Bonferroni post hoc test; * $p < 0.05$ vs. NTg mice, & $p < 0.05$ vs. NTg + CY-09 mice, ## $p < 0.01$ vs. 3×Tg-AD mice).

Author Contributions: Conceptualization, S.H. and Q.L.; data curation, Q.L.; formal analysis, S.H funding acquisition, J.N. and Q.L.; investigation, S.H.; methodology, S.H.; project administration, S.H K.Z. and Y.H.; resources, Z.H. and X.L.; software, X.H.; supervision, J.N. and Q.L.; writing—original draft, S.H.; writing—review and editing, P.X., Q.X. and Q.L. All authors have read and agreed to the published version of the manuscript.

Funding: This research was funded by the National Key Research and Development Program of China, grant No. 2018YFE0118900; the Shenzhen Science and Technology Innovation Commission, grant No. JCYJ20200109110001818; the Guangdong Provincial Key S&T Program, grant No. 2018B030336001, and the Shenzhen-Hong Kong Institute of Brain Science-Shenzhen Fundamental Research Institutions, grant No. 2022SHIBS0003.

Institutional Review Board Statement: The animal study protocol was approved by the Institutional Animal Care and Use Committee of Tongji Medical College of Huazhong University of Science and Technology (IACUC Number: 2390; approved date: 27 February 2018). All the animal experiments complied with the animal care protocol.

Informed Consent Statement: Not applicable.

Data Availability Statement: The data presented in this study are available on request from the corresponding author.

Acknowledgments: We would like to thank the Instrumental Analysis Center of Shenzhen University (Xili Campus) and the Central Research Facilities of the College of Life Science and Oceanography of Shenzhen University for providing access to the experimental instruments (laser scanning confocal microscope, LSM710, ZEISS, Oberkochen, Germany).

Conflicts of Interest: The authors declare the following financial interest or personal relationships that may be considered as potential competing interests: Q.X. reports that he is the chief scientist and co-founder of Suzhou RAYCAN Technology Co., Ltd. (Suzhou, China). S.H., Z.H., X.H., X.L., K.Z. Y.H., P.X., J.N. and Q.L. have nothing to disclose.

References

1. Butterfield, D.A.; Halliwell, B. Oxidative stress, dysfunctional glucose metabolism and Alzheimer disease. *Nat. Rev. Neurosci.* **2019**, *20*, 148–160. [CrossRef] [PubMed]
2. Scheltens, P.; Blennow, K.; Breteler, M.M.B.; de Strooper, B.; Frisoni, G.B.; Salloway, S.; Van der Flier, W.M. Alzheimer's disease *Lancet* **2016**, *388*, 505–517. [CrossRef] [PubMed]
3. Swanson, K.V.; Deng, M.; Ting, J.P. The NLRP3 inflammasome: Molecular activation and regulation to therapeutics. *Nat. Rev Immunol.* **2019**, *19*, 477–489. [CrossRef]
4. Yang, Y.; Wang, H.; Kouadir, M.; Song, H.; Shi, F. Recent advances in the mechanisms of NLRP3 inflammasome activation and its inhibitors. *Cell Death Dis.* **2019**, *10*, 128. [CrossRef] [PubMed]
5. Han, S.; He, Z.; Jacob, C.; Hu, X.; Liang, X.; Xiao, W.; Wan, L.; Xiao, P.; D'Ascenzo, N.; Ni, J.; et al. Effect of Increased IL-1beta on Expression of HK in Alzheimer's Disease. *Int. J. Mol. Sci.* **2021**, *22*, 1306. [CrossRef]
6. Wolf, A.J.; Reyes, C.N.; Liang, W.; Becker, C.; Shimada, K.; Wheeler, M.L.; Underhill, D.M. Hexokinase is an innate immune receptor for the detection of bacterial peptidoglycan. *Cell* **2016**, *166*, 624–636. [CrossRef] [PubMed]
7. Hughes, M.M.; O'Neill, L.A. Metabolic regulation of NLRP3. *Immunol. Rev.* **2018**, *281*, 88–98. [CrossRef] [PubMed]

8. Kelley, N.; Jeltema, D.; Duan, Y.; He, Y. The NLRP3 inflammasome: An overview of mechanisms of activation and regulation. *Int. J. Mol. Sci.* **2019**, *20*, 3328. [CrossRef]
9. Zhong, Z.; Umemura, A.; Sanchez-Lopez, E.; Liang, S.; Shalapour, S.; Wong, J.; Karin, M. NF-κB restricts inflammasome activation via elimination of damaged mitochondria. *Cell* **2016**, *164*, 896–910. [CrossRef]
10. Novoderezhkina, E.A.; Zhivotovsky, B.D.; Gogvadze, G.V. Induction of unspecific permeabilization of mitochondrial membrane and its role in cell death. *Mol. Biol.* **2016**, *50*, 43–58. [CrossRef]
11. GTakkinen, J.S.; Lopez-Picon, F.R.; Al Majidi, R.; Eskola, O.; Krzyczmonik, A.; Keller, T.; Haaparants-Solin, M. Brain energy metabolism and neuroinflammation in ageing APP/PS1-21 mice using longitudinal (18)F-FDG and (18)F-DPA-714 PET imaging. *J. Cereb. Blood. Flow Metab.* **2017**, *37*, 2870–2882. [CrossRef] [PubMed]
12. Cunnane, S.C.; Trushina, E.; Morland, C.; Prigione, A.; Casadesus, G.; Andrews, Z.B.; Beal, M.F.; Bergersen, L.H.; Brinton, R.D.; de la Monte, S.; et al. Brain energy rescue: An emerging therapeutic concept for neurodegenerative disorders of ageing. *Nat. Rev. Drug Discov.* **2020**, *19*, 609–633. [CrossRef] [PubMed]
13. Zhang, Y.H.; Yan, X.Z.; Xu, S.F.; Pang, Z.Q.; Li, L.B.; Yang, Y.; Fan, Y.G.; Wang, Z.; Yu, X.; Guo, C.; et al. alpha-Lipoic Acid Maintains Brain Glucose Metabolism via BDNF/TrkB/HIF-1alpha Signaling Pathway in P301S Mice. *Front. Aging Neurosci.* **2020**, *12*, 262. [CrossRef]
14. Harris, R.A.; Tindale, L.; Cumming, R.C. Age-dependent metabolic dysregulation in cancer and Alzheimer's disease. *Biogerontology* **2014**, *15*, 559–577. [CrossRef]
15. Ou, Y.N.; Xu, W.; Li, J.Q.; Guo, Y.; Cui, M.; Chen, K.L.; Huang, Y.Y.; Dong, Q.; Tan, L.; Yu, J.T.; et al. FDG-PET as an independent biomarker for Alzheimer's biological diagnosis: A longitudinal study. *Alzheimer Res. Ther.* **2019**, *11*, 57. [CrossRef] [PubMed]
16. McNay, E.C.; Pearson-Leary, J. GluT4: A central player in hippocampal memory and brain insulin resistance. *Exp. Neurol.* **2020**, *323*, 113076. [CrossRef] [PubMed]
17. Pearson-Leary, J.; McNay, E.C. Novel Roles for the Insulin-Regulated Glucose Transporter-4 in Hippocampally Dependent Memory. *J. Neurosci.* **2016**, *36*, 11851–11864. [CrossRef]
18. Marko, D.M.; Foran, G.; Vlavcheski, F.; Baron, D.C.; Hayward, G.C.; Baranowski, B.J.; Necakov, A.; Tsiani, E.; MacPherson, R.E.K. Interleukin-6 Treatment Results in GLUT4 Translocation and AMPK Phosphorylation in Neuronal SH-SY5Y Cells. *Cells* **2020**, *9*, 1114. [CrossRef]
19. Lauretti, E.; Li, J.G.; Di Meco, A.; Pratico, D. Glucose deficit triggers tau pathology and synaptic dysfunction in a tauopathy mouse model. *Transl. Psychiatry* **2017**, *7*, e1020. [CrossRef]
20. Tumminia, A.; Vinciguerra, F.; Parisi, M.; Frittitta, L. Type 2 Diabetes Mellitus and Alzheimer's Disease: Role of Insulin Signalling and Therapeutic Implications. *Int. J. Mol. Sci.* **2018**, *19*, 3306. [CrossRef]
21. Vigneri, R.; Goldfine, I.D.; Frittitta, L. Insulin, insulin receptors, and cancer. *J. Endocrinol. Investig.* **2016**, *39*, 1365–1376. [CrossRef] [PubMed]
22. Gabbouj, S.; Ryhanen, S.; Marttinen, M.; Wittrahm, R.; Takalo, M.; Kemppainen, S.; Martiskainen, H.; Tanila, H.; Haapasalo, A.; Hiltunen, M.; et al. Altered Insulin Signaling in Alzheimer's Disease Brain—Special Emphasis on PI3K-Akt Pathway. *Front. Neurosci.-Switz.* **2019**, *13*, 629. [CrossRef] [PubMed]
23. Ramm, G.; Larance, M.; Guilhaus, M.; James, D.E. A role for 14-3-3 in insulin-stimulated GLUT4 translocation through its interaction with the RabGAP AS160. *J. Biol. Chem.* **2006**, *281*, 29174–29180. [CrossRef]
24. Zaulkffali, A.S.; Razip, N.N.M.; Alwi, S.S.S.; Abd Jalil, A.; Abd Mutalib, M.S.; Gopalsamy, B.; Chang, S.K.; Zainal, Z.; Ibrahim, N.N.; Zakaria, Z.A.; et al. Vitamins D and E Stimulate the PI3K-AKT Signalling Pathway in Insulin-Resistant SK-N-SH Neuronal Cells. *Nutrients* **2019**, *11*, 2525. [CrossRef] [PubMed]
25. Blazhenets, G.; Ma, Y.; Sorensen, A.; Rucker, G.; Schiller, F.; Eidelberg, D.; Frings, L.; Meyer, P.T. Principal Components Analysis of Brain Metabolism Predicts Development of Alzheimer Dementia. *J. Nucl. Med.* **2019**, *60*, 837–843. [CrossRef] [PubMed]
26. Liu, C.C.; Hu, J.; Tsai, C.-W.; Yue, M.; Melrose, H.L.; Kanekiyo, T.; Bu, G. Neuronal LRP1 Regulates Glucose Metabolism and Insulin Signaling in the Brain. *J. Neurosci.* **2015**, *35*, 5851–5859. [CrossRef]
27. Bertrand, R.L. Iron Accumulation, Glutathione Depletion, and Lipid Peroxidation Must Occur Simultaneously during Ferroptosis and are Mutually Amplifying Events. *Med. Hypotheses* **2017**, *101*, 69–74. [CrossRef]
28. Fujimaki, M.; Furuya, N.; Saiki, S.; Amo, T.; Imamichi, Y.; Hattori, N. Iron Supply via NCOA4-Mediated Ferritin Degradation Maintains Mitochondrial Functions. *Mol. Cell. Biol.* **2019**, *39*, e00010–e00019. [CrossRef]
29. Butteerfield, D.A.; Kimball, D.B. Oxidative Stress, Amyloid-βPeptide, and Altered Key Molecular Pathways in the pathogenesis and Progression of Alzheimer's Disease. *J. Alzheimers Dis.* **2018**, *62*, 1345–1367. [CrossRef]
30. King, M.E.; Gamblin, T.C.; Kuret, J.; Binder, L.I. Differential Assembly of Human Tau Isoforms in the Presence of Arachidonic Acid. *J. Neurochem.* **2000**, *74*, 1749–1757. [CrossRef]
31. Kawarabayashi, T.; Shoji, M.; Younkin, L.H.; Lin, W.L.; Dickson, D.W.; Murakami, T.; Matsubara, E.; Abe, K.; Ashe, K.H.; Younkin, S.G. Dimeric Amyloid Protein Rapidly Accumulates in Lipid Rafts followed by Apolipoprotein E and Phosphorylated Tau Accumulation in the Tg2576 Mouse Model of Alzheimer's Disease. *Neurobiol. Dis.* **2004**, *24*, 3801–3809. [CrossRef]
32. Jiang, H.; He, H.; Chen, Y.; Huang, W.; Cheng, J.; Ye, J.; Wang, A.; Tao, J.; Wang, C.; Liu, Q.; et al. Identification of a selective and direct NLRP3 inhibitor to treat inflammatory disorders. *J. Exp. Med.* **2017**, *214*, 3219–3238. [CrossRef] [PubMed]
33. Yao, J.; Irwin, R.; Chen, S.H.; Hamilton, R.; Cadenas, E.; Brinton, R.D. Ovarian hormone loss induces bioenergetic deficits and mitochondrial beta-amyloid. *Neurobiol. Aging* **2012**, *33*, 1507–1521. [CrossRef] [PubMed]

34. Ding, F.; Yao, J.; Zhao, L.; Mao, Z.; Chen, S.; Brinton, R.D. Ovariectomy induces a shift in fuel availability and metabolism in the hippocampus of the female transgenic model of familial Alzheimer's. *PLoS ONE* **2013**, *8*, e59825. [CrossRef] [PubMed]
35. Fueger, B.J.; Czernin, J.; Hildebrandt, I.; Tran, C.; Halpern, B.S.; Stout, D.; Weber, W.A. Impact of animal handling on the results of F-18-FDG PET studies in mice. *J. Nucl. Med.* **2006**, *47*, 999–1006.
36. He, Z.; Li, X.; Wang, Z.; Tu, S.; Feng, J.; Du, X.; Liu, Q. Esculentoside A alleviates cognitive deficits and amyloid pathology through peroxisome proliferator-activated receptor gamma-dependent mechanism in an Alzheimer's disease model. *Phytomedicine* **2022**, *98*, 153956. [CrossRef]
37. Vorhees, C.V.; Williams, M.T. Morris water maze: Procedures for assessing spatial and related forms of learning and memory. *Nat. Protoc.* **2006**, *1*, 848–858. [CrossRef] [PubMed]
38. He, Z.; Han, S.; Wu, C.; Liu, L.; Zhu, H.; Liu, A.; Lu, Q.; Huang, J.; Du, X.; Li, N.; et al. Bis(ethylmaltolato)oxidovanadium(IV) inhibited the pathogenesis of Alezheimer's disease in triple transgenic model mice. *Metallomic* **2020**, *12*, 474–490. [CrossRef]
39. Zhang, H.; Zhao, Y.; Zhao, D.; Chen, X.; Khan, N.U.; Liu, X.; Zheng, Q.; Liang, Y.; Zhu, Y.; Iqbal, J.; et al. Potential biomarkers identified in plasma of patients with gestational diabetes mellitus. *Metabolomics* **2021**, *17*, 99. [CrossRef] [PubMed]
40. He, Z.; Han, S.; Zhu, H.; Hu, X.; Li, X.; Hou, C.; Wu, C.; Xie, Q.; Li, N.; Du, X.; et al. The Protective Effect of Vanadium on Cognitive Impairment and the Neuropathology of Alzheimer's Disease in APPSwe/PS1dE9 Mice. *Front. Mol. Neurosci.* **2020**, *13*, 21. [CrossRef]
41. Granov, A.; Tiutin, L.; Schwarz, T. Methodical Aspects of Using PET. In *Positron Emission Tomography*; Granov, A., Tiutin, L., Schwarz, T., Eds.; Springer: Berlin/Heidelberg, Germany, 2013; pp. 25–39.
42. Ising, C.; Venegas, C.; Zhang, S.; Scheiblich, H.; Schmidt, S.V.; Vieira-Saecker, A.; Schwartz, S.; Albasset, S.; McManus, R.M.; Tejera, D.; et al. NLRP3 inflammasome activation drives tau pathology. *Nature* **2019**, *575*, 669–673. [CrossRef] [PubMed]
43. Heneka, M.T.; Kummer, M.P.; Stutz, A.; Delekate, A.; Schwartz, S.; Vieira-Saecker, A.; Griep, A.; Axt, D.; Remus, A.; Tzeng, T.C.; et al. NLRP3 is activated in Alzheimer's disease and contributes to pathology in APP/PS1 mice. *Nature* **2013**, *493*, 674–678. [CrossRef] [PubMed]
44. Zimmer, E.R.; Parent, M.J.; Souza, D.G.; Leuzy, A.; Lecrux, C.; Kim, H.I.; Gauthier, S.; Pellerin, L.; Hamel, E.; Rosa-Neto, P. [(18)F]FDG PET signal is driven by astroglial glutamate transport. *Nat. Neurosci.* **2017**, *20*, 393–395. [CrossRef]
45. Bateman, R.J.; Xiong, C.; Benzinger, T.L.; Fagan, A.M.; Goate, A.; Fox, N.C.; Marcus, D.S.; Cairns, N.J.; Xie, X.; Blazey, T.M.; et al. Clinical and biomarker changes in dominantly inherited Alzheimer's disease. *N. Engl. J. Med.* **2012**, *367*, 795–804. [CrossRef] [PubMed]
46. Vlassenko, A.G.; Vaishnavi, S.N.; Couture, L.; Sacco, D.; Shannon, B.J.; Mach, R.H.; Morris, J.C.; Raichle, M.E.; Mintun, M.A. Spatial correlation between brain aerobic glycolysis and amyloid-β (Aβ) deposition. *Proc. Natl. Acad. Sci. USA* **2010**, *107*, 17763–17767. [CrossRef]
47. de Nazareth, A.M. Type 2 diabetes mellitus in the pathophysiology of Alzheimer's disease. *Dement. Neuropsychol.* **2017**, *11*, 105–113. [CrossRef] [PubMed]
48. Peng, W.; Tan, C.; Mo, L.; Jiang, J.; Zhou, W.; Du, J.; Zhou, X.; Liu, X.; Chen, L. Glucose transporter 3 in neuronal glucose metabolism: Health and diseases. *Metab. Clin. Exp.* **2021**, *123*, 154869. [CrossRef] [PubMed]
49. Szablewski, L. Glucose Transporters in Brain: In Health and in Alzheimer's Disease. *J. Alzheimers Dis.* **2017**, *55*, 1307–1320. [CrossRef] [PubMed]
50. Winkler, E.A.; Nishida, Y.; Sagare, A.P.; Rege, S.V.; Bell, R.D.; Perlmutter, D.; Sengillo, J.D.; Hillman, S.; Kong, P.; Nelson, A.R.; et al. GLUT1 reductions exacerbate Alzheimer's disease vasculo-neuronal dysfunction and degeneration. *Nat. Neurosci.* **2015**, *18*, 521–530. [CrossRef]
51. Duarte, A.I.; Santos, M.S.; Oliveira, C.R.; Moreira, P.I. Brain insulin signalling, glucose metabolism and females' reproductive aging: A dangerous triad in Alzheimer's disease. *Neuropharmacology* **2018**, *136*, 223–242. [CrossRef]
52. Saleh, R.A.; Eissa, T.F.; Abdallah, D.M.; Saad, M.A.; El-Abhar, H.S. Peganum harmala enhanced GLP-1 and restored insulin signaling to alleviate-AlCl3-induced Alzheimer-like pathology model. *Sci. Rep.* **2021**, *11*, 12040. [CrossRef] [PubMed]
53. Blazquez, E.; Velazquez, E.; Hurtado-Carneiro, V.; Ruiz-Albusac, J.M. Insulin in the brain: Its pathophysiological implications for States related with central insulin resistance, type 2 diabetes and Alzheimer's disease. *Front. Endocrinol.* **2014**, *5*, 161. [CrossRef] [PubMed]
54. Ratzmann, K.P.; Hampel, R. Glucose and insulin concentration patterns in cerebrospinal fluid following intravenous glucose injection in humans. *Endokrinologie* **1980**, *76*, 185–188.
55. Valvona, C.J.; Fillmore, H.L.; Nunn, P.B.; Pilkington, G.J. The Regulation and Function of Lactate Dehydrogenase A: Therapeutic Potential in Brain Tumor. *Brain Pathol.* **2016**, *26*, 3–17. [CrossRef] [PubMed]
56. Demetrius, L.A.; Eckert, A.; Grimm, A. Sex differences in Alzheimer's disease: Metabolic reprogramming and therapeutic intervention. *Trends Endocrin. Met.* **2021**, *32*, 963–979. [CrossRef] [PubMed]
57. Zheng, X.; Boyer, L.; Jin, M.; Mertens, J.; Kim, Y.; Ma, L.; Ma, L.; Hamm, M.; Gage, F.H.; Hunter, T. Metabolic reprogramming during neuronal differentiation from aerobic glycolysis to neuronal oxidative phosphorylation. *eLife* **2016**, *5*, e13374. [CrossRef] [PubMed]
58. Moon, J.S.; Hisata, S.; Park, M.A.; DeNicola, G.M.; Ryter, S.W.; Nakahira, K.; Choi, A.M.K. mTORC1-Induced HK1-Dependent Glycolysis Regulates NLRP3 Inflammasome Activation. *Cell Rep.* **2015**, *12*, 102–115. [CrossRef]

9. Zhong, Z.; Liang, S.; Sanchez-Lopez, E.; He, F.; Shalapour, S.; Lin, X.J.; Wong, J.; Ding, S.; Seki, E.; Schnabl, B. New mitochondrial DNA synthesis enables NLRP3 inflammasome activation. *Nature* **2018**, *560*, 198–203. [CrossRef]
0. Hasan, M.; Reddy, S.M.; Das, N.K. Ferritinophagy is not required for colon cancer cell growth. *Cell Biol. Int.* **2020**, *44*, 2307–2314. [CrossRef]
1. Seo, E.; Kang, H.; Choi, H.; Choi, W.; Jun, H.S. Reactive oxygen species—Induced changes in glucose and lipid metabolism contribute to the accumulation of cholesterol in the liver during aging. *Aging Cell* **2018**, *18*, e12895. [CrossRef]
2. Sagare, A.P.; Deane, R.; Zlokovic, B.V. Low-density lipoorotein receptor-related protein 1: A physiological Aβhomeostatic mechanism with multiple therapeutic opportunities. *Pharmacol. Ther.* **2012**, *136*, 94–105. [CrossRef] [PubMed]
3. Kanekiyo, T.; Liu, C.C.; Shinohara, M.; Li, J.; Bu, G.J. LRP1 in Brain Vascular Smooth Muscle Cells Mediates Local Clearance of Alzheimer's Amyloid-β. *J. Neurosci.* **2012**, *32*, 16459–16465. [CrossRef] [PubMed]
4. Cortes, V.L.; Costales, P.; Bernues, J.; Lopez, C.; Badimon, L. Sterol Regulatory Element-binding Protein-2 Negatively Regulates Low Density Lipoprotein Receptor-related Protein Transcription. *J. Mol. Biol.* **2006**, *359*, 950–960. [CrossRef] [PubMed]

antioxidants

MDPI

Article

CHIR99021 Maintenance of the Cell Stemness by Regulating Cellular Iron Metabolism

Yingying Han [1], Yong He [2], Xiaofang Jin [1], Jiayi Xie [3], Peng Yu [1], Guofen Gao [1], Shiyang Chang [2,*], Jianhua Zhang [1,*] and Yan-Zhong Chang [1,*]

1 Laboratory of Molecular Iron Metabolism, Key Laboratory of Animal Physiology, Biochemistry and Molecula Biology of Hebei Province, Ministry of Education Key Laboratory of Molecular and Cellular Biology, College of Life Science, Hebei Normal University, Shijiazhuang 050024, China
2 College of Basic Medical Sciences, Hebei Medical University, Shijiazhuang 050017, China
3 Department of Automatic, Tsinghua University, Beijing 100084, China
* Correspondence: changshiyangie@163.com (S.C.); jianhuazcell@163.com (J.Z.); yzchang@hebtu.edu.cn (Y.-Z.C.); Tel./Fax: +86-311-80787539 (Y.-Z.C.)

Abstract: CHIR99021 is an aminopyrimidine derivative, which can efficiently inhibit the activity o glycogen synthesis kinase 3α (GSK-3α) and GSK-3β. As an essential component of stem cell cultur medium, it plays an important role in maintaining cell stemness. However, the mechanism of its role i not fully understood. In the present study, we first found that removal of CHIR99021 from embryoni stem cell culture medium reduced iron storage in mouse embryonic stem cells (mESCs). CHIR9902 treated Neuro-2a cells led to an upregulation of ferritin expression and an increase in intracellular iro levels, along with GSK3β inhibition and Wnt/GSK-3β/β-catenin pathway activation. In additior iron treatment activated the classical Wnt pathway by affecting the expression of β-catenin in th Neuro-2a cells. Our data link the role of iron in the maintenance of cell stemness via the Wnt/GSK 3β/β-catenin signaling pathway, and identify intermediate molecules, including Steap1, Bola2, anc Kdm6bos, which may mediate the upregulation of ferritin expression by CHIR99021. These finding reveal novel mechanisms of the maintenance of cell stemness and differentiation and provide theoretical basis for the development of new strategies in stem cell treatment in disease.

Keywords: CHIR99021; glycogen synthesis kinase 3; classical Wnt signaling pathway; ferritin Neuro-2a; cell stemness

Citation: Han, Y.; He, Y.; Jin, X.; Xie, J.; Yu, P.; Gao, G.; Chang, S.; Zhang, J.; Chang, Y.-Z. CHIR99021 Maintenance of the Cell Stemness by Regulating Cellular Iron Metabolism. *Antioxidants* **2023**, *12*, 377. https://doi.org/10.3390/antiox12020377

Academic Editor: Dimitrios Galaris

Received: 8 December 2022
Revised: 19 January 2023
Accepted: 1 February 2023
Published: 4 February 2023

1. Introduction

Embryonic stem cells (ESCs), derived from early embryos or primary gonads, have the abilities of self-renewal and pluripotency [1,2]. Oct4, Sox2, and Nanog jointly regulate the pluripotency of ESCs [3]. The Wnt signaling pathway is a very important pathway to affect the stem cell behavior, which contains a complex protein interaction network [4] When the Wnt signal is absent, the scaffold proteins, Axin, adenomatous polyposis col (APC), GSK-3β, and casein kinase 1α (CK1α), form a protein destruction complex, which induces β-catenin phosphorylation and then degradation by the proteasome. Conversely the Wnt ligands bind to the Frizzled receptor and recruit the Dishevelled (DVL) protein resulting in the dissociation of the destruction complex. Subsequently, β-catenin binds to T-cell factor/lymphoid enhancer factor (TCF/LEF) to activate the target genes [5]. The non-canonical Wnt pathway does not require β-catenin. This alternate signaling cascade activates multiple intracellular targets. The most extensively studied are the Wnt/planar cell polarity (PCP) and Wnt/calcium (Ca^{2+}) pathways [6–8].

The canonical Wnt/β-catenin signaling pathway plays a key role in embryonic development and stem cell stemness. The activation of this pathway has been shown to increase the expression of pluripotent genes in mESCs. Besides, the activation of the pathway can significantly up-regulate the expression of prostate tumor stem cell markers B-cell-specific Moloney murine leukemia virus integration site 1 (BMI-1), Aldehyde dehydrogenase 1A1 (ALDH1A1), Cluster of differentiation-44 (CD44), aldehyde dehydrogenase 1 family, member A3 (ALDH1A3), and SOX2 mRNA [9]. Additionally, studies in colorectal cancer have found that transmembrane 4L6 family member 1 (TM4SF1) can maintain cancer cell stemness and epithelial-mesenchymal transformation (EMT) through the Wnt/β-catenin/c-Myc/SOX2 pathway [10].

Iron is an element essential to normal physiological functions and metabolic activities [11]. Most of the iron is liganded within functional proteins, including hemoproteins, iron-sulfur (Fe/S) proteins, and enzymes which contain non-heme, non-Fe/S iron. In addition, there is a fraction of uncommitted iron, called the labile iron pool (LIP). These different forms of iron can participate in numerous of important biological processes, such as oxygen transport, electron transport, gene expression, cell proliferation, and cell differentiation [12]. Both iron deficiency and excess can have harmful effects. Therefore, it is essential to maintain iron homeostasis to support the normal physiological activities, while protecting against the toxicity of the catalytic metal. Iron levels are tightly controlled by some proteins that regulate the absorption, storage, circulation, and utilization of iron [12,13]. Cellular iron uptake is mainly mediated by the plasma iron transport protein, transferrin (Tf), and its receptor, transferrin receptor 1 (TfR1). Tf-bound iron enters to the LIP, with some of this iron binding to cytosolic poly (rC) binding proteins PCBP1 and PCBP2, which can transport the metal to the ubiquitous iron storage protein, ferritin, or other non-heme iron-proteins. Excess iron can be transported out of cells by ferroportin 1 (FPN1) [14]. Cellular iron homeostasis is regulated by a post-transcriptional regulation mechanism mediated by iron regulatory proteins (IRPs). These two RNA-binding proteins specifically bind to the conserved motif, iron response element (IRE), in the untranslated regions (UTRs) of mRNAs encoding iron metabolism-related proteins [15].

The fate of stem cells is regulated by many factors [16], among which the endogenous factors mainly include the expression of genes, while the exogenous factors mainly include differentiation and inhibition between cells and the effects of external substances. As an exogenous factor, iron also affect the stemness and differentiation of stem cells. Metal ions, such as Mn, Co, Al and Fe, can regulate cell attachment and affect neuronal differentiation [17,18]. Besides, it was found that the neural differentiation of mESCs decreased upon addition of different concentrations of iron oxide nanoparticles to retinoic acid-induced mESCs [19]. In human dental pulp stem cells, the iron chelator, deferoxamine (DFO), can promote the expression of proteinaceous factors related to the stem cell characteristics [20–22]. Besides, at both cellular level and in clinical treatment, iron also play vital roles in the normal function of mesenchymal stem cells [23].

Both iron and the Wnt/GSK-3β/β-catenin pathway play a role in the maintenance of cell stemness. In fact, the iron and GSK-3β pathways may intersect. It has been found that Deferasirox (DFX) can trigger the proteolysis of cyclin D1 in mantle cell lymphoma (MCL), which requires the participation of GSK-3β [24]. In a human neuroblastoma cell line (SH-SY5Y), $FeSO_4$ can activate GSK-3β, and down-regulate the phosphorylation of GSK-3β [25]. In addition, in APP/PS1 transgenic mice, intranasal administration of DFO can abrogate the activity of GSK-3β, thus inhibiting tau protein phosphorylation [26]. In Parkinson's disease dementia (PDD), the neurotoxicity resulted by iron can promote the activation of GSK-3β-related pathways, thus playing an important role in the pathological synergism of α-syn, tau and Aβ [27]. These studies suggest that there may be a specific relationship between iron and GSK-3β, however it is unclear how GSK3 may affect iron homeostasis. In this study, we investigated the effects of CHIR99021 on cellular iron homeostasis. Our data demonstrate that iron has an effect on cell stemness and that this effect is associated with the Wnt/GSK-3β/β-catenin pathway. Our findings provide insight into hitherto unknown regulatory mechanisms controlling stem cell stemness maintenance and differentiation.

2. Materials and Methods

2.1. Cell Culture

The Neuro-2a cell line, a mouse neuroblastoma cells, was cultured in DMEM (GIBCO, Grand Island, New York, USA 8121382) with 10% fetal bovine serum (Biological Industries, Cromwell, CT, USA) and 1× Pen/Strep (GIBCO, 10378016). Cells were maintained at 37 °C in a humidified incubator containing 5% CO_2 (Thermo Fisher Scientific, Waltham, MA, USA).

2.2. Preparation of MEF Feeder Cells and Culture of mESCs

Healthy ICR male and female mice aged 2–3 months were selected to mate, and primary mouse embryonic fibroblast (MEF) feeder cells were obtained from pregnant mice at day 12.5 or 13.5 postfertilization. The embryonic day 12.5 (E12.5) or E13.5 fetal mice were isolated, and the head, limbs, tail, and internal organs were removed, leaving only the torso. The tissue fragments were then cut to <1 mm^3 pieces and digested into individual cells with 0.25% trypsin. Next, the cells were cultured in MEF feeder culture medium, which contained DMEM supplemented with 10% FBS, 1× nonessential amino acids (GIBCO, 11140), 1× L-glutamine (Specialty media, TMS-002-C) and 1× Pen/Strep. MEF feeder cells of generation 2–3 were incubated with MEF feeder culture medium with mitomycin C (10 μg/mL) for 2–3 h to inhibit proliferation. Afterwards, the cells were cryopreserved until subsequent use in embryonic stem cell culture experiments.

The mESCs were generated and banked in our laboratory. A 0.1% gelatin was added to cover the wells of 24-well plates and incubated in a tissue culture incubator for 2–3 h in advance of use. The MEF feeder cells were then resuscitated and seeded onto the gelatin-coated plates until the cells covered the bottom. Additionally, mESCs were then resuscitated and maintained in KnockOut™ DMEM (GIBCO, 10829018) supplemented with 15% KnockOut™ serum replacement (GIBCO, A3181502), 1× nucleosides for ES cells (Merck Millipore, ES-008-D), 1× nonessential amino acids, 1× L-glutamine, 1× 2-mercaptoethanol (GIBCO, 21985023), 1× Pen/Strep, 1000 U/mL LIF (GIBCO, A35933), 3 μM CHIR99021 (Selleck, Houston, TX, USA S2924), and 1 μM PD0325901 (Selleck, S1036).

2.3. Drug Treatment

Cells were treated with CHIR99021, deferoxamine (DFO, Sigma, D9533) or ferric ammonium citrate (FAC, Sigma, St. Louis, MO, USA F5879). The concentrations and treatment times were as indicated in the results and figures.

2.4. Western Blot Analysis

Cells of the control group and drug-treated group were lysed in RIPA lysis buffer containing 95% RIPA buffer (Solarbio, Beijing, China R0010), 2% protease inhibitors (Roche, 04693116001), 2% phosphatase inhibitors (Roche, Basel, Switzerland, 04906837001), and 1% PMSF (Solarbio, P0100) for 30 min, shaken every 5 min. The protein concentration was assessed by a BCA kit (Yeasen, Shanghai, China, 20201ES76). Protein samples were resolved by SDS-PAGE (10% or 12% acrylamide) at 25 μg per lane and then transferred onto nitrocellulose (NC) membranes, which were cut appropriately, and blocked in 5% skimmed milk for 1.5 h and then incubated with different primary antibodies overnight at 4 °C. The next day, after washing, the membranes were incubated with the corresponding secondary antibodies at room temperature for 1.5 h. Finally, we used an ECL kit (CWBIO, CW0049M) and a chemiluminescence imager (Bio-Rad, Hercules, CA, USA) for visualization. Quantitative analysis of the protein bands was normalized to β-actin, as an internal reference.

The primary and corresponding secondary antibodies we used were as follows: mouse anti-TfR1 (1:5000, 13-6890, Invitrogen, Waltham, MA, USA), rabbit anti-FPN1 (1:8000, MTP11-S, Alpha Diagnostic International, San Antonio, TX, USA), rabbit anti-ferritin heavy chain (1:5000, ab183781, Abcam, Cambridge, UK), rabbit anti-ferritin light chain (1:5000, ab109373, Abcam), rabbit anti-DMT1(+IRE) (1:5000, NRAMP21-S, Alpha Diagnostic International, USA), rabbit anti-DMT1(-IRE) (1:5000, NRAMP23-S, Alpha Diagnostic International, USA), mouse anti-GSK-3α/β (1:5000, sc-7291, Santa Cruz, Dallas, TX, USA), rabbit anti-p-GSK3β (Ser9) (1:5000, #9336, CST, Danver, MA, USA), rabbit anti-β-catenin (1:2000, 610153, BD), rabbit anti-NCOA4 (1:5000, Abbkine, Wuhan, China), rabbit anti-cyclin D1 (1:2000, #2922, CST, USA), rabbit anti-Ki67 (1:10,000, ab15580, Abcam), mouse anti-β-actin (1:10,000, CW0096, CWBIO, Beijing, China); goat anti-mouse lgG (H+L) (1:10,000, RS0001, Immunoway, Plano, TX, USA), and goat anti-rabbit lgG (H+L) (1:10,000, RS0002, Immunoway, USA).

2.5. Total RNA Extraction and qRT-PCR

Total RNA was extracted from neuro-2a cells or mESCs using TRIzol reagent (Invitrogen, 15596018). RNA was then reverse-transcribed to cDNA. Next, we mixed 10 μL SYBR Green PCR Master Mix (CWBIO, CW0957), 1μg cDNA, 0.4 μL forward primer, 0.4 μL reverse primer and 5.2 μL RNase-free water for the subsequent PCR reaction. qRT-PCR amplification was performed using a Bio-RAD CFX96 Connect Real-Time System. The primer sequences used are listed in Table 1.

Table 1. Primer sequences used for qRT-PCR.

Gene (Gene Accession Number)	Forward (5′-3′)	Reverse (5′-3′)
TfR1 (NM_001357298.1)	GAGTATCACTTCCTGTCGCCCTATG	GCTGAGAGAGTGTGAGAGCCAGAGC
FPN1 (NM_016917.2)	TTCCGCACTTTCCGAGATG	AGTCAAAGCCCAGGACTGTCA
FtH (NM_010239.2)	TGCCATCAACCGCCAGATCAAC	TCTTCAGAGCCACATCATCT CGGTC
FtL (NM_010240.2)	CAACCATCTGACCAACCTCCGCAG	AAAGAGATACTCGCCCAGAGATCC
Oct4 (NM_013633.3)	GAGGAGTCCCAGGACATGAA	AGATGGTGGTCTGGCTGAAC
Sox2 (NM_011443.4)	CTGCAGTACAACTCCATGACCAG	GGACTTGACCACAGAGCCCAT
Steap1 (NM_027399.3)	GGTCGCCATTACCCTCTTGG	GGTATGAGAGACTGTAAACAGCG
Bola2 (NM_175103.3)	GAACTCAGCGCCGATTACCTC	CAGTGGCTTTCCCTCGAACTT
Kdm6bos (NM_001017426.2)	AGTGAGGAAGCCGTATGCTG	AGCCCCATAGTTCCGTTTGTG
β-actin (NM_007393.5)	AGGCCCAGAGCAAGAGAGGTA	TCTCCATGTCGTCCCAGTTG

2.6. RNA-Sequencing (RNA-Seq) and Data Analysis

The Neuro-2a control group cells and CHIR99021-treatment group samples were collected in TRIzol (Invitrogen, 15596018) and sent to Majorbio (Shanghai, China). The original data and results of RNA-sequencing were analyzed by Majorbio.

2.7. Inductively Coupled Plasma Mass Spectrometry (ICP-MS)

Total cellular iron content was evaluated by ICP-MS. Cell samples were centrifuged and collected after 0.25% trypsin digestion. Then, the samples were digested in 500 μL 65% nitric acid at room temperature overnight. The nitric acid was evaporated by heating for 20 min in a metal bath at 90 °C. Afterwards, 500 μL 30% H_2O_2 was added and allowed to react at 70 °C for 15 min, and then continued to vaporize for at least 6 h at 100 °C until the liquid in the tube was almost fully volatilized. The digested samples were re-suspended and mixed in 1 mL ultra-pure water for subsequent ICP-MS.

2.8. Alkaline Phosphatase Staining

For mESCs, the BCIP/NBT Alkaline Phosphatase Color Development kit (Beyotime, Jiangsu, China, C3206) was used according to the manufacturer's instructions. Cells were carefully washed 3–5 times with PBS and fixed with an appropriate amount of 4% formaldehyde for 20 min. Next, cells were incubated with the BCIP/NBT working solution for 15 min in the dark. An appropriate amount of double-distilled water was added to the plates to wash the cells 1–2 times to stop the color reaction. The results were examined and imaged under a light microscope (OLYMPUS, FV300).

2.9. Electric Cell-Substrate Impedance Sensing (ECIS) Detection of Cell Proliferation

The proliferation of cells was detected using an ECIS instrument (Applied BioPhysics, Troy, NY, USA). The experimental materials and reagents prepared include: 0.22 μm filter, syringe, sterile water, and 100 mM cysteine. An 8W10E+ electrode plate was pre-treated with 400 μL cysteine solution at a final concentration of 10 mM in each well, and incubated at 37 °C overnight. The electrode plate was washed 3 times with sterile water the next day. The cells were then seeded onto the electrode plate with about 1.5×10^4 cells per well and left for 5 min. The real-time detection was performed after cells placed in the 37 °C incubator and connected to the ECIS instrument.

2.10. Profiles of Mitochondrial Respiration and Glycolysis

An Oxygraph-2k (O2k, TissueGnostics Asia Pacific Limited, Beijing, China) was used to assess the degree of oxidative phosphorylation and glycolysis of cell samples. The cells were digested, centrifuged, resuspended and counted, and then added into the chambers of the instrument. For the levels of oxidative phosphorylation, oxygen consumption rate (OCR) was recorded after injections of pyruvate (P), malic acid (M), ADP (D), oligomycin (Omy), the mitochondrial uncoupling agent, FCCP (U+), and antimycin A (Ama). For the level of glycolysis, the extracellular acidification rate (ECAR) was recorded, after injections of glucose (Glu), oligomycin (Omy), and 2-deoxy-glucose (2-DG), according to the manufacturer's instructions. The data were exported and analyzed by DatLab (New York, NY, USA, 7.4.0.4) software.

2.11. Statistical Analysis

All data are presented as the mean ± standard error (SEM). The statistical graphs were generated using GraphPad Prism 6 software. Differences between two groups were compared by non-paired *t*-tests. Differences were considered statistically significant when $p < 0.05$.

3. Results

3.1. The Effects of CHIR99021 and Iron on Cell Stemness in mESCs

The small molecule compound, CHIR99021, is an aminopyrimidine derivative which can efficiently inhibit GSK-3α and GSK-3β. It is also commonly used to culture ESCs and induced pluripotent stem cells (iPSCs). To explore the effects of CHIR99021 on iron in mESCs, we used qRT-PCR to examine the transcriptional levels of H-ferritin (FtH) and L-ferritin (FtL), which make up the ubiquitous intracellular iron storage protein. The expression levels of *FtH* mRNA were significantly decreased after the removal of CHIR99021 alone or with simultaneous removal of LIF, PD0325901, and CHIR99021 (Figure 1A). Likewise, the expression levels of *FtL* mRNA were also significantly diminished after removal of CHIR99021 alone (Figure 1B). There was no significant difference in the expression levels of *Oct4* mRNA after removing CHIR99021 alone or removing LIF, PD0325901 and CHIR99021 (Figure S1A), while the levels of *Sox2* mRNA were significantly decreased (Figure S1B). These results demonstrated that removal of CHIR99021 reduced iron storage in mESCs.

To further examine the relationship between CHIR99021, iron, and cell stemness in mESCs, we supplemented normal embryonic stem cell medium and embryonic stem cell differentiation medium, in which LIF, PD0325901, and CHIR99021 were simultaneously removed, with 25 μM ferric ammonium citrate (FAC) and examined the mRNA levels of the pluripotency genes, *Oct4* and *Sox2*. Supplementing the embryonic stem cell culture medium with iron did not affect the expression of these mRNAs. However, the levels of *Sox2* mRNA were significantly increased after supplementing the differentiation medium with 25 μM FAC (Figure 1C,D). By alkaline phosphatase staining, we also found that the stemness was weakened in mESCs cultured with differentiation medium, with the cells exhibiting a differentiated, spreading morphology. After supplementing the differentiation medium with 25 μM, 50 μM, or 100 μM FAC, although the mESCs also appeared to differentiate, the color of the cells was darker than that of the DV- group (Figure 1E,F). Together, these results suggested that removal of CHIR99021 decreases iron levels and the stemness of mESCs; supplementation with iron may be beneficial to the maintenance of stemness.

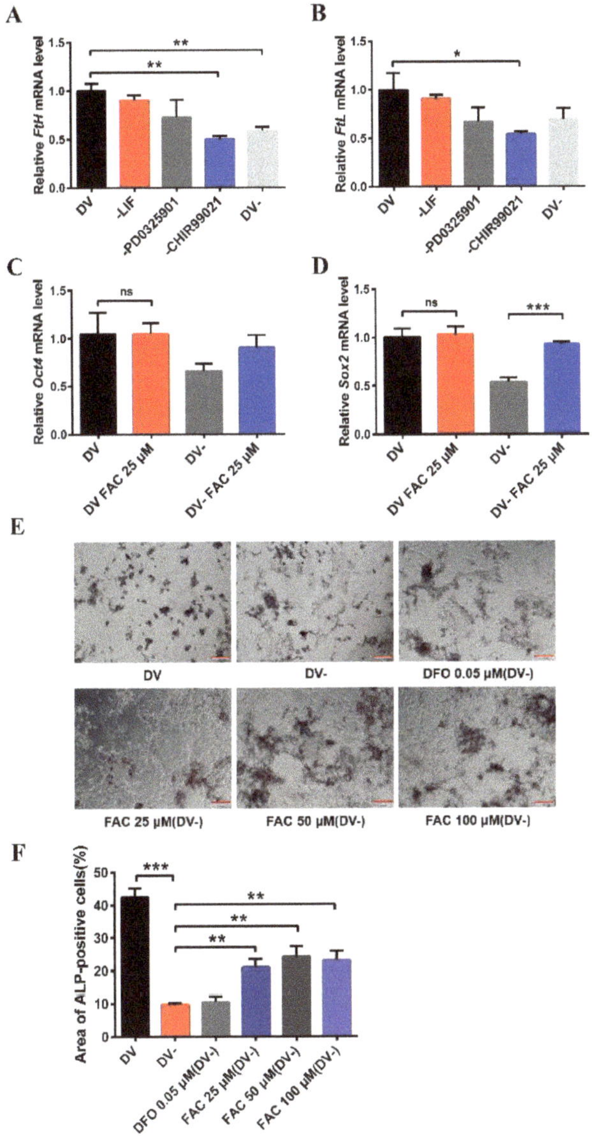

Figure 1. Effects of removal CHIR99021 and iron on the cell stemness in mESCs. (**A**,**B**) qRT-PCR analysis of *FtH* mRNA levels (**A**) and *FtL* mRNA levels (**B**) after incubation for 72 h in full derivation medium (DV) or without LIF, PD0325901, CHIR99021 or all three combined (DV-). (**C**,**D**) qRT-PCR analysis of pluripotency genes *Oct4* (**C**) and *Sox2* (**D**) expression after incubation for 72 h in full derivation medium or without LIF, PD0325901 and CHIR99021 (DV-), without or with the addition of 25 μM FAC. The mRNA levels are normalized to GAPDH mRNA levels. (**E**) Alkaline phosphatase (ALP) staining of mESCs treated for 72 h in full derivation medium (DV) or without LIF, PD0325901 and CHIR99021 (DV-), without or with the indicated amounts of FAC. Scale bar = 100 μm. (**F**) Quantification of area of ALP-positive cells (%). The data are expressed as the mean ± SEM. * $p < 0.05$, ** $p < 0.01$, *** $p < 0.001$, ns, not significant.

3.2. CHIR99021 Increasing the Expression of Ferritins, Cellular Iron Levels and Inhibiting the Expression of FPN1 in Neuro-2a Cells

In order to further explore the effects of CHIR99021 on iron metabolism, we treated Neuro-2a cells with CHIR99021 and assessed the expression of iron metabolism related proteins, including FtH, FtL, TfR1, FPN1, and DMT1 (+/-IRE) (divalent metal transporter 1). The protein levels of FtH and FtL increased significantly in CHIR99021-treated Neuro-2a cells (Figure 2A,B). The levels of *FtH* and *FtL* mRNA were consistent with the increased protein levels (Figure 2C). As ferritin is the main intracellular iron storage protein, we next investigated whether total intracellular iron levels were changed by ICP-MS; total intracellular iron levels increased significantly after CHIR99021 treatment (Figure 2D). We also evaluated the levels of TfR1 protein and mRNA and found no significant differences (Figure 2E–G). However, the protein and mRNA levels of FPN1 [28] decreased significantly (Figure 2E,F,H). Additionally, the protein level of DMT1 decreased significantly (Figure S2). These results indicated that iron metabolism has been affected, and iron export decreased, which is expected to increase the amount of uncommitted iron in cells and, in turn, stimulate an increase in ferritin expression.

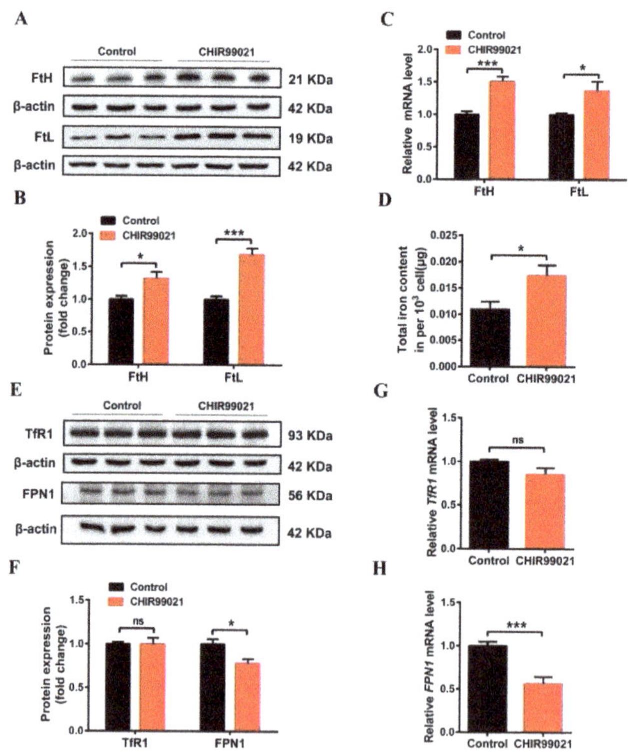

Figure 2. Changes in iron metabolism in Neuro-2a cells treated with CHIR99021. (**A**,**E**) Western blot analysis of FtH and FtL (**A**), TfR1 and FPN1 (**E**) protein after the cells were treated with or without 3 μM CHIR99021 [29,30] for 72 h. (**B**,**F**) Quantification of Western blot data of FtH and FtL (**B**), TfR1 and FPN1 (**F**). (**C**,**G**,**H**) qRT-PCR analysis of *FtH* and *FtL* (**C**), *TfR1* (**G**), and *FPN1* (**H**) mRNA levels. (**D**) ICP-MS detection of total cellular iron levels in Neuro-2a cells treated with or without CHIR99021. The relative expression levels are normalized to β-actin and then expressed as the fold of control group. The data are presented as the mean ± SEM. * $p < 0.05$, *** $p < 0.001$, ns, not significant.

3.3. DFO Abrogates the Increase of Ferritin Induced by CHIR99021

In order to further confirm that CHIR99021 increases cellular iron levels, we co-treated cells with CHIR99021 and the iron chelator, DFO. We initially performed a concentration-dependence experiment, using 0–40 μM DFO. We found that the protein levels of FtH and FtL were decreased in the presence of 5 μM DFO (Figure 3A–D). Therefore, we proceeded to treat cells with 5 μM DFO and 3 μM CHIR99021. As previously shown, CHIR99021 treatment significantly increased the levels of FtH and FtL, while the expression of these proteins significantly decreased when DFO was included in the incubation (Figure 3E–H). These results suggested that the sequestration of iron could alleviate the increase of ferritin expression induced by CHIR99021.

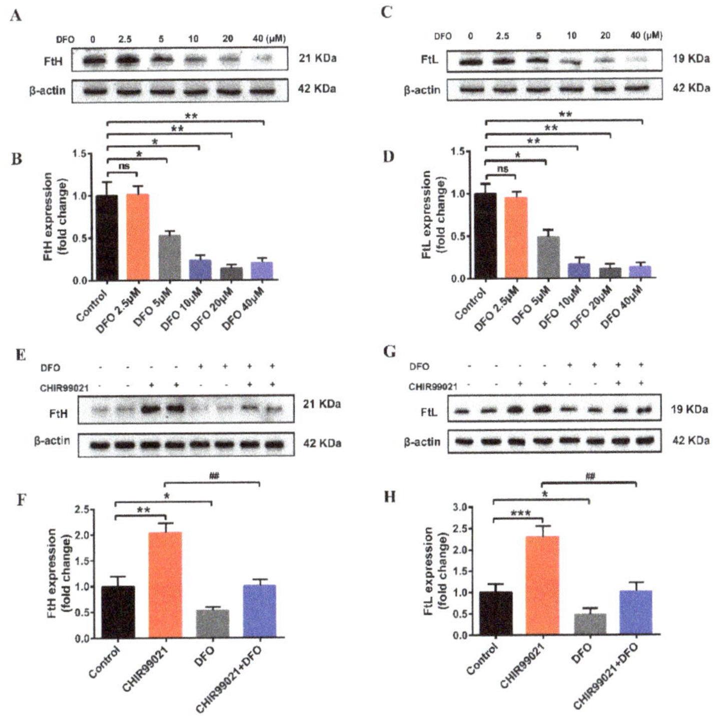

Figure 3. DFO prevents the increase of ferritin expression induced by CHIR99021. (**A–D**) Western blot analysis of FtH and FtL protein after treatment with the indicated concentrations of DFO. (**E**,**F**) Western blot analysis of FtH levels after treatment with or without CHIR99021 and/or DFO. (**G**,**H**) Western blot analysis of FtL levels after treatment with or without CHIR99021 and/or DFO. The relative expression levels are normalized to β-actin. The data are expressed as the mean ± SEM. * $p < 0.05$, ** $p < 0.01$, *** $p < 0.001$, ## $p < 0.01$, ns, not significant.

3.4. *CHIR99021 Stimulates the Expression of Steap1, Bola2 and Kdm6bos Increasing in Neuro-2a Cells*

In order to further explore the effect of CHIR99021 treatment on iron homeostasis and the mechanism of increased ferritin expression, we performed RNA-seq analysis in cells treated with or without CHIR99021. As shown in Figure 4A, we identified 1364 differentially expressed genes between the two groups, of which 580 genes were up-regulated and 784 genes were down-regulated. Gene Ontology (GO) enrichment analysis categorizes groupings of genes into three general areas: biological process, cellular component, and molecular function. From the results of this type of analysis, we found most differentially expressed genes to be concentrated in cellular processes: biological regulations and metabolic processes of biological processes; the composition of cells and organelles in cellular components; and molecular binding parts in molecular functions (Figure 4B). The Kyoto Encyclopedia of Genes and Genomes (KEGG) database divides biological metabolic pathways into six areas, including metabolism, genetic information processing, environmental information processing, cellular processes, biological systems, and human disease We found that the differentially expressed genes were more heavily distributed in th process of translation and signal transduction (Figure 4C). These results helped us to hon in on the genes targeted by CHIR99021.

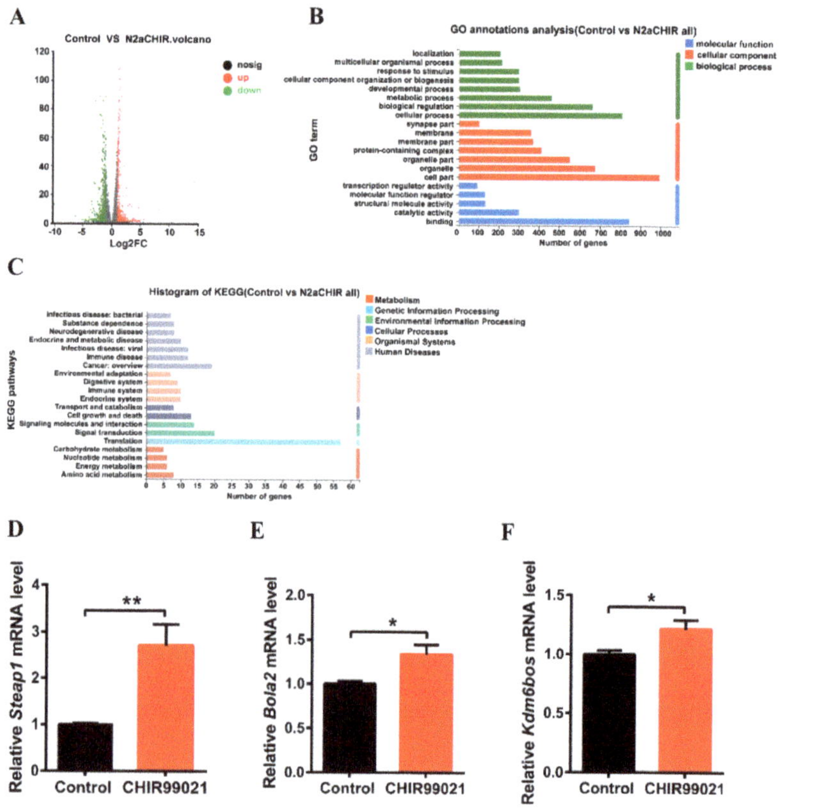

Figure 4. Comparison of RNA-seq results and verification of the differentially expressed genes between control and CHIR99021-treated cells. (**A**) Volcano plot of mRNA expression differences (**B**) GO enrichment analysis between control and CHIR99021-treated Neuro-2a cells. (**C**) Statistical classification analysis of KEGG pathways between control and CHIR99021-treated Neuro-2a cells (**D–F**) qRT-PCR analysis of *Steap1* (**D**), *Bola2* (**E**) and *Kdm6bos* (**F**) mRNA levels with or without CHIR99021 treatment. The mRNA levels are normalized to β-actin mRNA levels and then expressed as the fold of control group. The data are expressed as the mean ± SEM. * $p < 0.05$, ** $p < 0.01$.

Next, based on the results of sequencing and alignment, we determined the genes which both $|\log2^{FC}| \geq 1$ and were related to iron or metal ion binding from 1364 differentially expressed genes. Six-segment transmembrane epithelial antigen of prostate 1 (STEAP1) has oxidoreductase activity and metal ion binding function in the molecular functional grouping enriched by GO, and also plays a key role in iron homeostasis [31]. BolA-like protein (Bola2) participates in the assembly of iron-sulfur clusters on biological process grouping [32,33]. KDM6B is a KDM1 lysine-specific demethylase 6B, which has methyltransferase activity and metal ion binding function in the molecular function grouping. By qRT-PCR, we confirmed that the mRNA levels of *Steap1*, *Bola2* and *Kdm6bos* increased significantly upon CHIR99021 treatment (Figure 4D–F).

3.5. Both CHIR99021 and Iron Promote the Expression of β-Catenin

The small molecular compound, CHIR99021, is a highly effective inhibitor of GSK-3 α/β, while β-catenin is a downstream target gene of GSK-3β in the classical Wnt pathway [34]. To verify the effect of CHIR99021 on the Wnt/GSK-3β/β-catenin pathway, we examined the related proteins. We found that CHIR99021 significantly downregulated the expression of GSK-3α/β, concomitantly upregulating the levels of p-GSK3β and β-catenin (Figure 5A,B,D,E). Thus, the ratio of p-GSK3β/GSK-3β increased significantly with the treatment of CHIR99021 (Figure 5C). In addition, we found that 25 μM and 50 μM FAC significantly increased the expression of β-catenin (Figure 5F–K). In consideration of these and the previous results, we hypothesized that there may be a relationship between β-catenin and iron level, particularly on the context of CHIR99021 exposure.

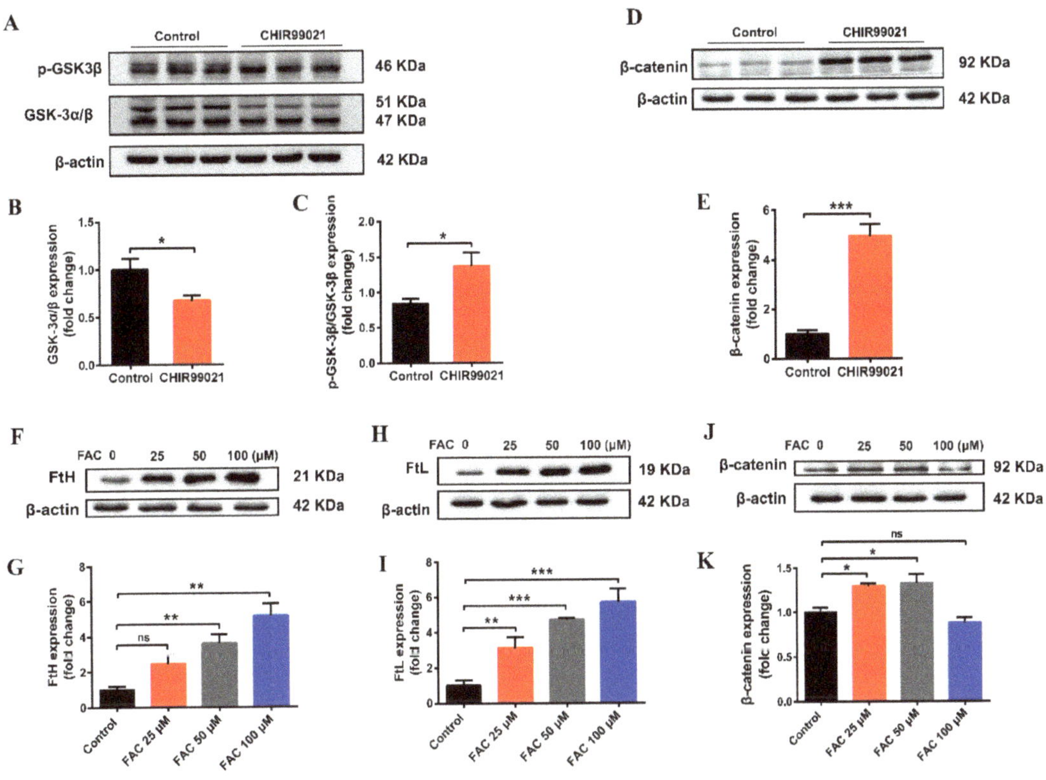

Figure 5. Both CHIR99021 and iron stimulate the expression of β-catenin in Neuro-2a cells. (**A**,**D**) Western blot analysis of GSK-3 α/β, p-GSK3β and β-catenin protein in Neuro-2a cells treated with or without CHIR99021. (**B**,**C**,**E**) Quantification of the Western blot analysis of GSK-3 α/β (**B**), p-GSK3β/GSK-3β (**C**), and β-catenin (**E**). (**F**,**H**,**J**) Western blot analysis of FtH (**F**), FtL (**H**) and β-catenin (**J**) protein after treatment with the indicated concentrations of FAC. (**G**,**I**,**K**) Quantification of the Western blot analysis of FtH (**G**), FtL (**I**) and β-catenin (**K**). The relative expression levels are normalized to β-actin and then expressed as the fold of control group. The data are expressed as the mean ± SEM. * $p < 0.05$, ** $p < 0.01$, *** $p < 0.001$, ns, not significant.

3.6. CHIR99021 Affects the Growth of Neuro-2a Cells

Our above results demonstrate that CHIR99021 treatment altered intracellular iron homeostasis, including changes in iron metabolism-related molecules, as well as significant increases in intracellular iron levels. Among its many roles in enzymes and other proteins, iron participates in the synthesis and replication of DNA, thus playing important roles in the process of cell growth [35]. We therefore proceeded to examine the expression of cyclin D1 and Ki67 proteins. Treatment of cells with CHIR99021 led to significant decreases in both of these proteins (Figure 6A–C). In addition, we found by electric cell-substrate impedance sensing that CHIR99021 inhibits cell proliferation (Figure 6D).

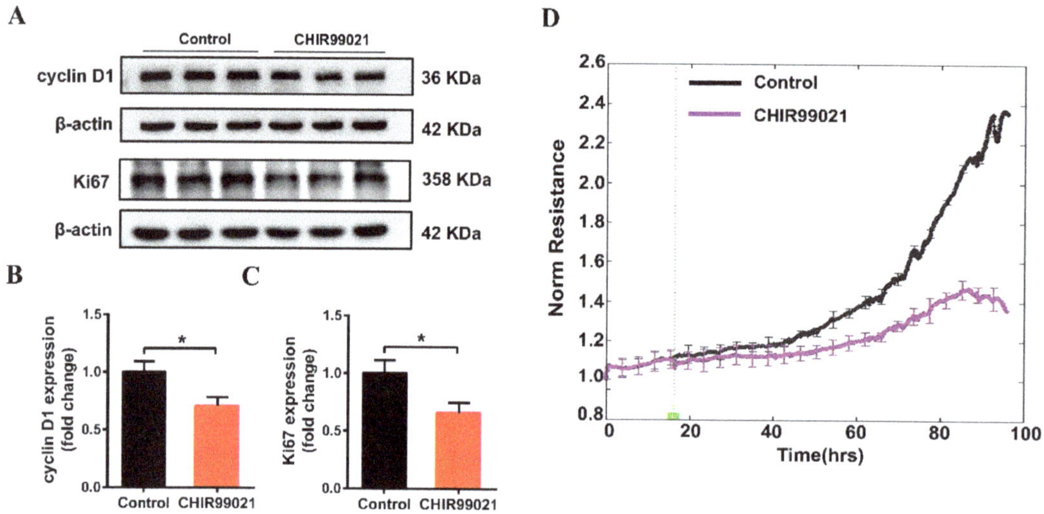

Figure 6. CHIR99021 inhibits the growth of Neuro-2a cells. (**A–C**) Western blot analysis and quantification of the expression of cyclin D1 and Ki67 protein. The relative expression levels are normalized to β-actin and then expressed as the fold of control group. (**D**) Detection of cell proliferation by electric cell-substrate impedance sensing. The data are expressed as the mean ± SEM. * $p < 0.05$.

3.7. CHIR99021 Treatment Decreases Mitochondrial Oxidative Phosphorylation and Enhances Glycolytic Capacity in Neuro-2a Cells

Oxidative phosphorylation (OXPHOS) and glycolysis are two major pathways of energy metabolism in mammalian cells. Normally, these two pathways are regulated to adapt to changes of external environment. As described above, we found that the GSK3 inhibitor, CHIR99021, can not only alter cellular iron levels and iron metabolism-related proteins, but also activate the Wnt/GSK-3β/β-catenin pathway. Both iron and this Wnt pathway could affect the maintenance of cell stemness [9,23,36]. When stem cells are in an undifferentiated state, mitochondria are immature and perinuclearly distributed, with lower OXPHOS capacity and higher glycolytic activity. Importantly, cellular energy metabolism patterns can also regulate cell fate through epigenetic mechanisms [37,38]. Therefore, we used the cellular energy metabolism analyzer, O2k, to evaluate cellular oxygen consumption and glycolytic capacity. The overall mitochondrial metabolic capacity decreased in cells treated with CHIR99021 (Figure 7A,B); basal respiration, ATP production-related oxygen consumption, and maximal respiration were all significantly decreased. Spare respiratory capacity, which represents the potential responsiveness of cells to energy demands, showed a downward trend (Figure 7C). In contrast, the overall glycolytic capacity of the cells was increased after CHIR99021 treatment (Figure 7D), while the maximum glycolytic capacity and glycolysis reserve capacity were all significantly increased (Figure 7E). These results suggest that CHIR99021 treatment diminishes cellular OXPHOS levels and enhances cellular glycolytic capacity, which is consistent with the characteristic that cells in a state of stemness mainly rely on glycolysis rather than OXPHOS.

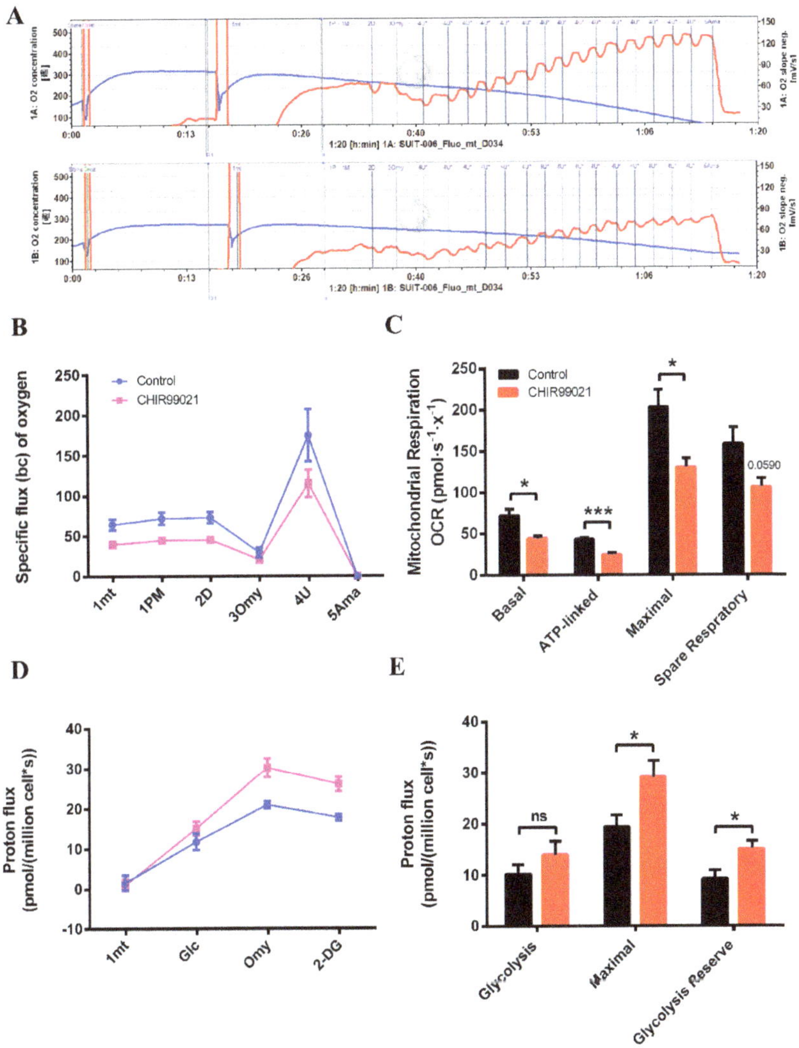

Figure 7. CHIR99021 treatment decreases mitochondrial oxidative phosphorylation and enhances glycolytic capacity in Neuro-2a cells. (**A**) Overall oxygen consumption capacity of cells, as measured by O2k. Top panel, control group; bottom panel, CHIR99021-treated group. (**B**) Oxygen flux of cells after adding the indicated reagents. (**C**) Basal respiration, ATP-linked respiration, maximal respiration and spare respiratory capacity, as indicated by oxygen consumption rate (OCR) in Neuro-2a cells treated as shown. (**D**) Proton flux of cells after adding the indicated reagents. (**E**) Basal glycolysis, maximal glycolytic capacity and glycolysis reserve capacity in cells treated as shown. The data are expressed as the mean ± SEM. * $p < 0.05$, *** $p < 0.001$, ns, not significant.

4. Discussion

A multifunctional protein, GSK3, is involved in a variety of cellular processes, including proliferation, metabolism and embryonic development [39]. It contains GSK-3α and GSK-3β. The latter is one of the main members of the canonical Wnt pathway. Recent studies have revealed that this pathway plays an important role in the maintenance of stemness [9,10,40]. As an essential element, iron also has many biological functions. Among these, iron has been identified as a key factor in the maintenance of cancer stem cell (CSC) stemness. The iron chelation can inhibit the expression of the stem cell marker, Nanog, thu revealing a new treatment strategy for cancer [41]. However, it is not clear whether there i a relationship between GSK3 and iron, and, if so, what that mechanism may entail.

CHIR99021, a small molecule compound, can efficiently inhibit both GSK3. Th compound is also commonly used in the culture of stem cells. In this study, we foun that the expression of stemness-related genes and the transcription level of FtH and FtL i mESCs were significantly decreased upon removal of CHIR99021 from mouse embryoni stem cell culture medium. On the other hand, iron supplementation of embryonic sten cell differentiation medium was beneficial to the maintenance of mESCs stemness. In fac we found that CHIR99021 can affect cell stemness by affecting intracellular iron levels. T further explore the relationship between GSK3 and cellular iron homeostasis, we treate Neuro-2a cells with CHIR99021, which led to increased expression of FtH and FtL at bot the mRNA and protein levels, as well as increased intracellular iron content. Cellular iro homeostasis is regulated at the post transcriptional level by the IRE/IRP system. Whe intracellular iron levels increase, IRPs are unable to bind to the IREs in the 5′-UTR of th mRNAs encoding ferritin and FPN1, thus permitting their translation, while binding t the 3′-UTR in the mRNA encoding TfR1 is required to stabilize the message, so TfR1 level decrease under elevated cellular iron conditions. Regulation occurs in the opposite directio upon depletion of cellular iron [42]. Our finding of increased ferritin, with concomitan decreases in FPN1 and unchanged TfR1 after treatment with CHIR99021, indicates that th changes in iron homeostasis induced by CHIR99021 may not be regulated by the syster However, treatment of cells with iron chelation and CHIR99021 alleviated the increase o ferritin expression induced by CHIR99021. Thus, we conclude that CHIR99021 indee affects cellular iron homeostasis.

To further explore the mechanism of ferritin up-regulation by CHIR99021, we use RNA-seq and narrowed the differentially regulated genes based on iron or metal iror binding, finally identifying three genes of interest: *Steap1*, *Bola2* and *Kdm6bos*. The mRN expression of all three of these genes increased significantly after CHIR99021 treatmen Interestingly, STEAP1 is highly expressed in prostate cancer [43]. In 2016, Kim et a reported the presence of a single heme b prosthetic group in STEAP1, which can reduc metal ion complexes and oxygen. Diferric transferrin (Fe_2Tf) binds to TfR1, which mediate iron uptake into cells. Then, Fe^{3+} is released from Tf and must be reduced prior to expor from the endosome by the Fe^{2+} transporter, DMT1. This reduction can be mediated b STEAP1. After export from the endosome, iron can be incorporated into functional protein or stored in ferritin [31]. Therefore, the increase of ferritin expression induced by CHIR9902 treatment may be a consequence of increased levels of STEAP1. Bola2 has been shown t bind to the multi-functional binding protein PCBP1 to form an intermediate iron chaperon complex and participate in the assembly of [2Fe-2S] clusters, while PCBP1 may bind iror in the LIP to transfer iron to ferritin to form a PCBP1-Fe-GSH-BolA2 complex [32,33 Therefore, there may be a connection between ferritin and BolA2. KDM6B is a KDM lysine-specific demethylase with methyltransferase activity and metal ion binding functior KDM6B requires Fe^{2+} as a cofactor [44]. Together, our data provide support for a lin between GSK3 activity and ferritin synthesis, possibly via the proteins encoded by th above genes, however, the specific mechanisms of these processes require further study.

As one of the main members of Wnt/β-catenin pathway, GSK-3β can phosphorylat β-catenin and cause it to be degraded by proteasome. When the phosphorylation of GSK 3β at Ser9 is increased, the activity of GSK-3β decreases [34]. Our results show that, wit CHIR99021, the levels of GSK-3α/β were significantly decreased, while the p-GSK3β/GSK 3β ratio was significantly increased, ultimately leading to promotion of the expression o the downstream target molecule β-catenin. Thus, the Wnt/GSK-3β/β-catenin pathwa was significantly activated. We therefore asked, since iron levels increased upon CHIR9902 treatment, could iron also affect the expression of β-catenin? To this end, we treated cell with FAC, which promoted the expression of β-catenin. The finding is consistent with a report demonstrating a positive correlation between FtL and β-catenin in glioma cells [45

Iron can involve in the regulation of cell growth. Both iron deficiency and excess can affect the cell cycle, which may then affect a series of critical organismal processes [46,47]. Our data that cell proliferation was inhibited with the treatment of CHIR99021 is consistent with these properties of iron in mammals.

Iron has also been found to play a biological role in maintaining the stemness of mesenchymal stem cells and cancer stem cells [23,41]. The activation of the classical Wnt pathway has been reported to inhibit the differentiation of stem cells [9,10,36]. Additionally, in stem cells, the mode of energy metabolism mainly depends on glycolysis rather than oxidative phosphorylation. On the other hand, the mode of cell energy metabolism can also determine the fate of stem cells, with respect to stemness or differentiation, through epigenetic regulation [37,38]. We found that oxidative phosphorylation decreased, along with increased glycolysis, upon treatment with CHIR99021.

5. Conclusions

In summary, our study uncovers a relationship between GSK3 and ferritin, along with alterations in cellular iron metabolism that affect cells stemness and proliferation. These findings not only open new avenues for further exploration of stemness maintenance and differentiation, but also provide an insight for the treatment of stem cells in disease.

Supplementary Materials: The following supporting information can be downloaded at: https://www.mdpi.com/article/10.3390/antiox12020377/s1, Figure S1: Effects of removal CHIR99021 on the expression of Oct4 and Sox2 genes. Figure S2: Expression of DMT1 in Neuro-2a cells treated with CHIR99021.

Author Contributions: Formal analysis, Y.-Z.C., S.C., J.Z., P.Y. and G.G.; writing—review and editing, Y.-Z.C., S.C. and J.Z.; supervision, Y.-Z.C., S.C. and J.Z.; writing—original draft preparation, Y.H. (Yingying Han); methodology, Y.H. (Yong He), X.J. and J.X. All authors have read and agreed to the published version of the manuscript.

Funding: This research was funded by the National Natural Science Foundation of China (grant numbers 31520103908, 32070962).

Institutional Review Board Statement: This study does not involve humans or animals.

Informed Consent Statement: Not applicable.

Data Availability Statement: All of the data is contained within the article.

Acknowledgments: The work was supported by the National Natural Science Foundation of China (grant numbers 31520103908, 32070962).

Conflicts of Interest: The authors declare no conflict of interest.

References

1. Evans, M.J.; Kaufman, M.H. Establishment in culture of pluripotential cells from mouse embryos. *Nature* **1981**, *292*, 154–156. [CrossRef] [PubMed]
2. Martin, G.R. Isolation of a pluripotent cell line from early mouse embryos cultured in medium conditioned by teratocarcinoma stem cells. *Proc. Natl. Acad. Sci. USA* **1981**, *78*, 7634–7638. [CrossRef] [PubMed]
3. Chambers, I.; Smith, A. Self-renewal of teratocarcinoma and embryonic stem cells. *Oncogene* **2004**, *23*, 7150–7160. [CrossRef]
4. Nusse, R.; Clevers, H. Wnt/beta-Catenin Signaling, Disease, and Emerging Therapeutic Modalities. *Cell* **2017**, *169*, 985–999. [CrossRef] [PubMed]
5. Huang, P.; Yan, R.; Zhang, X.; Wang, L.; Ke, S.; Qu, Y. Activating Wnt/beta-catenin signaling pathway for disease therapy: Challenges and opportunities. *Pharmacol. Ther.* **2019**, *196*, 79–90. [CrossRef] [PubMed]
6. Corda, G.; Sala, A. Non-canonical WNT/PCP signalling in cancer: Fzd6 takes centre stage. *Oncogenesis* **2017**, *6*, e364. [CrossRef]
7. Lang CM, R.; Chan, C.K.; Veltri, A.; Lien, W.H. Wnt Signaling Pathways in Keratinocyte Carcinomas. *Cancers* **2019**, *11*, 1216. [CrossRef] [PubMed]
8. Wen, X.; Wu, Y.; Awadasseid, A.; Tanaka, Y.; Zhang, W. New Advances in Canonical Wnt/beta-Catenin Signaling in Cancer. *Cancer Manag. Res.* **2020**, *12*, 6987–6998. [CrossRef]
9. Wang, F.; Liu, L.; Gao, W.Q. Activation of Wnt/ß-catenin signaling enhanced stemness of prostate cancer cells. *Tumor* **2019**, *39*, 525–533+567.

10. Tang, Q.; Chen, J.; Di, Z.; Yuan, W.; Zhou, Z.; Liu, Z.; Han, S.; Liu, Y.; Ying, G.; Shu, X.; et al. TM4SF1 promotes EMT and cancer stemness via the Wnt/beta-catenin/SOX2 pathway in colorectal cancer. *J. Exp. Clin. Cancer Res.* **2020**, *39*, 232. [CrossRef]
11. Niu, Y.M.; Yang, T.L. A number of important biochemical function of trace metal elements and its relationship to human health. *Stud. Trace Elem. Health* **2014**, *31*, 78–80.
12. Hentze, M.W.; Muckenthaler, M.U.; Galy, B.; Camaschella, C. Two to tango: Regulation of Mammalian iron metabolism. *Cell* **2010**, *142*, 24–38. [CrossRef] [PubMed]
13. Pantopoulos, K.; Porwal, S.K.; Tartakoff, A.; Devireddy, L. Mechanisms of mammalian iron homeostasis. *Biochemistry* **2012**, *51*, 5705–5724. [CrossRef] [PubMed]
14. Philpott, C.C. Coming into view: Eukaryotic iron chaperones and intracellular iron delivery. *J. Biol. Chem.* **2012**, *287*, 13518–13523. [CrossRef]
15. Zhang, D.L.; Ghosh, M.C.; Rouault, T.A. The physiological functions of iron regulatory proteins in iron homeostasis—An update. *Front. Pharmacol.* **2014**, *5*, 124. [CrossRef]
16. Zhu, J.; Wu, X.Y. Modulating factors in the differentiation of stem cells: Present research and progress. *J. Clin. Rehabil. Tissue Eng. Res.* **2009**, *13*, 4.
17. Lein, P.; Gallagher, P.J.; Amodeo, J.; Howie, H.; Roth, J.A. Manganese induces neurite outgrowth in PC12 cells via upregulation of alpha(v) integrins. *Brain Res.* **2000**, *885*, 220–230. [CrossRef]
18. Hong, J.-H.; Noh, K.-M.; Yoo, Y.-E.; Choi, S.-Y.; Park, S.-Y.; Kim, Y.-H.; Chung, J.-M. Iron promotes the survival and neurite extension of serum-starved PC12 cells in the presence of NGF by enhancing cell attachment. *Mol. Cells* **2003**, *15*, 10–19. [PubMed]
19. Rostami, A.A.; Kouchesfahani, H.M.; Kiani, S.; Fakheri, R. Iron Oxide Nanoparticles Reduced Retinoic Acid Induced-neuronal Differentiation of Mouse Embryonic Stem Cells By ROS Generation. *Arch. Iran. Med.* **2015**, *18*, 586–590.
20. Zhou, Y.; Zhang, H.; Liao, S.; Hu, F.Q.; Yi, J.; Liu, Y.B.; Jin, J.D. Immunomodulatory effects of deferoxamine and interferon gamma on human dental pulp stem cells. *Chin. J. Tissue Eng. Res.* **2022**, *26*, 8.
21. Hu, K.; Olsen, B.R. The roles of vascular endothelial growth factor in bone repair and regeneration. *Bone* **2016**, *91*, 30–38. [CrossRef]
22. Pricola, K.L.; Kuhn, N.Z.; Haleem-Smith, H.; Song, Y.; Tuan, R.S. Interleukin-6 maintains bone marrow-derived mesenchymal stem cell stemness by an ERK1/2-dependent mechanism. *J. Cell. Biochem.* **2009**, *108*, 577–588. [CrossRef]
23. Mehta, K.J. Role of iron and iron-related proteins in mesenchymal stem cells: Cellular and clinical aspects. *J. Cell. Physiol.* **2021**, *236*, 7266–7289. [CrossRef]
24. Samara, A.; Shapira, S.; Lubin, I.; Shpilberg, O.; Avigad, S.; Granot, G.; Raanani, P. Deferasirox induces cyclin D1 degradation and apoptosis in mantle cell lymphoma in a reactive oxygen species- and GSK3beta-dependent mechanism. *Br. J. Haematol.* **2021**, *192*, 747–760. [CrossRef]
25. Zhu, J.L.; Lu, Z.S.; Wu, Y.; Xu, Y.J.; Yao, J.; Wang, X.X.; Wang, D.C. Copper, iron, zinc and aluminum induce the apoptosis of SH-SY5Y cells through Akt/GSK-3β signaling pathways. *Pract. Geriatr.* **2017**, *31*, 723–727.
26. Guo, C.; Wang, P.; Zhong, M.-L.; Wang, T.; Huang, X.-S.; Li, J.-Y.; Wang, Z.-Y. Deferoxamine inhibits iron induced hippocampal tau phosphorylation in the Alzheimer transgenic mouse brain. *Neurochem. Int.* **2013**, *62*, 165–172. [CrossRef] [PubMed]
27. Han, J.; Fan, Y.; Wu, P.; Huang, Z.; Li, X.; Zhao, L.; Ji, Y.; Zhu, M. Parkinson's Disease Dementia: Synergistic Effects of Alpha-Synuclein, Tau, Beta-Amyloid, and Iron. *Front. Aging Neurosci.* **2021**, *13*, 743754. [CrossRef] [PubMed]
28. Ganz, T. Cellular iron: Ferroportin is the only way out. *Cell Metab.* **2005**, *1*, 155–157. [CrossRef]
29. Wu, Y.; Liu, W.; Chen, J.; Liu, S.; Wang, M.; Yang, L.; Chen, C.; Qi, M.; Xu, Y.; Qiao, Z.; et al. Nuclear Exosome Targeting Complex Core Factor Zcchc8 Regulates the Degradation of LINE1 RNA in Early Embryos and Embryonic Stem Cells. *Cell Rep.* **2019**, *29*, 2461–2472. [CrossRef]
30. Ai, Z.; Shao, J.; Wu, Y.; Yu, M.; Du, J.; Shi, X.; Shi, X.; Zhang, Y.; Guo, Z. CHIR99021 enhances Klf4 Expression through beta-Catenin Signaling and miR-7a Regulation in J1 Mouse Embryonic Stem Cells. *PLoS ONE* **2016**, *11*, e0150936. [CrossRef]
31. Kim, K.; Mitra, S.; Wu, G.; Berka, V.; Song, J.; Yu, Y.; Poget, S.; Wang, D.-N.; Tsai, A.-L.; Zhou, M. Six-Transmembrane Epithelial Antigen of Prostate 1 (STEAP1) Has a Single b Heme and Is Capable of Reducing Metal Ion Complexes and Oxygen. *Biochemistry* **2016**, *55*, 6673–6684. [CrossRef] [PubMed]
32. Patel, S.J.; Frey, A.G.; Palenchar, D.J.; Achar, S.; Bullough, K.Z.; Vashisht, A.; Wohlschlegel, J.A.; Philpott, C.C. A PCBP1-BolA2 chaperone complex delivers iron for cytosolic [2Fe-2S] cluster assembly. *Nat. Chem. Biol.* **2019**, *15*, 872–881. [CrossRef]
33. Patel, S.J.; Protchenko, O.; Shakoury-Elizeh, M.; Baratz, Z.; Jadhav, S.; Philpott, C.C. The iron chaperone and nucleic acid–binding activities of poly(rC)-binding protein 1 are separable and independently essential. *Proc. Natl. Acad. Sci. USA* **2021**, *118*, e2104666118. [CrossRef]
34. He, X.J.; Zheng, W.W.; Jia, L.L.; Li, H.; Xu, L.M.; Zeng, J.W.; Xu, H.F.; Zheng, C.S.; Ye, J.X.; Wu, G.W.; et al. Study on the Mechanism of Tiaogu Tablet Regulating Wnt/β-catenin Signaling Pathway to Delay Articular Cartilage Degeneration in Osteoarthritis. *Rheum. Arthritis* **2019**, *8*, 5–9.
35. Bartels, P.L.; Stodola, J.L.; Burgers, P.M.J.; Barton, J.K. A Redox Role for the [4Fe4S] Cluster of Yeast DNA Polymerase delta. *J. Am. Chem. Soc.* **2017**, *139*, 18339–18348. [CrossRef]
36. Aulicino, F.; Pedone, E.; Sottile, F.; Lluis, F.; Marucci, L.; Cosma, M.P. Canonical Wnt Pathway Controls mESC Self-Renewal Through Inhibition of Spontaneous Differentiation via β-Catenin/TCF/LEF Functions. *Stem Cell Rep.* **2020**, *15*, 646–661. [CrossRef] [PubMed]

7. Rafalski, V.A.; Mancini, E.; Brunet, A. Energy metabolism and energy-sensing pathways in mammalian embryonic and adult stem cell fate. *J. Cell Sci.* **2012**, *125 Pt 23*, 5597–5608. [CrossRef] [PubMed]
8. Shyh-Chang, N.; Ng, H.H. The metabolic programming of stem cells. *Genes Dev.* **2017**, *31*, 336–346. [CrossRef]
9. Nagini, S.; Sophia, J.; Mishra, R. Glycogen synthase kinases: Moonlighting proteins with theranostic potential in cancer. *Semin. Cancer Biol.* **2019**, *56*, 25–36. [CrossRef] [PubMed]
0. Liu, X.; Su, K.; Sun, X.; Jiang, Y.; Wang, L.; Hu, C.; Zhang, C.; Lu, M.; Du, X.; Xing, B. Sec62 promotes stemness and chemoresistance of human colorectal cancer through activating Wnt/beta-catenin pathway. *J. Exp. Clin. Cancer Res.* **2021**, *40*, 132. [CrossRef]
1. Narusaka, T.; Ohara, T.; Noma, K.; Nishiwaki, N.; Katsura, Y.; Kato, T.; Sato, H.; Tomono, Y.; Kikuchi, S.; Tazawa, H.; et al. Nanog is a promising chemoresistant stemness marker and therapeutic target by iron chelators for esophageal cancer. *Int. J. Cancer* **2021**, *149*, 347–357. [CrossRef] [PubMed]
2. Joshi, R.S.; Moran, E.; Sanchez, M. Cellular Iron Metabolism—The IRP/IRE Regulatory Network. *Iron Metab.* **2012**, *2*, 25–58.
3. Hubert, R.S.; Vivanco, I.; Chen, E.; Rastegar, S.; Leong, K.; Mitchell, S.C.; Madraswala, R.; Zhou, Y.; Kuo, J.; Raitano, A.B.; et al. STEAP: A prostate-specific cell-surface antigen highly expressed in human prostate tumors. *Proc. Natl. Acad. Sci. USA* **1999**, *96*, 14523–14528. [CrossRef]
4. Lagunas-Rangel, F.A. KDM6B (JMJD3) and its dual role in cancer. *Biochimie* **2021**, *184*, 63–71. [CrossRef]
5. Liu, J.; Gao, L.; Zhan, N.; Xu, P.; Yang, J.; Yuan, F.; Xu, Y.; Cai, Q.; Geng, R.; Chen, Q. Hypoxia induced ferritin light chain (FTL) promoted epithelia mesenchymal transition and chemoresistance of glioma. *J. Exp. Clin. Cancer Res.* **2020**, *39*, 137. [CrossRef] [PubMed]
6. Khan, A.; Singh, P.; Srivastava, A. Iron: Key player in cancer and cell cycle? *J. Trace Elem. Med. Biol.* **2020**, *62*, 126582. [CrossRef]
7. Troadec, M.-B.; Courselaud, B.; Détivaud, L.; Haziza-Pigeon, C.; Leroyer, P.; Brissot, P.; Loréal, O. Iron overload promotes Cyclin D1 expression and alters cell cycle in mouse hepatocytes. *J. Hepatol.* **2006**, *44*, 391–399. [CrossRef]

antioxidants

MDPI

Article

Stimulation of Hepatic Ferritinophagy Mitigates *Irp2* Depletion-Induced Anemia

Yutong Liu [1,2], Yuxuan Li [1], Liu Yang [1], Jiaqi Shen [1], Hongting Zhao [1], Weichen Dong [1,3], Yanzhong Chang [4], Tong Qiao [2] and Kuanyu Li [1,2,*]

1 Jiangsu Key Laboratory of Molecular Medicine, Medical School, Nanjing University, Nanjing 210093, China
2 Department of Vascular Surgery, Affiliated Drum Tower Hospital, Medical School, Nanjing University, Nanjing 210093, China
3 Department of Neurology, Affiliated Jinling Hospital, Medical School, Nanjing University, Nanjing 210002, China
4 College of Life Science, Hebei Normal University, Shijiazhuang 050024, China
* Correspondence: likuanyu@nju.edu.cn

Abstract: Background: Iron regulatory proteins (IRPs) maintain cellular iron homeostasis. Due to aberrant tissue-iron distribution, *Irp2*-deficient mice suffer microcytic anemia and neurodegeneration while iron overload occurs in the liver and intestine. We previously found that *Irp2* deficiency-induced Hif2 plays an important role in neurodegeneration. Methods: To test the role of Hif2 in *Irp2* deficiency-induced anemia, we used *Irp2* global knockout mice. Following Hif2 inhibition, routine blood tests, iron availability in bone marrow, histological assays, and biochemical analysis were performed to assess anemia improvement and tissue iron distribution. Results: We found that Hif2 inhibition improved anemia. The increased iron bioavailability for erythropoiesis was mainly derived from hepatic iron release, and secondly from enhanced intestinal absorption. We further demonstrate that nuclear receptor coactivator 4 (Ncoa4) was upregulated for iron release via the process of ferritinophagy. The released iron was utilized not only for intracellular Fe-S biogenesis but also for erythropoiesis after being exported from the liver to circulation. The hepatic iron export reduced hepcidin expression to further support iron absorption through the hepcidin-ferroportin axis to alleviate intestinal iron overload. Conclusion: Irp2 not only regulates cellular iron homeostasis but also tissue iron distribution by managing the involvement of Hif2-Ncoa4.

Keywords: IRP2; Hif2 inhibition; Fe-S clusters; ferritinophagy; microcytic anemia

Citation: Liu, Y.; Li, Y.; Yang, L.; Shen, J.; Zhao, H.; Dong, W.; Chang, Y.; Qiao, T.; Li, K. Stimulation of Hepatic Ferritinophagy Mitigates *Irp2* Depletion-Induced Anemia. *Antioxidants* **2023**, *12*, 566. https://doi.org/10.3390/antiox12030566

Academic Editor: Kostas Pantopoulos

Received: 15 January 2023
Revised: 13 February 2023
Accepted: 16 February 2023
Published: 24 February 2023

1. Introduction

Iron is essential for almost all living organisms. It functions as a cofactor in the form of Fe-S clusters or heme, or by itself, participating in numerous vital physiological processes, including mitochondrial respiration, oxygen transfer, DNA repair, and enzymatic catalysis [1–3]. While iron deficiency can lead to cognitive defects in children and anemia in adults, excess iron is also detrimental because it stimulates oxidative stress, subsequently causing tissue injury and disease [4]. As a result, it is imperative to maintain the body's iron levels within an acceptable range.

To maintain iron homeostasis at the cellular level, iron regulatory proteins (IRP1 and IRP2) post-transcriptionally regulate the expression of iron-related proteins via the iron-responsive element (IRE) (see review in [5]). When iron is deficient, IRPs bind to the IRE in the 5′-untranslated region (5′-UTR) of L- and H-ferritin and ferroportin 1 (FPN1) mRNA to inhibit the translation of these genes to limit iron storage and efflux. Meanwhile, IRPs can stabilize mRNA by binding to the IRE in the 3′-UTR of divalent metal transporter 1 (DMT1) and transferrin receptor 1 (TfR1), which promotes iron uptake and ultimately alleviates iron deficiency [5,6]. In contrast, when iron is redundant, IRP1 binds to [4Fe-4S] clusters and loses its ability to bind IREs [7]; meanwhile, F-box and leucine-rich repeats protein 5

(FBXL5), a subunit of ubiquitin ligase complex, specifically recognizes IRP2 and promotes its degradation by gaining oxygen-responsive [2Fe-2S] clusters [8–10]. This, in turn, inhibits iron uptake, boosts iron storage, or exports excess iron.

Systemic deficiency of *Irp2* in mice results in microcytic anemia, erythropoietic protoporphyria [11], neurodegeneration [12], and diabetes [13,14], which are clinically manifested by phenotypes of patients with bi-allelic loss-of-function variants in *IREB2* [15]. *Irp2* depletion-induced microcytic anemia is not simply derived from global iron deficiency but is accompanied by iron overload in the liver and intestine [16]. This abnormal iron distribution might injure hepatocytes and enterocytes for liver metabolism and nutrient absorption.

Recently, we found that *Irp2*-null mutation causes downregulation of frataxin (Fxn) and IscU, two of the core components in the Fe-S cluster biogenesis machinery [17]. Consequently, mitochondrial dysfunction occurs, which shifts energy metabolism from oxidative phosphorylation (OXPHOS) to glycolysis in $Irp2^{-/-}$ murine embryonic fibroblasts (MEFs) [18]. We further demonstrated that *Irp2* deficiency induces the expression of hypoxia-inducible factors Hif1α and Hif2α. Hif2, but not Hif1, suppresses mitochondrial Fe-S biosynthesis and OXPHOS in *Irp2*-deficient cells [18]. These results have been confirmed in *Irp2* knockout mice, showing that an increase in Hif2α switches energy metabolism from OXPHOS to glycolysis, and the inhibition of Hif2 by PT–2385, a selective Hif2 inhibitor, may alleviate the neurological disorder by improving mitochondrial function [19], which is accomplished by an increase of Fe-S biogenesis [18,19].

Mitochondria not only act as the powerhouse of the cell but also as Fe-S- and heme-houses. Two key enzymes in the second and last steps of heme synthesis (5′-aminolevulinate dehydratase (ALAD) in cytosol and ferrochelatase (FECH) in mitochondria, respectively) require Fe-S clusters as prosthetic groups to exert enzymatic activities [20,21]. We speculated that Hif2 inhibition might also improve *Irp2* deletion-induced microcytic anemia by enhancing Fe-S biogenesis. However, iron deficiency is induced by *Irp2* depletion in bone marrow [11,16]. Then, our question was where the erythroblasts in bone marrow would obtain enough iron.

In this study, we found that *Irp2* depletion-induced abnormal iron distribution was corrected by PT–2385 administration, indicated by a mitigation of iron accumulation in the liver and an increased iron content in hematopoietic tissue and serum. Further, we demonstrated that the iron release from ferritin in the liver is accomplished by nuclear receptor coactivator 4 (Ncoa4)-mediated ferritinophagy. Therefore, the symptoms of anemia were improved.

2. Materials and Methods

2.1. Animals and Tissue Collection

Irp2 homozygous global knockout (*Irp2* KO) mice and WT mice in the C57BL/6J background were generated in our previous study [19]. Briefly, the *Irp2* KO and WT mice used in the experiment were the descendants of *Irp2* heterozygous knockout mice purchased from MMRRC at UC Davis (cat. no. 030490-MU, Davis, CA, USA). Mice were group-housed in standard housing conditions under a 12 h light-dark cycle at 25 °C. All experiments were approved by the Animal Investigation Ethics Committee of Nanjing University and were performed according to the Guidelines for the Care and Use of Laboratory Animals of the National Institutes of Health, USA.

The animals were weighed and euthanized after drug treatment. Blood was collected for subsequent routine blood and biochemical examinations. After saline perfusion, tissues including liver, kidney, and intestine were immediately frozen in liquid nitrogen or transferred to paraformaldehyde (4%) to be used as needed. Bone marrow was collected from the hind limbs of mice.

2.2. Drug Treatment

PT–2385 (MedChemExpress, Shanghai, China) dissolved in dimethylsulfoxide (DMSO) and diluted with saline was administered at 0.4 mg/kg bw by intraperitoneal injection qod for 1 month. The same DMSO volume was diluted with saline as the vehicle control Six-month-old mice were randomly divided into four groups: WT mice with vehicle treatment (WT), WT mice with PT–2385 treatment (WT+PT–2385), *Irp2* KO mice with vehicle treatment (*Irp2* KO), and *Irp2* KO mice with PT–2385 treatment (*Irp2* KO+PT–2385).

2.3. Routine Blood Examinations

The red blood cell (RBC) count, hemoglobin (HGB) concentration, and mean corpuscular volume (MCV) (n = 8) were determined by a Mindray automatic hematology analyzer (BC-2800vet, Shenzhen, China).

2.4. Blood Biochemical Examinations

Blood was collected in heparinized tubes and centrifugated at 1200× g for 15 min at 4 °C. Serum samples were prepared for determination of alanine aminotransferase (ALT), aspartate aminotransferase (AST), and total bilirubin (TBil) by an auto-chemical analyzer (Beckman Coulter AU5421, Brea, CA, USA).

2.5. Western Blot Analysis

Mouse tissues were lysed by RIPA lysis buffer (Wuhan Servicebio Technology Co., Ltd., Wuhan, China) and homogenized by a KZ-II 2100 rpm High-Speed Tissue Homogenizer (Wuhan Servicebio Technology Co., Ltd., Wuhan, China). The protein concentration was quantified by Bradford buffer. Total proteins were prepared (20–40 µg/lane), run in SDS-PAGE gels at 100 V, transferred to nitrocellulose membranes, and incubated with primary and secondary antibodies for analysis.

The information for primary antibodies is as follows: anti-Hif2α (rabbit, 1:1000; cat# 109616), anti-ferritin light chain (rabbit, 1:1000; cat# 69090), anti-Ncoa4 (rabbit, 1:1000; cat# ab86707), and anti-Sdhb (rabbit, 1:2000; at# 178423) from Abcam (Cambridge, MA, USA). Anti-Ndufs1 (rabbit, 1:1000; cat# 12444-1-AP), anti-Uqcrfs1 (rabbit, 1:2000; cat# 1843-1-AP), anti-Atg5 (rabbit, 1:2000; cat# 10181-2-AP), anti-Glut1 (rabbit, 1:1000; cat# 21829-1-AP), anti-GAPDH (mouse, 1:5000; cat# 60004-1-Ig), and anti-Lamp1 (mouse, 1:1000′ cat#67300-1-Ig) from Proteintech (Chicago, IL, USA). Anti-LC3A/B (rabbit, 1:1000; cat# 4108) from Cell Signaling Technology (Danvers, MA, USA), anti-β-Actin (rabbit, 1:10,000; cat# AP0060) from Bioworld (Nanjing, China), anti-Fpn1 (mouse, 1:1000; cat# MTP11-A) from Alpha Diagnostic (San Antonio, TX, USA), anti-p62 (rabbit, 1:1000 cat# A7758) from ABclonal (Wuhan, China), and anti-Fxn, Iscu, Irp1, and Irp2 (polyclonal, self-made, raised from rabbits). All self-made antibodies were validated in previous studies [17,18]. The validation of the antibody against Fth is shown in Figure S1. The secondary antibodies were anti-rabbit IgG (HRP) (1:50,000; cat# 111-035-144) and anti-mouse IgG (HRP) (1:50,000; cat# 115-035-146) from Jackson ImmunoResearch Laboratories (West Grove, PA, USA). Quantification of the density of Western bands was performed with ImageJ software. Each experiment was repeated at least 3 times, independently.

2.6. Quantitative Real-Time PCR

Total RNA was isolated from tissues or cells using the RNA isolator Total RNA Extraction Reagent (Vazyme Biotech, Nanjing, China) and reverse-transcribed to cDNA by using HiScript® III RT SuperMix for qPCR (+gDNA wiper) (Vazyme Biotech, Nanjing, China), according to the manufacturer's instructions. RT-qPCR experiments were performed with ChamQ Universal SYBR qPCR Master Mix (Vazyme Biotech, Nanjing, China) in an Applied Biosystems ViiA™ 7 system. The following primers were used: for *Actin*, forward primer 5′-GCCACTGCCGCATCCTCTTC-3′ and reverse primer 5′-AGCCTCAGGGCATCGGAACC-3′; for *Epo*, forward primer 5′-AGTTGCCTTCTTGGGACTGA-3′ and reverse primer 5′-GCCACTCCTTCTGTGACTCC-3′; for *Hamp*, forward primer 5′-CTCCTGCTTCTCCTCCTTGC-3′ and reverse primer 5′-GCAATGTCTGCCCTGCTTTC-3′. *Actin* was used as a control.

2.7. Histological Staining

Peripheral blood smears were prepared by retro-orbital bleeding, then the smears were fixed in methanol for 10 min and stained with Wright-Giemsa stain (Wuhan Servicebio Technology Co., Ltd., Wuhan, China) at room temperature, following the manufacturer's protocol.

For Prussian blue staining, tissue sections underwent dewaxing and hydration steps: dewaxing for 10 min in xylene twice, hydration for 5 min in 100–50% ethanol gradient buffers, rinsing for 5 min in running water at room temperature, and incubation in 1% potassium ferrocyanide in 0.12 M HCl for 30 min to 4 h. For liver slides and bone marrow smears, before being dehydrated by gradual ethyl alcohol solutions, sections were stained in Nuclear Fast Red Staining Solution (Wuhan Servicebio Technology Co., Ltd.) for 5–10 min; for small intestine slides, sections were color-rendered by 3, 3′-diaminobenzidin (DAB) (cat. no. D5905-1007AB; Sigma-Aldrich, Shanghai, China) for 40 min, and then dehydrated according to the same procedure as above.

Images were viewed through a biological microscope (BX43, Olympus, Tokyo, Japan) and captured using a digital camera.

2.8. Ferrozine Assays

The iron content of tissues was detected by ferrozine assays, as previously described [19]. Briefly, 11 µL of concentrated HCl (11.6 M) was added to 50 µL of tissue sample lysate or serum (250 µg total protein). The samples were heated at 95 °C for 20 min and centrifuged at the highest speed for 10 min. Then, 45 µL of supernatant was removed to a new tube. Next, 18 µL of ascorbate (75 mM), 18 µL of ferrozine (10 mM), and 36 µL of saturated ammonium acetate (NH_4Ac) were sequentially added to each tube, with incubation for 2 min at room temperature between each step. The absorbance was read at 570 nm (BioTek ELx800, Shanghai, China).

2.9. ELISA

The serum EPO and interleukin-6 (IL-6) levels were quantified using specific ELISA kits according to the manufacturer's instructions (Invitrogen, Waltham, MA, USA).

2.10. Ferritin Iron Staining in Gels

In-gel ferritin iron staining was performed as described previously [22].Briefly, tissue samples of the lysate (200 µg total protein in 20 µL) were heated at 70 °C for 10 min and centrifuged at 12,000 rpm (~13,400× *g*) and 4 °C for 10 min. All supernatant was removed to a new tube and 4× loading buffer without β-mercaptoethanol (BME) and SDS was added. All samples were run in pre-cooled non-denaturing gels at 100 V for 4 h, at 4 °C. Finally, the gels were stained in Prussian blue staining buffer (1% potassium ferrocyanide in 0.12 M HCl) overnight. The data were collected by photography and quantified by ImageJ.

2.11. Statistical Analysis

All data are presented as mean ± SD. One-way ANOVA was performed using GraphPad Prism version 8.2.1 (La Jolla, CA, USA). p-value < 0.05 was considered significant.

3. Results

3.1. Hif2 Inhibition Ameliorates Anemia in Irp2 Knockout Mice

To test the role of Hif2 in *Irp2* null-induced anemia in mice, six-month-old mice were randomly divided into four groups: wild-type (WT), WT+PT–2385, *Irp2* KO, and *Irp2* KO +PT–2385. PT–2385, as a selective Hif2 inhibitor, was administered (0.4 mg/kg bw i.p. qod) for 1 month. First, by monitoring the changes in body weight pre- and post-administration, we verified that intraperitoneal administration of PT–2385 did not affect mouse growth (Figure S2A). Physiological tolerance was assessed by measuring aspartate aminotransferase, alanine aminotransferase, and bilirubin levels before and after administration (Figure S2B–D), indicating that the dose of PT–2385 was within a safe range. Following that, we confirmed the presence of small hypochromic red blood cells in peripheral blood smears of *Irp2* KO mice and found the reverse phenotype following PT–2385 administration (Figure 1A). Other hematological indexes proved the significant rescue of *Irp2* KO-induced anemia by PT–2385 treatment in terms of red blood cell count and hemoglobin level (Figure 1B,C), with one exception: mean corpuscular volume (MCV) was not reversed by PT–2385 (Figure 1D), likely because one month of treatment was not enough time to improve all phenotypes.

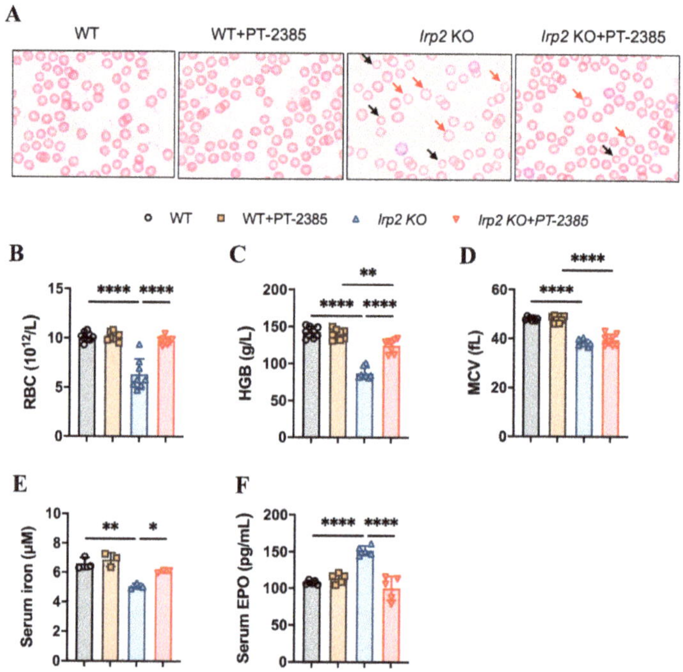

Figure 1. Hif2 inhibition by PT–2385 ameliorates anemia in *Irp2* KO mice. Six-month-old mice were divided into four groups: WT (DMSO as vehicle), WT+PT–2385 (0.4 mg/kg i.p. qod), *Irp2* KO (vehicle), and *Irp2* KO+PT–2385. (**A**) Improved hypochromic anemia by PT–2385 in *Irp2* KO mice on peripheral smear determined by Wright-Giemsa staining. Red arrows point to hypochromic red blood cells, black arrows point to small red blood cells. Routine blood examinations: (**B**) number of red blood cells (RBCs), (**C**) hemoglobin (HGB) concentration, and (**D**) mean corpuscular volume (MCV). (**E**) Serum iron content detected by ferrozine assays. (**F**) Serum EPO levels detected by enzyme-linked immunosorbent assays (ELISA). (**E**) $n = 3$, (**B–D**,**F**) $n = 5–8$. Results are shown as mean ± SD. One-way ANOVA was used for significance. * $p < 0.05$, ** $p < 0.01$, **** $p < 0.0001$.

It is known that Hif2 mediates the transcriptional activation of erythropoietin (*EPO*), a gene encoding a master hormone regulator of erythropoiesis, by binding 5′ hypoxia responsive element (HRE), thereby contributing to the changes in tissue oxygen concentration for adaptive adjustment [23]. Our previous study [19] and others [11] have found that deletion of *Irp2* in mice increases Hif2α, leading to increased EPO, but this increase may be ineffective due to limited iron content in hematopoietic tissues. We wonder whether the improvement in anemia results from the increased iron availability when Hif2 is inhibited by PT–2385. To this end, we measured serum non-heme iron and EPO levels before and after PT–2385 administration. Consistent with previous studies [11], serum iron levels in *Irp2* KO mice were significantly lower than in WT mice, while serum EPO levels were higher. However, both changes remarkably returned following treatment with PT–2385 (Figure 1E,F).

3.2. Enhanced Fe-S Cluster Biogenesis by Hif2 Inhibition Rescues Iron Deficiency to Recover Hematopoietic Function in Irp2 Knockout Mice

Bone marrow is the vital hematopoietic tissue, and iron availability is one of the main factors affecting hematopoiesis. Previous studies have shown that *Irp2* KO-induced microcytic anemia in mice is derived from insufficient iron reserves in the bone marrow [11]. We expected that the therapeutic effect of PT–2385 on anemia would be proven due to the increasing iron availability in serum (Figure 1E). Therefore, we detected the iron content in bone marrow samples by ferrozine assays and in bone marrow smears by Prussian blue staining. Consistent with previous studies [11,16], *Irp2* deletion did cause iron deprivation in bone marrow, and PT–2385 treatment significantly reversed it to WT levels (Figure 2A,B). The biochemical data from Western blotting also provided evidence that Hif2α expression was significantly increased in *Irp2* KO mice but decreased after drug treatment, and that expression of ferritin H and L subunits (Fth and Ftl) increased with PT–2385 administration; in particular, Ftl significantly increased (Figure 2C, quantified in Figure 2D), indicating the rescued iron availability for erythropoiesis.

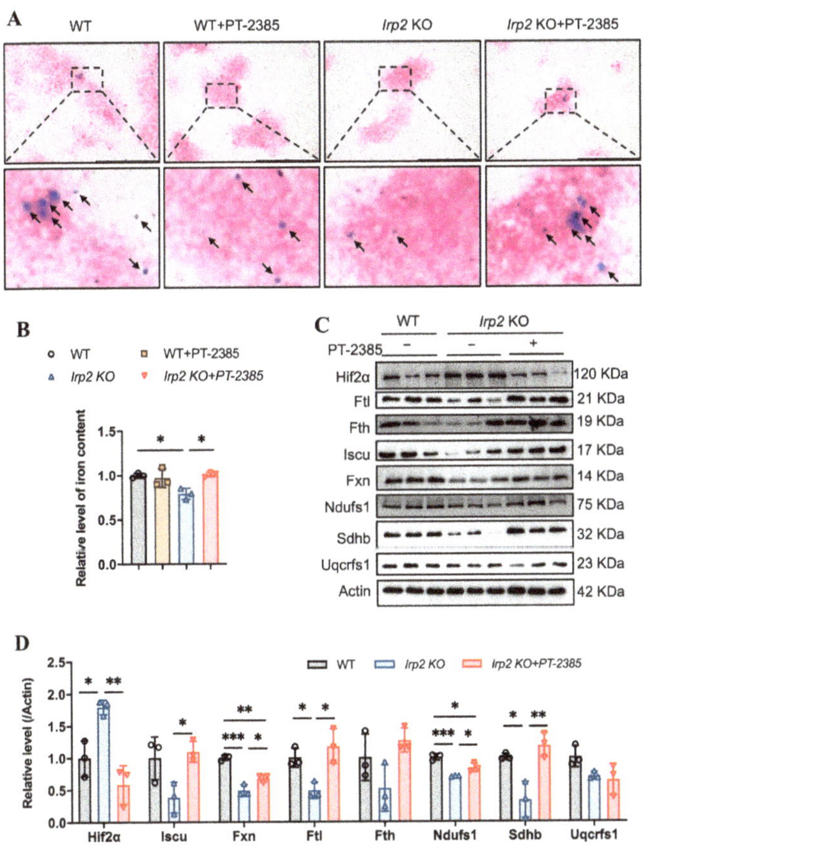

Figure 2. Hif2 inhibition increases available iron content and Fe-S cluster synthesis to maintain hematopoietic function in bone marrow. (**A**) Prussian blue staining of bone marrow smears. Arrows point to iron-stained positive areas. (**B**) Iron content in bone marrow determined by ferrozine assays. (**C**) Western blot analysis of expression of Hif2, iron-related proteins (Fth and Ftl), Fe-S cluster synthesis-related proteins (Fxn and Iscu), and respiratory chain-related proteins (Ndufs1, Sdhb, and Uqcrfs1) in bone marrow. Actin was used as an internal control. (**D**) Quantification of data in (C) to show significance; $n = 3$. Results are shown as mean ± SD. One-way ANOVA was used for significance. * $p < 0.05$, ** $p < 0.01$, *** $p < 0.001$.

Our previous studies revealed that *Irp2* KO-induced functional iron deficiency downregulated the synthesis of mitochondrial Fe-S clusters [18], which are essential for mitochondrial heme biogenesis and mitochondrial function. Then, we asked whether mitochondria could use the available iron. Therefore, we detected the expression of proteins related to Fe-S cluster synthesis and mitochondrial complex in the bone marrow. The results showed that Fe-S cluster synthesis-related proteins Fxn and Iscu and mitochondrial complex I- and II-related subunits Ndufs1 and Sdhb more or less decreased, but significantly, in *Irp2* KO mice, except complex III subunit Uqcrfs1. However, the changed proteins recovered after PT–2385 administration (Figure 2C, quantified in Figure 2D), further in favor of the notion that *Irp2* KO-induced upregulation of Hif2 negatively modulates the mitochondrial OXPHOS [18,19].

On the other hand, inhibition of Hif2 might reduce the production of EPO, which is a direct target of Hif2 [23], to suppress hematopoiesis. EPO is primarily produced in the kidney rather than the bone marrow. Therefore, we detected the Hif2α protein and *Epo* mRNA levels in the kidneys, and the results confirm that the *Irp2* KO-induced upregulation of Hif2α (Figure 3A,B) and increased mRNA of *Epo* returned to the proper level after administration of PT–2385 (Figure 3C), accompanying an improvement in anemia in *Irp2* KO mice (Figure 1). Along with the mild increase of renal iron content (Figure 3D), the proteins related to Fe-S cluster biogenesis and mitochondrial function in the kidney recovered the most, comparable to WT, after administration (Figure 3E, quantified in Figure 3F), further supporting that mitochondrial dysfunction, more specifically Fe-S cluster insufficiency, is triggered by *Irp2* KO-induced upregulation of Hif2α [17,18].

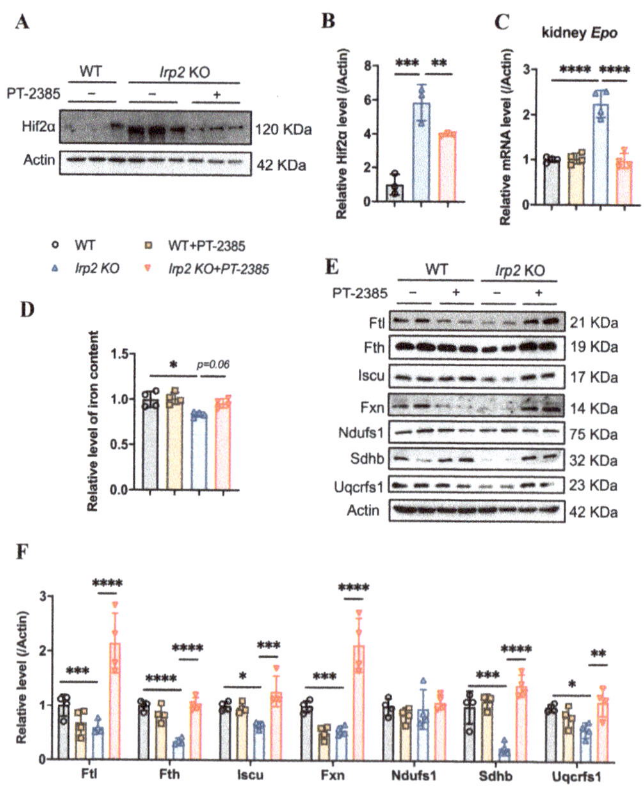

Figure 3. Hif2 inhibition reduces abnormally high EPO levels and rescues mitochondrial iron metabolism in kidneys. (**A**) Protein levels of Hif2α revealed by Western blot analysis. Actin was used as an internal control. (**B**) Quantification of Hif2α to show significance. (**C**) Relative mRNA levels of

Epo in kidney, detected by quantitative real-time PCR. (**D**) Iron content in kidneys determined by ferrozine assays. (**E**) Western blot analysis of iron-related proteins (Fth and Ftl), Fe-S cluster synthesis-related proteins (Fxn and Iscu), and respiratory chain-related proteins (Ndufs1, Sdhb, and Uqcrfs1) in kidney. Actin was used as an internal control. (**F**) Quantification of data in (**E**) to show significance; $n \geq 3$. Results are shown as mean ± SD. One-way ANOVA was used for significance. * $p < 0.05$, ** $p < 0.01$, *** $p < 0.001$, **** $p < 0.0001$.

3.3. PT–2385 Treatment Modulated Intestinal Iron Absorption and Hepatic Iron Release in Irp2 KO Mice

While iron deficiency has been found in bone marrow and kidneys in *Irp2* KO mice, iron overload has been found in the intestine and liver [16]. We wondered whether and how the available iron came from the intestine for more absorption or from the liver for more iron release. Based on this question, we first determined the changes in iron content in the small intestine. The results showed that iron accumulation was inhibited in the small intestine of *Irp2* KO mice after PT–2385 treatment, as demonstrated by decreased total iron content (Figure 4A), fewer iron-stained positive areas (Figure 4B), and reduced ferritin expression (Figure 4C). This inhibition was correlated with mildly increased intestinal Fpn1 (Figure 4C, quantified in Figure 4D) and remarkably decreased liver hepcidin (*Hamp*) expression (Figure 4E). Liver is the primary site of synthesis and secretion of hepcidin. Its expression is regulated by inflammation and iron, allowing us to detect the expression of inflammatory factor IL-6. The ELISA results showed an increase in the level of serum IL-6 in *Irp2* KO mice, which decreased after PT–2385 administration (Figure 4F). Other inflammatory markers (interleukin-1β and C-reactive protein) were not found to be changed in *Irp2* KO mice.

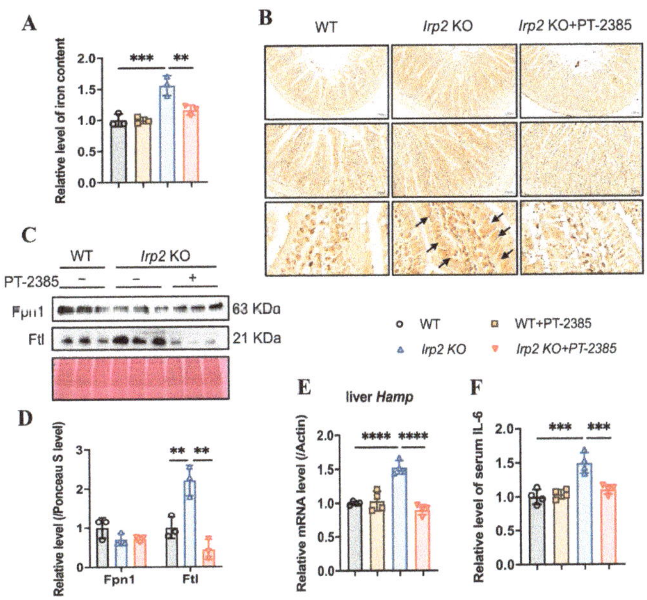

Figure 4. Abnormal iron accumulation is suppressed in the small intestine and hepcidin levels are downregulated in the liver of *Irp2* KO mice by PT–2385 treatment. (**A**) Iron content in the small intestine determined by ferrozine assays. (**B**) Representative images of DAB-enhanced Prussian blue staining for iron in the small intestine. Arrows indicate iron-stained epithelial cells. (**C**) Protein levels of Fpn1 and Ftl revealed by Western blot analysis. Ponceau S stain was used as an internal control. (**D**) Quantification of (**C**) to show significance. (**E**) Relative mRNA levels of Hepcidin in the liver, detected by RT-PCR. (**F**) Relative serum IL-6 levels, detected by ELISA. $n \geq 3$. Results are shown as mean ± SD. One-way ANOVA was used for significance. ** $p < 0.01$, *** $p < 0.001$, **** $p < 0.0001$.

Hepcidin regulates iron efflux by mediating Fpn1 endocytosis and ensuring tha hepatic iron can be mobilized to maintain iron homeostasis [24]. We expected that th iron accumulation would be alleviated due to the diminished *Hamp*. First, we detecte the effect of PT–2385 on Hif2α expression. As in other tissues, Hif2α was induced by *irp* ablation and reduced by PT–2385 treatment, and the Hif2α-targeted *Glut*1 gene showe a similar pattern (Figure 5A,B). Indeed, we found that PT–2385 could effectively reduc the accumulation of hepatic iron caused by *Irp2* deletion, verified by ferrozine assays an Prussian blue iron staining (Figure 5C,D). Biochemically, iron-related proteins Fpn1, Fth and Ftl all changed to present reduced iron accumulation (Figure 5E,F).

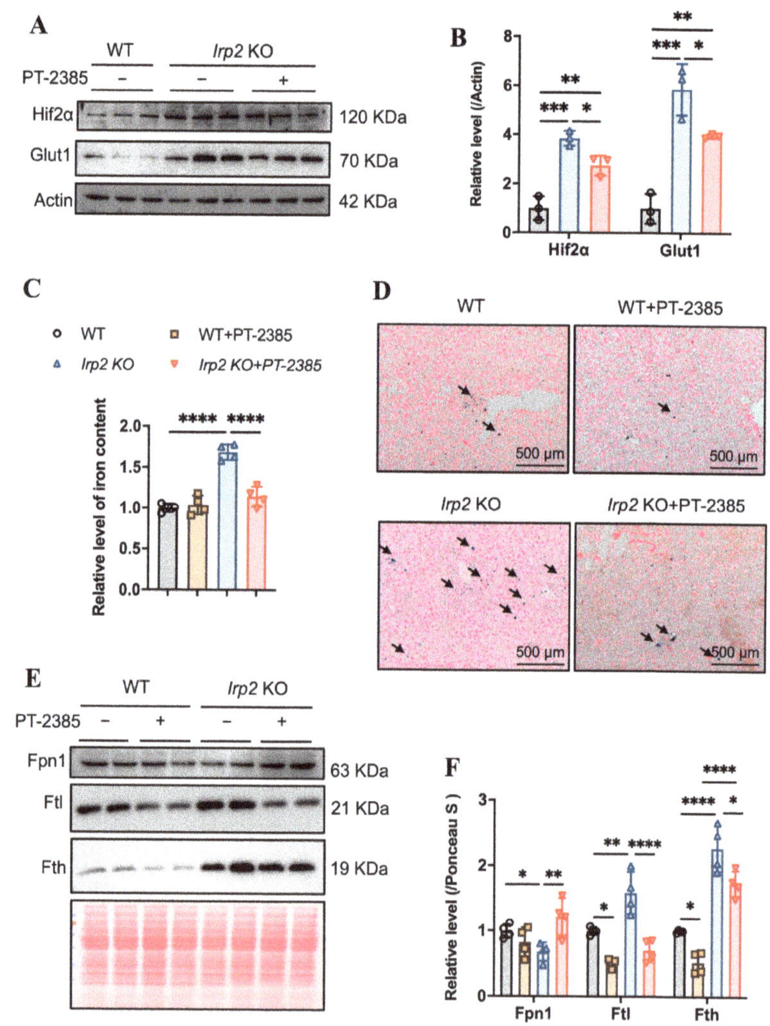

Figure 5. PT–2385 treatment inhibits iron accumulation in the liver. (**A**) Protein levels of Hif2α anc Glut1 revealed by Western blot analysis. Actin was used as internal control. (**B**) Quantification o (**A**) to show significance. (**C**) Ferrozine-based iron content in the liver. (**D**) Prussian blue staining o liver sections. Arrows indicate stained iron. (**E**) Iron-related proteins (Fpn1, Fth, and Ftl) detected by Western blot analysis. Ponceau S stain was used as an internal control. (**F**) Quantification o (**E**) to show significance. $n \geq 3$. Results are shown as mean ± SD. One-way ANOVA was used fo significance. * $p < 0.05$, ** $p < 0.01$, *** $p < 0.001$, **** $p < 0.0001$.

Collectively, these results indicate that PT–2385 administration could enhance iron absorption in the gut and iron release from the liver to rebuild iron homeostasis and proper hematopoiesis in *Irp2* KO mice.

3.4. Hif2 Inhibition Reconstitutes Tissue Iron Homeostasis by Initiating Ncoa4-Mediated Ferritinophagy in Liver

We wondered what mediates the remobilization of iron from the liver after Hif2 inhibition. In addition to maintaining systemic iron homeostasis through the hepcidin-FPN1 axis and cellular iron homeostasis through the IRP-IRE system, another regulatory strategy has been revealed in some tissues or cells, such as red blood cells and macrophages, as Ncoa4 mediates ferritinophagy, a process that maintains intracellular iron homeostasis to sustain erythropoiesis [25,26].

Considering that Ncoa4 is a critical cargo receptor for autophagic degradation of ferritin and the subsequent release of iron, we speculated that PT–2385 treatment could induce Ncoa4-mediated ferritinophagy based on the decline in ferritin. Therefore, we examined the expression level of Ncoa4. It turned out that Ncoa4 expression increased (Figure 6A, quantified in Figure 6B), in agreement with the mitigated iron accumulation and lower ferritin levels in the liver, comparable to WT mice (Figure 5C,E). Subsequently, we examined the autophagic elongation component Atg5, the autophagic cargo selection component Lc3, and the classical autophagy receptor p62. Atg5 forms a constitutive complex with Atg12 to promote the transfer of Lc3-I to phosphatidylethanolamine (PE). Lc3-I is then lipidated and converted to Lc3-II by ATG4 during autophagy activation, and p62 engages with autophagic substrates and carries them to autophagosomes for degradation (see review in [27]). As a result, the Lc3-II/Lc3-I ratio and decreased p62 levels are widely used as markers of autophagy activation. Consistent with the increased Ncoa4 levels, the expression of Atg5 significantly increased, the Lc3-II/Lc3-I ratio slightly increased, and p62 levels decreased in *Irp2* KO mice after treatment with PT–2385 compared with the vehicle group (Figure 6C,D), indicating an increase in the autophagic process in the liver. Autophagy is a major intracellular system that derives its degradative abilities from the lysosome. We then detected the lysosomal-associated membrane protein 1 (Lamp1). The results showed that Lamp1 protein levels were significantly lower in *Irp2* KO mice, and were increased after treatment, comparable to WT levels (Figure 6E,F). This suggests that *Irp2* deletion may have a direct effect on lysosomal biogenesis and function and, ultimately, autophagic flow, explaining why ferritinophagy efficiency is reduced in irp2 KO mice despite a decrease in p62 levels. In addition, to measure loaded iron in ferritin, we stained the iron by in-gel assays [22] and found significantly less iron in the liver after PT–2385 treatment (Figure 6G,H), correlating with the ferritin levels (Figure 6A).

In summary, Hif2 inhibition by PT–2385 stimulated the release of iron from hepatic ferritin storage by commencing Ncoa4-mediated ferritinophagy, and ultimately corrected the *Irp2* KO-induced aberrant iron distribution in tissues for appropriate erythropoiesis.

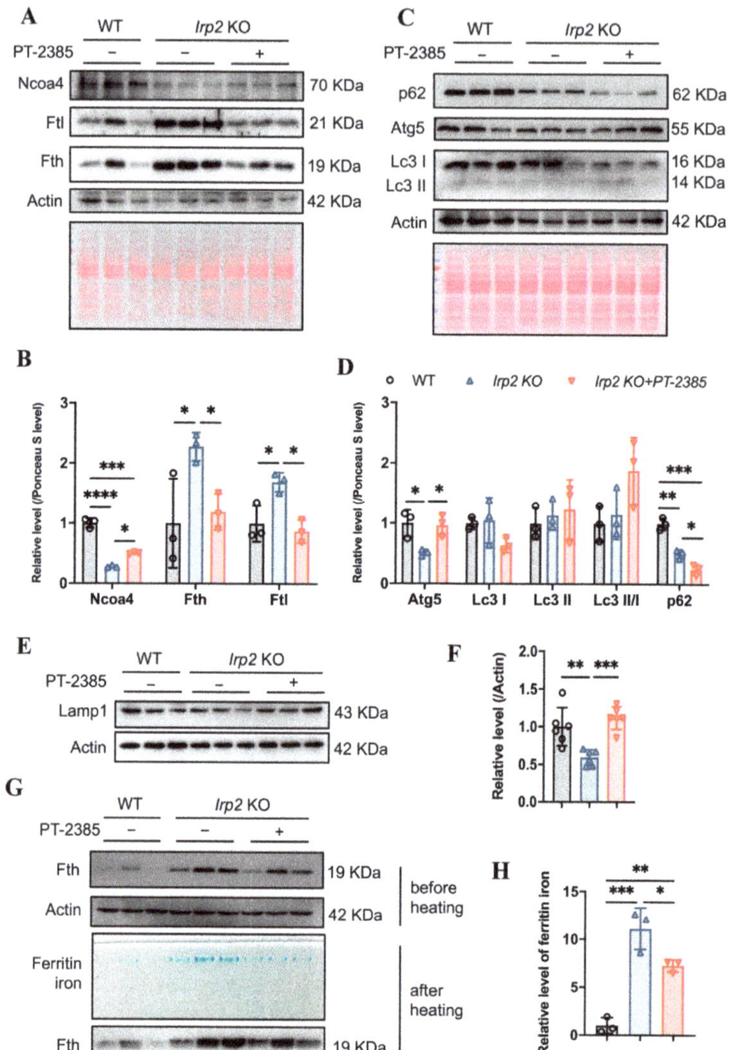

Figure 6. Hif2 inhibition stimulates Ncoa4-mediated ferritinophagy for iron release from ferritin in *Irp2* KO liver. (**A**) Expression of Ncoa4 and iron-related proteins (Fth and Ftl) after PT–2385 treatment in liver of *Irp2* KO and WT mice. Actin and Ponceau S stain were used as internal controls. (**B**) Quantification of (**A**) to show significance. (**C**) Expression of autophagy-related proteins (Atg5, Lc3, and p62). Actin and Ponceau S stain were used as internal controls. (**D**) Quantification of (**C**) to show significance of Atg5, Lc3, and p62 changes. (**E**) Lamp1 expression. Actin was used as an internal control. (**F**) Quantification of Lamp1 to show significance. (**G**,**H**) Result of ferritin iron staining in gels. $n \geq 3$. Results are shown as mean ± SD. One-way ANOVA was used for significance. * $p < 0.05$, ** $p < 0.01$, *** $p < 0.001$, **** $p < 0.0001$.

4. Discussion

Here, we observed that inhibition of Hif2 by PT–2385 could alleviate *Irp2* depletion-induced anemia (Figure 7). Mechanically, it was revealed that upregulation of Ncoa4 initiated ferritinophagy in the liver to release iron into the serum for erythropoiesis in bone marrow. Concurrent with the reduced liver iron, liver *Hamp* expression decreased, which systemically affected iron absorption and recycling. Liver Fpn1 was found to be significantly upregulated though intestinal Fpn1 only slightly, and it is expected that the efficiency of iron transportation was increased in the intestine and liver due to the reduced hepcidin after PT–2385 administration. Therefore, anemia is cured, and aberrant iron distribution is corrected (Figure 7).

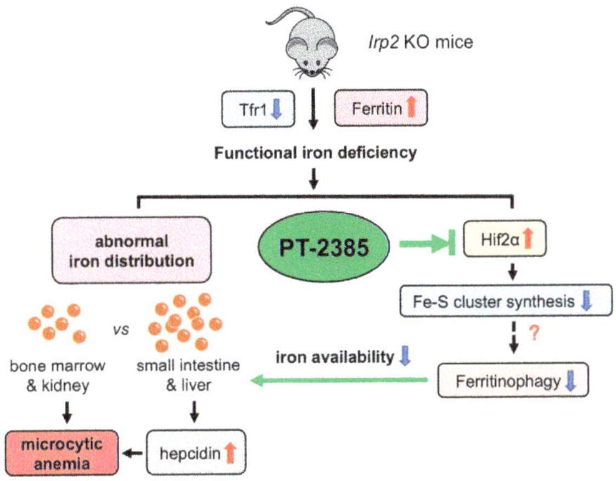

Figure 7. Working model of the role of *Irp2* KO-induced upregulation of Hif2 in microcytic anemia. *Irp2* deficiency causes functional iron deprivation by reducing cellular iron uptake (downregulated transferrin receptor 1, Tfr1) and increasing cellular iron storage capacity (upregulated ferritin), eventually leading to the development of microcytic anemia symptoms in *Irp2* KO mice. Iron accumulation was observed in the intestine and liver, but iron deficiency was seen in erythropoiesis-related organs such as bone marrow and kidney. At the molecular level, *Irp2* deficiency increases Hif2α expression, interferes with Fe-S cluster synthesis, and may affect ferritinophagy, resulting in decreased iron availability. Furthermore, elevated levels of hepcidin secreted by the liver aggravate the abnormal iron distribution. Hif2α inhibition by PT–2385 alleviated symptoms of anemia in *Irp2* KO mice by increasing ferritinophagy levels.

Dietary iron acquisition in mammals is mediated by intestinal absorption. In the proximal gut, ferric iron (Fe^{3+}) is reduced to ferrous iron (Fe^{2+}) by duodenal cytochrome b (Dcytb), a ferric reductase [28], and then transported across the apical membrane by divalent metal transporter 1 (Dmt1) [29]. According to the body's need for iron, iron can be either kept in ferritin or exported via the only known iron exporter, Fpn1, over the basolateral membrane and into the circulation [30]. Three major intestinal proteins (Dcytb, Dmt1, and Fpn1) have been identified as direct targets of HIF2α, and oral administration of HIF2α inhibitors (1,3-diaminopropane, DAP, and reuterin) significantly suppressed iron uptake via the intestine, implying that HIF2α is required to maintain intestinal iron absorption [31]. Interestingly, in our study, when *Irp2* KO mice were administered the HIF2α inhibitor PT–2385 intraperitoneally, there was no significant change in body weight. In addition, no significant change of intestinal Fpn1 expression was observed, while the excessive iron accumulation in the small intestine was significantly alleviated after Hif2α inhibition. This discrepancy suggests efficient iron transport via Fpn1 over the basolateral membrane and, to some extent, intraperitoneally administrated PT–2385 being bypassed from the gut. How *Irp2* ablation induces Hif2α upregulation in the liver even though iron accumulation was observed there remains a question. This phenomenon was observed previously in a study showing that the Hif2α protein level remained constantly high even under iron repletion conditions in *Irp2*-ablated cells [18]. Probably, the iron- and oxygen-dependent degradation process is disturbed when *Irp2* is depleted. These results also support the finding that besides IRP1 [32–34], IRP2 also regulates Hif2α via an unknown mechanism.

As a primary circulating regulator of iron, hepcidin regulates three significant iro fluxes into plasma: dietary iron absorption in the intestine, iron recycling by macrophag phagocytosis, and iron mobilization from hepatocytes (see review in [35]). A recent stud showed that when erythropoiesis intensifies, EPO increases erythroferrone (ERFE) produ tion in the bone marrow and spleen erythroblasts in a JAK2/STAT5-dependent manne and ERFE is secreted into the bloodstream and acts directly on the liver to suppres hepcidin [36]. Even though EPO levels are significantly increased in *Irp2* KO mice, hep cidin is not suppressed. On the contrary, hepcidin is expressed more than in wild-typ mice, suggesting that the increase in EPO is invalid [11] and is the consequence of iro deficiency in the bone marrow. In addition, inflammatory signals in the body may induc hepcidin expression. For instance, interleukin-6 upregulates hepcidin expression in viv through the IL-6/JAK2/STAT3 pathway [37], and other cytokines, such as IL-1β, that rel on BMP/SMAD signaling also increase hepcidin expression [38]. In this study and ou previous study [19], we found an increase in serum IL-6, not IL-1β, in *Irp2* KO mice and decrease in IL-6 after PT–2385 administration. The decreased Fpn1 in the intestine and live of *Irp2* KO mice and the increase in the liver after Hif2 inhibition confirm that hepcidin is master regulator of Fpn1, though *Fpn1* is a member of the Hif2 regulons.

Iron mobilization from hepatocytes through Fpn1, we think, plays a very importan role in correcting the aberrant iron distribution in *Irp2* KO mice. Iron, if not utilized o exported, is stored in ferritin nanocages consisting of 24 subunits of FTH1 and FTL pe cage. The cage is able to hold up to 4500 iron atoms [1]. What causes iron to release from ferritin after Hif2 inhibition? Ferritin is regulated at least at three levels, transcriptionall by NFκB [39], post-transcriptionally by the IRE-IRP system [40], and post-translationall by Ncoa4-mediated ferritinophagy [41]. In this study, we found a decrease in the level o ferritin protein, not mRNA, in *Irp2* KO mice after PT–2385 administration, which led u to hypothesize that PT–2385 treatment activates Ncoa4-mediated ferritinophagy. Ncoa interacts with the ferritin heavy chain and mediates ferritin transport to the lysosome vi the autophagosome for degradation, which is essential for maintaining iron homeostasis especially in erythrocytes [26,42]. In agreement with these findings, PT–2385 treatmen increased Ncoa4 expression and autophagic flux in the liver, which led to a massive efflu of hepatic stored iron via Fpn1 and, consequently, to circulation for iron redistribution and erythropoiesis in bone marrow. Though Ncoa4 is regulated by Hif at the transcriptiona level and mediates the mobilization of murine hepatic iron stores after blood loss [43], we found that inhibition of Hif2 increased Ncoa4 expression, suggesting post-transcriptional regulation of Ncoa4. Very recently, Ncoa4 has been demonstrated to be selectively tar geted for the management of iron overload disorders [44]. This explains our finding that increased Ncoa4 alleviated liver iron overload to improve systemic iron homeostasis. In addition, Ncoa4 has been revealed to be regulated by iron-dependent HERC2-mediated proteolysis [42]. We wonder if a dependence on Fe-S clusters, besides iron, is also possible because Hif2 inhibition by PT–2385 shifts cellular energy metabolism from glycolysis to oxidative phosphorylation by enhancing Fe-S cluster biogenesis and mitochondrial func tion to improve *Irp2* KO-induced neurodegeneration [18,19]. This possibility needs to be further investigated.

Overall, we validated the role of Hif2 in *Irp2* KO-induced anemia. Combining our find ings and others' previous findings, we have highlighted that Hif2α-Ncoa4 is a complex axis that contributes to iron metabolic disorders, including neurodegeneration, iron-overload disorder, and anemia.

Supplementary Materials: The following supporting information can be downloaded at https //www.mdpi.com/article/10.3390/antiox12030566/s1: Figure S1. The validation of self-made antibody against Fth. Figure S2. Body weight and blood biochemical indices.

Author Contributions: Conceptualization, Y.L. (Yutong Liu), Y.L. (Yuxuan Li) and K.L.; methodology Y.L. (Yutong Liu), Y.L. (Yuxuan Li) and H.Z.; software, Y.L. (Yutong Liu), Y.L. (Yuxuan Li) and L.Y. validation, Y.L. (Yutong Liu) and L.Y.; formal analysis, Y.L. (Yutong Liu), H.Z. and Y.L. (Yuxuan Li) investigation, Y.L. (Yutong Liu), Y.L. (Yuxuan Li), H.Z., L.Y. (Liu Yang) and J.S.; resources, K.L. and Y.C.; data curation, K.L. and Y.L. (Yutong Liu); writing—original draft preparation, Y.L. (Yutong Liu) and Y.L. (Yuxuan Li); writing—review and editing, Y.L. (Yutong Liu), W.D., T.Q. and K.L.; visualization, K.L.; supervision, K.L. and Y.C.; project administration, K.L.; funding acquisition, K.L. and Y.L. (Yutong Liu). All authors have read and agreed to the published version of the manuscript.

Funding: This research was funded by the Natural Science Foundation of China, grant number 32271221.

Institutional Review Board Statement: The animal study protocol was approved by the Laboratory Animal Welfare and Ethical Committee of Nanjing University (approval number: IACUC-2003118, approval date: 17 March 2020).

Informed Consent Statement: Not applicable.

Data Availability Statement: Data is contained within the article and supplementary material.

Acknowledgments: Y.L. (Yutong Liu) is partially supported by the 2021 Graduate Research and Practice Innovation Program at Medical School, Nanjing University.

Conflicts of Interest: The authors declare no conflict of interest.

References

1. Muckenthaler, M.U.; Rivella, S.; Hentze, M.W.; Galy, B. A Red Carpet for Iron Metabolism. *Cell* **2017**, *168*, 344–361. [CrossRef] [PubMed]
2. Maio, N.; Rouault, T.A. Outlining the Complex Pathway of Mammalian Fe-S Cluster Biogenesis. *Trends Biochem. Sci.* **2020**, *45*, 411–426. [CrossRef] [PubMed]
3. Paul, V.D.; Lill, R. Biogenesis of cytosolic and nuclear iron-sulfur proteins and their role in genome stability. *Biochim. Biophys. Acta* **2015**, *1853*, 1528–1539. [CrossRef] [PubMed]
4. Andrews, N.C.; Schmidt, P.J. Iron homeostasis. *Annu. Rev. Physiol.* **2007**, *69*, 69–85. [CrossRef] [PubMed]
5. Rouault, T.A. The role of iron regulatory proteins in mammalian iron homeostasis and disease. *Nat. Chem. Biol.* **2006**, *2*, 406–414. [CrossRef] [PubMed]
6. Hentze, M.W.; Muckenthaler, M.U.; Galy, B.; Camaschella, C. Two to tango: Regulation of Mammalian iron metabolism. *Cell* **2010**, *142*, 24–38. [CrossRef] [PubMed]
7. Wallander, M.L.; Leibold, E.A.; Eisenstein, R.S. Molecular control of vertebrate iron homeostasis by iron regulatory proteins. *Biochim. Biophys. Acta* **2006**, *1763*, 668–689. [CrossRef]
8. Salahudeen, A.A.; Thompson, J.W.; Ruiz, J.C.; Ma, H.W.; Kinch, L.N.; Li, Q.; Grishin, N.V.; Bruick, R.K. An E3 ligase possessing an iron-responsive hemerythrin domain is a regulator of iron homeostasis. *Science* **2009**, *326*, 722–726. [CrossRef]
9. Vashisht, A.A.; Zumbrennen, K.B.; Huang, X.; Powers, D.N.; Durazo, A.; Sun, D.; Bhaskaran, N.; Persson, A.; Uhlen, M.; Sangfelt, O.; et al. Control of iron homeostasis by an iron-regulated ubiquitin ligase. *Science* **2009**, *326*, 718–721. [CrossRef]
10. Wang, H.; Shi, H.; Rajan, M.; Canarie, E.R.; Hong, S.; Simoneschi, D.; Pagano, M.; Bush, M.F.; Stoll, S.; Leibold, E.A.; et al. FBXL5 Regulates IRP2 Stability in Iron Homeostasis via an Oxygen-Responsive [2Fe2S] Cluster. *Mol. Cell* **2020**, *78*, 31–41.e5. [CrossRef]
11. Cooperman, S.S.; Meyron-Holtz, E.G.; Olivierre-Wilson, H.; Ghosh, M.C.; McConnell, J.P.; Rouault, T.A. Microcytic anemia, erythropoietic protoporphyria, and neurodegeneration in mice with targeted deletion of iron-regulatory protein 2. *Blood* **2005**, *106*, 1084–1091. [CrossRef]
12. LaVaute, T.; Smith, S.; Cooperman, S.; Iwai, K.; Land, W.; Meyron-Holtz, E.; Drake, S.K.; Miller, G.; Abu-Asab, M.; Tsokos, M.; et al. Targeted deletion of the gene encoding iron regulatory protein-2 causes misregulation of iron metabolism and neurodegenerative disease in mice. *Nat. Genet.* **2001**, *27*, 209–214. [CrossRef]
13. Zhou, Y.; Wu, W.; Xu, Z.; Liu, Y.; Chang, H.; Yu, P.; Zhang, X.; Yang, Y.; Liu, K.; Chang, Y. Iron regulatory protein 2 deficiency may correlate with insulin resistance. *Biochem. Biophys. Res. Commun.* **2019**, *510*, 191–197. [CrossRef]
14. Santos, M.; Anderson, C.P.; Neschen, S.; Zumbrennen-Bullough, K.B.; Romney, S.J.; Kahle-Stephan, M.; Rathkolb, B.; Gailus-Durner, V.; Fuchs, H.; Wolf, E.; et al. Irp2 regulates insulin production through iron-mediated Cdkal1-catalyzed tRNA modification. *Nat. Commun.* **2020**, *11*, 296. [CrossRef] [PubMed]
15. Costain, G.; Ghosh, M.C.; Maio, N.; Carnevale, A.; Si, Y.C.; Rouault, T.A.; Yoon, G. Absence of iron-responsive element-binding protein 2 causes a novel neurodegenerative syndrome. *Brain* **2019**, *142*, 1195–1202. [CrossRef] [PubMed]
16. Galy, B.; Ferring, D.; Minana, B.; Bell, O.; Janser, H.G.; Muckenthaler, M.; Schumann, K.; Hentze, M.W. Altered body iron distribution and microcytosis in mice deficient in iron regulatory protein 2 (IRP2). *Blood* **2005**, *106*, 2580–2589. [CrossRef]
17. Li, H.; Zhao, H.; Hao, S.; Shang, L.; Wu, J.; Song, C.; Meyron-Holtz, E.G.; Qiao, T.; Li, K. Iron regulatory protein deficiency compromises mitochondrial function in murine embryonic fibroblasts. *Sci. Rep.* **2018**, *8*, 5118. [CrossRef] [PubMed]
18. Li, H.; Liu, Y.; Shang, L.; Cai, J.; Wu, J.; Zhang, W.; Pu, X.; Dong, W.; Qiao, T.; Li, K. Iron regulatory protein 2 modulates the switch from aerobic glycolysis to oxidative phosphorylation in mouse embryonic fibroblasts. *Proc. Natl. Acad. Sci. USA* **2019**, *116*, 9871–9876. [CrossRef]
19. Shen, J.; Xu, L.; Li, Y.; Dong, W.; Cai, J.; Liu, Y.; Zhao, H.; Xu, T.; Holtz, E.M.; Chang, Y.; et al. Protective Effects of Hif2 Inhibitor PT-2385 on a Neurological Disorder Induced by Deficiency of Irp2. *Front. Neurosci.* **2021**, *15*, 715222. [CrossRef]
20. Liu, G.; Sil, D.; Maio, N.; Tong, W.H.; Bollinger, J.M., Jr.; Krebs, C.; Rouault, T.A. Heme biosynthesis depends on previously unrecognized acquisition of iron-sulfur cofactors in human amino-levulinic acid dehydratase. *Nat. Commun.* **2020**, *11*, 6310. [CrossRef]

21. Dailey, H.A.; Finnegan, M.G.; Johnson, M.K. Human ferrochelatase is an iron-sulfur protein. *Biochemistry* **1994**, *33*, 403–407 [CrossRef] [PubMed]
22. Tang, X.; Zhou, B. Ferritin is the key to dietary iron absorption and tissue iron detoxification in Drosophila melanogaster. *FASEB J* **2013**, *27*, 288–298. [CrossRef] [PubMed]
23. Storti, F.; Santambrogio, S.; Crowther, L.M.; Otto, T.; Abreu-Rodriguez, I.; Kaufmann, M.; Hu, C.J.; Dame, C.; Fandrey, J.; Wenger R.H.; et al. A novel distal upstream hypoxia response element regulating oxygen-dependent erythropoietin gene expression *Haematologica* **2014**, *99*, e45–e48. [CrossRef] [PubMed]
24. Nemeth, E.; Tuttle, M.S.; Powelson, J.; Vaughn, M.B.; Donovan, A.; Ward, D.M.; Ganz, T.; Kaplan, J. Hepcidin regulates cellular iron efflux by binding to ferroportin and inducing its internalization. *Science* **2004**, *306*, 2090–2093. [CrossRef]
25. Nai, A.; Lidonnici, M.R.; Federico, G.; Pettinato, M.; Olivari, V.; Carrillo, F.; Geninatti Crich, S.; Ferrari, G.; Camaschella, C. Silvestri, L.; et al. NCOA4-mediated ferritinophagy in macrophages is crucial to sustain erythropoiesis in mice. *Haematologica* **2021**, *106*, 795–805. [CrossRef] [PubMed]
26. Mancias, J.D.; Wang, X.; Gygi, S.P.; Harper, J.W.; Kimmelman, A.C. Quantitative proteomics identifies NCOA4 as the cargo receptor mediating ferritinophagy. *Nature* **2014**, *509*, 105–109. [CrossRef]
27. Galluzzi, L.; Green, D.R. Autophagy-Independent Functions of the Autophagy Machinery. *Cell* **2019**, *177*, 1682–1699. [CrossRef]
28. McKie, A.T.; Barrow, D.; Latunde-Dada, G.O.; Rolfs, A.; Sager, G.; Mudaly, E.; Mudaly, M.; Richardson, C.; Barlow, D.; Bomford, A.; et al. An iron-regulated ferric reductase associated with the absorption of dietary iron. *Science* **2001**, *291*, 1755–1759. [CrossRef]
29. Gunshin, H.; Mackenzie, B.; Berger, U.V.; Gunshin, Y.; Romero, M.F.; Boron, W.F.; Nussberger, S.; Gollan, J.L.; Hediger, M.A Cloning and characterization of a mammalian proton-coupled metal-ion transporter. *Nature* **1997**, *388*, 482–488. [CrossRef]
30. Abboud, S.; Haile, D.J. A novel mammalian iron-regulated protein involved in intracellular iron metabolism. *J. Biol. Chem.* **2000**, *275*, 19906–19912. [CrossRef]
31. Das, N.K.; Schwartz, A.J.; Barthel, G.; Inohara, N.; Liu, Q.; Sankar, A.; Hill, D.R.; Ma, X.; Lamberg, O.; Schnizlein, M.K.; et al. Microbial Metabolite Signaling Is Required for Systemic Iron Homeostasis. *Cell Metab.* **2020**, *31*, 115–130.e6. [CrossRef]
32. Anderson, S.A.; Nizzi, C.P.; Chang, Y.I.; Deck, K.M.; Schmidt, P.J.; Galy, B.; Damnernsawad, A.; Broman, A.T.; Kendziorski, C.; Hentze, M.W.; et al. The IRP1-HIF-2alpha axis coordinates iron and oxygen sensing with erythropoiesis and iron absorption. *Cell Metab.* **2013**, *17*, 282–290. [CrossRef]
33. Wilkinson, N.; Pantopoulos, K. IRP1 regulates erythropoiesis and systemic iron homeostasis by controlling HIF2alpha mRNA translation. *Blood* **2013**, *122*, 1658–1668. [CrossRef] [PubMed]
34. Ghosh, M.C.; Zhang, D.L.; Jeong, S.Y.; Kovtunovych, G.; Ollivierre-Wilson, H.; Noguchi, A.; Tu, T.; Senecal, T.; Robinson, G.; Crooks, D.R.; et al. Deletion of iron regulatory protein 1 causes polycythemia and pulmonary hypertension in mice through translational derepression of HIF2alpha. *Cell Metab.* **2013**, *17*, 271–281. [CrossRef] [PubMed]
35. Nemeth, E.; Ganz, T. Hepcidin and Iron in Health and Disease. *Annu. Rev. Med.* **2022**, *74*, 261–277. [CrossRef] [PubMed]
36. Kautz, L.; Jung, G.; Valore, E.V.; Rivella, S.; Nemeth, E.; Ganz, T. Identification of erythroferrone as an erythroid regulator of iron metabolism. *Nat. Genet.* **2014**, *46*, 678–684. [CrossRef] [PubMed]
37. Nemeth, E.; Rivera, S.; Gabayan, V.; Keller, C.; Taudorf, S.; Pedersen, B.K.; Ganz, T. IL-6 mediates hypoferremia of inflammation by inducing the synthesis of the iron regulatory hormone hepcidin. *J. Clin. Investig.* **2004**, *113*, 1271–1276. [CrossRef]
38. Shanmugam, N.K.; Chen, K.; Cherayil, B.J. Commensal Bacteria-induced Interleukin 1beta (IL-1beta) Secreted by Macrophages Up-regulates Hepcidin Expression in Hepatocytes by Activating the Bone Morphogenetic Protein Signaling Pathway. *J. Biol. Chem.* **2015**, *290*, 30637–30647. [CrossRef]
39. Pham, C.G.; Bubici, C.; Zazzeroni, F.; Papa, S.; Jones, J.; Alvarez, K.; Jayawardena, S.; De Smaele, E.; Cong, R.; Beaumont, C.; et al. Ferritin heavy chain upregulation by NF-kappaB inhibits TNFalpha-induced apoptosis by suppressing reactive oxygen species. *Cell* **2004**, *119*, 529–542. [CrossRef]
40. Yanatori, I.; Richardson, D.R.; Dhekne, H.S.; Toyokuni, S.; Kishi, F. CD63 is regulated by iron via the IRE-IRP system and is important for ferritin secretion by extracellular vesicles. *Blood* **2021**, *138*, 1490–1503. [CrossRef]
41. Kuno, S.; Fujita, H.; Tanaka, Y.K.; Ogra, Y.; Iwai, K. Iron-induced NCOA4 condensation regulates ferritin fate and iron homeostasis. *EMBO Rep.* **2022**, *23*, e54278. [CrossRef] [PubMed]
42. Mancias, J.D.; Pontano Vaites, L.; Nissim, S.; Biancur, D.E.; Kim, A.J.; Wang, X.; Liu, Y.; Goessling, W.; Kimmelman, A.C.; Harper, J.W. Ferritinophagy via NCOA4 is required for erythropoiesis and is regulated by iron dependent HERC2-mediated proteolysis. *Elife* **2015**, *4*, e10308. [CrossRef]
43. Li, X.; Lozovatsky, L.; Sukumaran, A.; Gonzalez, L.; Jain, A.; Liu, D.; Ayala-Lopez, N.; Finberg, K.E. NCOA4 is regulated by HIF and mediates mobilization of murine hepatic iron stores after blood loss. *Blood* **2020**, *136*, 2691–2702. [CrossRef] [PubMed]
44. Das, N.K.; Jain, C.; Sankar, A.; Schwartz, A.J.; Santana-Codina, N.; Solanki, S.; Zhang, Z.; Ma, X.; Parimi, S.; Rui, L.; et al. Modulation of the HIF2alpha-NCOA4 axis in enterocytes attenuates iron loading in a mouse model of hemochromatosis. *Blood* **2022**, *139*, 2547–2552. [CrossRef] [PubMed]

antioxidants

MDPI

rticle

Mitochondrial-Targeted Antioxidant MitoQ-Mediated Autophagy: A Novel Strategy for Precise Radiation Protection

ingting Bao [1,2,3,4,5], Xiongxiong Liu [1,3,4,5], Qingfeng Wu [1], Fei Ye [1,3,4,5], Zheng Shi [1,2,3,4,5], Dan Xu [1],
nhua Zhang [1,2,3,4,5], Zhihui Dou [1,2,3,4,5], Guomin Huang [1,2,3,4,5], Hong Zhang [1,2,3,4,5,*] and Chao Sun [1,2,3,4,5,*]

1 Department of Medical Physics, Institute of Modern Physics, Chinese Academy of Sciences, Lanzhou 730000, China
2 Advanced Energy Science and Technology Guangdong Laboratory, Huizhou 516000, China
3 Key Laboratory of Heavy Ion Radiation Medicine of Gansu Province, Lanzhou 730000, China
4 Key Laboratory of Heavy Ion Radiation Biology and Medicine of Chinese Academy of Sciences, Lanzhou 730000, China
5 College of Life Sciences, University of Chinese Academy of Sciences, Beijing 101408, China
* Correspondence: zhangh@impcas.ac.cn (H.Z.); sunchao@impcas.ac.cn (C.S.); Tel.: +86-(931)-519-6126 (H.Z.); +86-(931)-519-6027 (C.S.)

Abstract: Radiotherapy (RT) is one of the most effective cancer treatments. However, successful radiation protection for normal tissue is a clinical challenge. Our previous study observed that MitoQ, a mitochondria-targeted antioxidant, was adsorbed to the inner mitochondrial membrane and remained the cationic moiety in the intermembrane space. The positive charges in MitoQ restrained the activity of respiratory chain complexes and decreased proton production. Therefore, a pseudo-mitochondrial membrane potential (PMMP) was developed via maintenance of exogenous positive charges. This study identified that PMMP constructed by MitoQ could effectively inhibit mitochondrial respiration within normal cells, disrupt energy metabolism, and activate adenosine 5′-monophosphate (AMP)-activated protein kinase (AMPK) signaling to induce autophagy. As such, it could not lead to starvation-induced autophagy among tumor cells due to the different energy phenotypes between normal and tumor cells (normal cells depend on mitochondrial respiration for energy supply, while tumor cells rely on aerobic glycolysis). Therefore, we successfully protected the normal cells from radiation-induced damage without affecting the tumor-killing efficacy of radiation by utilizing selective autophagy. MitoQ-constructed PMMP provides a new therapeutic strategy for specific radiation protection.

Keywords: MitoQ; PMMP; autophagy; radioprotection; energy phenotype

itation: Bao, X.; Liu, X.; Wu, Q.; Ye, ; Shi, Z.; Xu, D.; Zhang, J.; Dou, Z.; uang, G.; Zhang, H.; et al. litochondrial-Targeted Antioxidant litoQ-Mediated Autophagy: A lovel Strategy for Precise Radiation rotection. *Antioxidants* **2023**, *12*, 453. ttps://doi.org/10.3390/ ntiox12020453

cademic Editor: Yan-Zhong Chang

eceived: 14 December 2022
evised: 8 February 2023
ccepted: 9 February 2023
ublished: 10 February 2023

1. Introduction

RT is one of the current first-line treatment options for cancers, with over 50% of all cancer patients being treated with RT [1,2]. However, RT can induce short-term and long-term toxicity. For instance, the exposure of brain tissue to therapeutic radiation is related to various adverse outcomes, including long-term neurocognitive sequelae, cognitive impairment, and endocrine dysfunction [3,4]. Thus, eliminating tumors while selectively protecting normal tissues is challenging for tumor radiotherapy. However, no radio-protective agents are clinically available for brain tumors, except for memantine hydrochloride, which is in clinical recruiting status [5]. Additionally, the only US Food and Drug Administration (FDA)-approved drug, amifostine, provides radioprotection in head and neck cancer and is unsuitable due to side effects and incomplete protection [6]. Many antioxidants ameliorate or prevent the side effects of RT, but some even potentially promote cancer development and metastasis in mice models [7–10]. Thus, unmet medical demand for safer and more effective radio-protective agents persists.

MitoQ, a mitochondria-targeted antioxidant, is developed to penetrate the blood–brai barrier (BBB) and neuronal membranes and has a good neuroprotective effect [11,12]. I our previous study, MitoQ induced autophagy by affecting mitochondrial respiration b constructing PMMP [13]. Autophagy is an intracellular "self-digestion" physiologica process that relies on lysosomes to degrade cytoplasmic unnecessary or dysfunctiona components [14]. Under normal physiological states, the binding of adenosine triphosphat (ATP) to AMPK suppresses the kinase activity. When energy starvation appears in cells, th ATP levels decrease, and the AMP levels increase. An important "energy sensor"—AMPI in eukaryotic cells is phosphorylated. The activated AMPK causes defective mammalia target of rapamycin (mTOR) signaling by inhibiting the phosphorylation of mTOR, boostin the formation of autophagosomes and autophagy flux [15–17]. Nature Chemical Biolog reported that inducing autophagy within ischemic brain cells by small molecule compound can reduce the expression of apoptotic factors, clear damaged organelles in time, an degrade and reuse accumulated long-lived proteins to supply energy for ischemic brai cells quickly, which is conducive to maintaining cell homeostasis and protecting cells [18 The α1-antitrypsin mutation can lead to hepatic endoplasmic reticulum stress, sever inflammatory response, and liver damage and carcinogenesis. It is reported that autophag effectively removes the mutated α1-antitrypsin in liver cells, relieves the stress state o the liver, and has a potent liver protection effect [19]. Therefore, autophagy ensures cel integrity, maintains effective cell function, helps cells survive the crisis, and promote cell survival in extreme cases [20]. Thus, if autophagy can be selectively induced amon radiated normal cells, normal tissue can be effectively protected during RT.

Eukaryotic cells primarily rely on oxidative phosphorylation (OXPHOS) of the mito chondrial respiratory chain for energy supply [21]. The electrons carried by nicotinamid adenine dinucleotide (NADH) and flavin adenine dinucleotide ($FADH_2$) are utilized as fue and offered for the molecular oxygen by the mitochondrial inner membrane respirator chain. This generates a proton gradient with proton pumping into the mitochondria intermembrane space through the respiratory chain complexes I, III, and IV. The dissipatior of the established proton gradient using complex V generates ATP [22]. Thus, the energ supply in normal cells originates from mitochondria, and the proton gradient inside th mitochondrial intermembrane space is the driving force maintaining the energy synthesis o cells. However, in the 1950s, Otto Warburg discovered a characteristic of energy metabolism distinguishing tumor cells from normal cells through high levels of glycolysis. Warbur believed that tumorigenesis led to mitochondrial respiratory dysfunction in tumor cell Tumor cells reprogram their energy metabolism by initiating mitochondria-independen energy supply pathway: aerobic glycolysis called the "Warburg effect" [23–25] to maintair normal intracellular ATP and NADH levels to ensure the need for malignant prolifera tion. This fact has been applied in positron emission tomography (PET) depending or 18F-fluorodeoxyglucose (FDG). The degree of glucose uptake of a malignancy as image by FDG-PET is associated with histologic measures of tumor differentiation [24]. In sum mary, significant differences in energy metabolism exist between normal and tumor cell Therefore, if the mitochondria-dependent energy supply pathway is targeted, leading tc normal cell starvation, autophagy can be selectively induced within normal cells. Howeve it is ineffective among tumor cells independent of mitochondria and relies primarily on gly colysis for energy supply. Here, we investigated the protective effect of MitoQ-constructe PMMP on normal cells during irradiation by selectively inducing autophagy in norma cells. In contrast, tumor cells are not protected due to the absence of autophagy.

2. Methods and Materials

2.1. Cell Culture

Human Astrocytes (HA) were purchased from ScienCell^{TM} and cultured in Astrocyte Medium (ScienCell^{TM}, Cat No.1801, San Diego, CA, USA) with 2% fetal bovine serum (ScienCell^{TM}, 0010), 1% penicillin/streptomycin (ScienCell^{TM}, 0503), and 1% astrocyte growth supplement (ScienCell^{TM}, 1852), in 5% CO_2 and at 37 °C. The culture vessels were coated with poly-L-lysine stock solution (ScienCell™, 0413) at 2 μg/cm^2. HA was utilized within 10 generations, authenticated using short tandem repeat (STR, ScienCell™), and routinely tested for *Mycoplasma*. Human glioblastoma cell line A172 cells were obtained from BeNa Culture Collection (BNCC) and cultured in RPMI 1640 medium (Meilunbio, MA0215-2, Liaoning, China) using 10% fetal bovine serum (TransGen Biotechnology, FS101-02, Beijing, China), in 5% CO_2 and at 37 °C. A172 cells were used within 15 passages from frozen stocks, authenticated using STR (BNCC), and were routinely evaluated for *Mycoplasma*.

2.2. Animals

Six-week-old male BALB/c nude mice were utilized for high performance liquid chromatography–mass spectrometry (HPLC-MS) and matrix-assisted laser desorption/ionization time-of-flight mass spectrometry (MALDI-TOF-MS) imaging. The mice weighed approximately 20 g and were purchased from Beijing Vital River Laboratory Animal Technology Co., Ltd. (Beijing, China). In addition, six-week-old male BALB/c nude mice bearing luciferase-positive U87MG orthotopic brain tumors, weighing approximately 20 g, were procured from Wuhan Servicebio Biotechnology Co., Ltd. (Wuhan, China). All the mice were housed under specific pathogen-free (SPF) conditions in facilities having freely available food and water. All animal procedures strictly followed the animal experimentation regulations in China and were approved by the Animal Ethics Committee of the Institute of Modern Physics of the Chinese Academy of Science (No. 2022(06)).

2.3. Polymerase Chain Reaction (PCR) Array Expression Profiling

The Human Signal Transduction Pathway Finder™ RT^2 Profiler™ PCR Array (Qiagen) determined the expression levels of 50 genes associated with glycolysis and the tricarboxylic acid (TCA) cycle within HA and A172 cells. Microarray data were normalized against the house-keeping genes by evaluating the ΔCt for each gene representative of glycolysis and TCA in the plate. Data were analyzed using the RT^2 PCR array data analysis web portal version 3.5.

2.4. Agilent Seahorse XF Technology

The Agilent Seahorse XF technology was used to analyze the cell energy phenotype of HA and A172 cells using the Agilent Seahorse XFp Cell Energy Phenotype Test Kit (Agilent, 103275-100, Santa Clara, CA, USA) based on the instructions. HA and A172 cells were grown in the Agilent Seahorse XFp Cell Culture Miniplate and cultured overnight. The Agilent Seahorse XFp Sensor Cartridge was hydrated using Agilent Seahorse XF Calibrant at 37 °C in a non-CO_2 incubator overnight. The Agilent Seahorse Sensor Cartridge, Calibrant, and Miniplate were obtained from the Agilent Seahorse XFp FluxPak (Agilent, 103022-100). The assay medium was prepared by supplementing Agilent Seahorse XF Base Medium (Agilent, 102353-100) with 1 mM pyruvate (Sigma, S8636, St. Louis, MO, USA), 2 mM glutamine (Sigma, G8540), and 10 mM glucose (Sigma, G8769). The cell culture medium of HA and A172 were replaced with the assay medium and cultured in a non-CO_2 incubator for 1 h. Oligomycin and cyanide p-trifluoromethoxylphenyl-hydrazone (FCCP) from the Agilent Seahorse XFp Cell Energy Phenotype Test Kit were combined to develop a stressor mix loaded into every port A of the hydrated sensor cartridge. The Agilent Seahorse XF Cell Energy Phenotype test was run using the Agilent Seahorse XFp Analyzer (Agilent,102745-100). The data were analyzed with the Agilent Seahorse XF Cell Energy Phenotype Test Report Generator.

2.5. HPLC-MS Analysis

The enrichment of MitoQ (Vosun Chemical, 444890-41-9) in mitochondria in vitro and the concentrations in brain and blood in vivo were determined using an EVOQ Qube LC-TQ system (Bruker, Germany) as in our previous study [13]. For measuring the MitoQ enrichment inside mitochondria in vitro, the cells were harvested and divided into two equal parts after treatment with MitoQ for 2 h, one for the homogenate of the whole cell and the other for the homogenate of the mitochondria. A mitochondrial extraction kit (Beijing Solarbio, SM0020, Beijing, China) was used to isolate the mitochondria. All the homogenates were extracted twice utilizing the mixture of methylene chloride and methanol at a 2:1 volume ratio, including the 2 mM butylated hydroxytoluene (Sigma-Aldrich, W218405). After drying, the residue was dissolved in cold methanol with 2 mM butylated hydroxytoluene.

For measuring the MitoQ in brain and blood in vivo, six-week-old male BALB/c nude mice were randomly divided into three groups (n = 5): control, intraperitoneal injection (i.p.), and intragastric administration (i.g.) groups. The i.p. and i.g. groups were administered intraperitoneally and intragastrically with MitoQ (5 mg/kg) for three days, respectively. Mice in the control group were administered intraperitoneally with saline for three days. Mice were sacrificed after treatment with MitoQ for 1 h on the last day, and blood and brain samples were collected to determine MitoQ. All the blood and brain samples were homogenized and processed as described above. Moreover, the gradient elution time was 6 min per sample, and MitoQ depicted a retention time close to 3 min.

2.6. Determination of Mitochondrial Membrane Potential (MMP)

Changes in the MMP were identified using JC-1 Mitochondrial Membrane Potential Assay Kit (Beyotime, C2006). The treated cells were incubated with a JC-1 working solution for 20 min at a 37 °C, 5% CO_2 incubator. Then, the cells were washed twice with 1×JC-1 staining buffer and analyzed by a CytoFLEX flow cytometer (Beckman, South Kraemer Boulevard Brea, CA, USA).

2.7. Detection of Respiratory Chain Complex Activities

HA and A172 cells were treated using MitoQ for 2 h and collected for isolating the mitochondria through a Mitochondrial extraction kit. After estimating the protein concentration of isolated mitochondrial samples, the respiratory chain complex activities were detected using the Mitochondrial Complex I Activity Colorimetric Assay Kit (Abcam, ab287847, Cambridge, UK) and the Mitochondrial Complex III Activity Assay Kit (Abcam, ab287844) based on the instructions provided with the reagent kits.

2.8. Measurement of H^+/2e

Mitochondria were extracted from HA and A172 cells through the Mitochondrial extraction kit. Isolated mitochondria were suspended within a solution with 0.25 M sucrose (Sigma, S9378), 1 mM ethylenebis (oxyethylenenitrilo) tetraacetic acid (EGTA), and 5 mM Tris, pH 7.4 at 0 °C. The proton pump rate (PPR) underwent the $K_3Fe(CN)_6$ pulse method as in our previous study [13]. Electron transport rates and proton ejection were determined using 557 double-beam spectrophotometers (PerkinElmer, Waltham, MA, USA) and PHM84 fast-responding pH electrode system (Radiometer Medical, Copenhagen, Denmark).

2.9. Untargeted Metabolomics Based on Ultra-High Performance Liquid Chromatography-Tandem Time-of-Flight Mass Spectrometry (UHPLC-Q-TOF MS)

The cells were treated using MitoQ, and after collection, the cells were washed with pre-cooled phosphate buffered saline (PBS) three times and removed from the supernatant. Pre-cooled methanol: acetonitrile: water (2:2:1, *v/v*) solution was added to each sample to precipitate protein. The samples were pulverized by the ultrasonic wave at 0 °C for 30 min, incubated at −20 °C for 10 min, and then centrifuged at 4 °C, 14,000× *g*, for 20 min. The supernatants were transferred into new 1.5 mL tubes and dried inside a centrifugal vacuum evaporator, then reconstituted using a 100 µL acetonitrile: water (1:1, *v/v*) mixture for further analysis. The metabolites in the samples were detected with Shanghai Applied Protein Technology Co., Ltd. (Shanghai, China) using UHPLC-Q-TOF MS. which is a 1290 Infinity Ultra-high Performance Liquid Chromatography system

(Agilent, Santa Clara, CA, USA) coupled with AB Triple TOF 6600 Mass Spectrometer (AB SCIEX, Radio Road Redwood City, CA, USA).

2.10. Western Blot Analysis

The cells were lysed in RIPA buffer (Solarbio, R0010). The protein concentration was determined using the BCA Protein Assay Kit (Meilunbio, MA0082). Equivalent amounts of protein were separated through SDS-polyacrylamide gels and electroblotted onto PVDF membranes (Bio-Rad, 1620177). After blocking using a 5% BSA Blocking Buffer (Solarbio, SW3015), the membranes were incubated with the primary antibody at 4 °C overnight. The antibodies used were AMPKα Antibody (Cell Signaling Technology, 2532, Danvers, MA, USA), Phospho-AMPKα Rabbit mAb (Cell Signaling Technology, 2535), mTOR Antibody (Cell Signaling Technology, 2972), Phospho-mTOR Antibody (Cell Signaling Technology, 2971), LC3B (E5Q2K) Mouse mAb (Cell Signaling Technology, 83506), SQSTM1/p62 Antibody (Cell Signaling Technology, 5114), and β-Actin Antibody (Cell Signaling Technology, 4967). Finally, the relevant protein was visualized through staining with an appropriate HRP-conjugated secondary antibody for 1 h and then enhanced with chemiluminescence. The expression of the protein was quantified using ImageJ software.

2.11. Transmission Electron Microscopy (TEM)

The ultrastructural analysis of autophagy was observed using TEM. The samples were fixed using a 3% glutaraldehyde solution (Sigma-Aldrich, G5882) for 24 h at 4 °C and post-fixed in 1% osmium tetroxide (Sigma-Aldrich, 75632) at 4 °C for 1 h. The cells were washed twice, dehydrated in successive ethanol baths, treated with two baths of propylene oxide, and progressively impregnated and embedded within Epon-Araldite resin (Electron Microscopy Sciences, 14900, Hatfield, PA, USA). Ultrathin sections of 50 nm thickness were acquired through a Leica Ultracut (Leica Microsystems, Wetzlar, Germany; EM UC6) and stained using uranyl acetate and lead citrate. The autophagosomes were observed through JEM-1220 TEM (JEOL, Tokyo, Japan).

2.12. Observation of Autophagosome and Autophagy Flux

The high content imaging system was used to visualize autophagic vacuoles within live cells. HA and A172 cells were grown in ViewPlate-96 Black, clear bottom, TC-treated (PerkinElmer, 6005182), and cultured overnight. The treated cells were stained using Cell Meter™ Autophagy Assay Kit (AAT Bioquest, 23002) and Hoechst 33,342 (Meilunbio, MA0126) based on the manufacturer's instructions. The images were acquired and analyzed through the Operetta CLSTM High content imaging system (PerkinElmer, Waltham, MA, USA).

Confocal microscopy was used to visualize and monitor autophagy flux in live cells. HA and A172 cells were seeded into a 35 mm confocal dish (NEST, 801001) and cultured overnight. Then, the cells were treated with MitoQ for 2 h. After that, cells were treated using the nuclear dye Hoechst 33,342, Autophagy Assay Kit (green), and LysoBrite Red (AAT Bioquest, 22645). The confocal images were acquired through Laser confocal microscopy (LSM700, Zeiss, Germany).

2.13. In Vitro Cell Proliferation Analysis

The Cell Counting Kit-8 (CCK8, Meilunbio, Dalian, China, MA0218) and 5-ethynyl-2′-deoxyuridine (EdU, Meilunbio, MA0425) labeling assay were used to measure cell proliferation based on the manufacturer's protocols. In brief, for the CCK8 assay, HA and A172 cells were seeded inside the 96-well microplates and cultured overnight. After incubation with MitoQ for 2 h, the cells were radiated using X-rays at doses of 4 Gy or carbon ion at 2 Gy. Then, 10 μL of CCK-8 reagent was added to each well after 24 h and incubated for 2 h. The optical density (OD) was determined at 450 nm through a multi-plate reader (Tecan Infinite M200, Männedorf, Switzerland).

For the EdU labeling assay, HA and A172 cells were seeded into a 35 mm confoca dish and cultured overnight. After treatment using MitoQ for 2 h, the cells were radiate with 4 Gy X-rays or 2 Gy carbon ion. Based on the manufacturer's instructions, the EdU k was utilized for assessing cell proliferative ability after irradiating for 24 h. Hoechst 33,34 (nuclear staining) was utilized to counterstain cells. The images were captured with Lase confocal microscopy.

2.14. MALDI-TOF-MS Imaging Mass Spectrometry Analysis

MALDI-TOF-MS imaging is used for label-free bioanalysis of the spatial distribution c pharmaceuticals, biomolecules, and other molecules from a tissue section [26]. The autofle speed™ MALDI-TOF-MS imaging (Bruker, Bremen, Germany) was utilized to revea the quantitative distribution of MitoQ across a tissue section (8 μm) having high spatia resolution from BALB/c nude mice injected with 5 mg/kg/day MitoQ intraperitoneally fc three consecutive days. 2,5-Dihydroxybenzoic acid (DHB, Bruker, 8201346, 30 g/L) wit 1% trifluoroacetic acid was utilized as the matrix for MALDI-TOF-MS imaging analysis.

2.15. In Vivo Imaging of Orthotopic Glioma Using Bioluminescence Imaging (BLI) and Magneti Resonance Imaging (MRI)

The tumor luciferase expression determines the tumor size through an in vivo imagin system (PerkinElmer IVIS Lumina LT Series III, Waltham, MA, USA). The mice were injecte intraperitoneally using 150 mg/kg of D-luciferin (PerkinElmer, Waltham, MA, USA) an anesthetized with vaporized isoflurane. The mice were stratified into sham, X-ray, an X-ray + MitoQ groups depending on the tumor luciferase expression (n = 5). The shan and X-ray groups received an i.p. injection of 200 μL saline for four consecutive days. Th X-ray + MitoQ group was intraperitoneally injected using MitoQ (10 mg/kg/day) for fou days. The X-ray and X-ray + MitoQ group received 16 Gy X-ray whole brain radiation o the third day 2 h after administration.

For MR imaging, T2-TSEI and T2-FLAIR MR images were acquired 24 h and seve days post-X-ray radiation, respectively, through a 1.0-tesla small animal magnetic resc nance scanner (XGY OPER 1.0). The coronal images were obtained with the followin parameters: FOV (field of view) = 4.0 × 4.0 cm, TR/TE = 4000/91 ms for T2-TSEI im ages, and TR/TE = 7000/104 ms for T2-FLAIR images, Matrix = 160 × 154, and Slic Thickness = 1.3 mm. The sagittal and horizontal images were obtained with the follow ing parameters: FOV = 5.0 × 5.0 cm, TR/TE = 5000/104 ms for T2-TSEI images, an TR/TE = 7500/101 ms for T2-FLAIR images, Matrix = 192 × 182 for T2-TSEI images an Matrix = 192 × 198 for T2-FLAIR, with Slice Thickness = 1.2 mm.

2.16. Histomorphological and Terminal dUTP Nicked End LABELING (TUNEL) Assessments of Brain

We used hematoxylin and eosin (H&E) staining to analyze damages caused by X-ray to normal brain tissue to determine the protective effects of MitoQ on normal brain tissu during X-ray radiotherapy. All the mice were sacrificed by cervical dislocation, and ther the brain tissues were embedded inside Tissue-Tek optimal cutting temperature (OCT Compound (SAKURA, 4583) and sectioned (12 μm). The 12 μm cryosections were staine using H&E and examined to observe histomorphological features through the Pannorami 250 Flash digital microscopes (P250 Flash digital microscopes; 3DHISTECH, Budapes Hungary). TUNEL assay was utilized to detect cell death-associated DNA fragmentatior (3′-OH DNA termini) on the tissue sections [27]. The tissue sections were stained base on the instructions provided by the Fluorescein (FITC) Tunel Cell Apoptosis Detection Ki (Servicebio, G1501). Then, the slides were incubated with DAPI (Invitrogen) for 10 mir and observed under P250 Flash digital microscopes.

2.17. Statistical Analysis

Statistical analysis was conducted using GraphPad Prism. The statistical analyses performed for different data are demonstrated in each figure legend, and the data were considered statistically significant if $p < 0.05$.

3. Results

3.1. Cell Energy Phenotype of HA and A172 Cells

In normoxia, fully differentiated tissues utilize OXPHOS. However, the most common metabolic hallmark of malignant tumors, i.e., the "Warburg effect," maintains a malignant tumor phenotype (Figure 1E) [28]. A heat map revealed that most glycolysis genes were up-regulated in A172 cells compared to HA cells, and genes associated with the TCA cycle were down-regulated (Figure 1A). Agilent Seahorse XF technology was further used to evaluate the cell energy phenotype of HA and A172 cells [29]. Based on the instructions, metabolic potential represented the percentage increase in stressed oxygen consumption rate (OCR) relative to the baseline OCR and stressed extracellular acidification rate (ECAR) relative to the baseline ECAR. It measures the ability of the cells to meet energy demand through mitochondrial respiration or aerobic glycolysis. Under the stress conditions induced by oligomycin and FCCP, ECAR increased by 40 mpH/min and OCR decreased by 4 pmol/min in A172 cells, while ECAR increased by 14 mpH/min and OCR did by 10 pmol/min in HA cells (Figure 1B,C). Compared to HA, the stressed ECAR/baseline ECAR ratio was significantly increased and the stressed OCR/baseline OCR ratio was significantly decreased in A172 cells (Figure 1D). The results depicted that HA cells meet energy demand primarily by mitochondrial respiration, while A172 cells meet energy needs primarily through aerobic glycolysis. Our results are consistent with the previous studies [28,30].

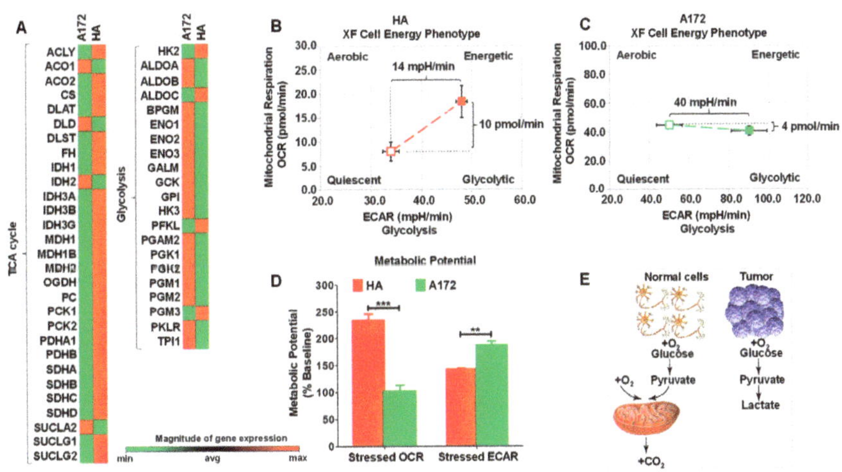

Figure 1. Cell energy phenotype of HA and A172 cells. (**A**) Human Signal Transduction Pathway Finder PCR Array was used to evaluate the expression of 50 essential genes related to glycolysis and the TCA cycle. Red indicates high expression levels, whereas green indicates low expression levels ($n = 3$). The Agilent Seahorse XFp Cell Energy Phenotype Test Kit was used to assess that normal astrocyte HA tends to supply energy through mitochondrial respiration (**B**). In contrast, glioma cell A172 tends to provide energy through glycolysis (**C**). (**D**) The metabolic Potential of HA and A172 cells is the ability of the cell to meet energy demand through respiration or glycolysis. (**E**) Schematic diagram of the different cellular energy phenotypes of normal and tumor cells. All the data were presented as mean ± SEM. Error bars represent SEM, and statistical significance between groups was analyzed using an unpaired *t*-test. ** $p < 0.01$; *** $p < 0.001$.

3.2. MitoQ Was Selectively Enriched in Mitochondria in HA Cells Higher than That in A172 Cells

It is hypothesized that defects in mitochondrial function could be the reason why cancer cells rely on aerobic glycolysis for energy supply [31]. The mitochondrial structure and functions of cancerous cells differ from those of normal cells. Moreover, MitoQ targeting mitochondria is based on MMP and lipophilicity [32]. Therefore, the enrichment of MitoQ within cancerous and normal cells could be different. The concentrations of MitoQ in isolated mitochondria and whole cells of HA and A172 cells were determined using HPLC-MS to evaluate the mitochondria-targeted ability of MitoQ. The HPLC-MS calibration curve ($y = 3446.7x - 9869.8$, $R^2 = 0.9999$) for MitoQ had been established at concentrations of 1, 10, 100, and 1000 ng/mL and subsequently became feasible and linear ($R^2 > 0.99$) (Figure 2A,B). The concentrations of MitoQ in whole cells and the isolated mitochondria of HA cells were higher than those in A172 cells, demonstrating that MitoQ was more easily enriched in normal cells (Figure 2C,D). Treatment of HA with MitoQ for 2 h resulted in a >80% enrichment of MitoQ in mitochondria, indicating that MitoQ could be effectively enriched within mitochondria of normal cells (Figure 2C).

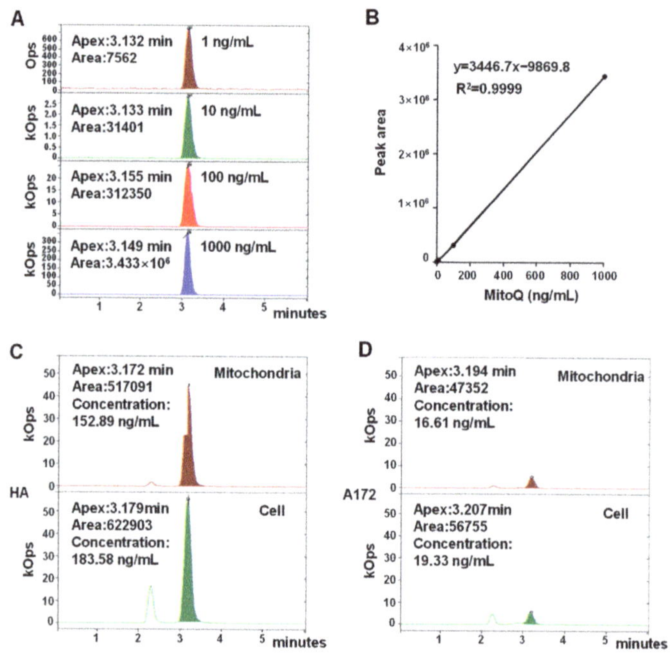

Figure 2. Quantitative analysis of MitoQ enrichment in mitochondria among HA and A172 cells. (**A**) HPLC−MS chromatograms of MitoQ at different concentrations. (**B**) The calibration curve of MitoQ within the concentration range of 1−1000 ng/mL. The concentrations of MitoQ in whole cells and the isolated mitochondria of HA (**C**) and A172 cells (**D**).

3.3. MitoQ-Induced PMMP in Normal Cells Was Higher than That in Tumor Cells

HA and A172 cells treated with MitoQ were stained using JC-1 and analyzed with flow cytometry to determine changes in MMP. Through treatment with MitoQ for 15 min, the MMP of HA cells increased by 13.15%, whereas that of A172 cells only increased by 5.58% (Figure 3A). The increase could be associated with the HPLC-MS results that showed that MitoQ was enriched in the mitochondria of HA cells at a much higher level than A172 cells (Figure 2C,D). In the meantime, the activities of respiratory chain complexes I and III and the PPR were restrained by MitoQ (Figure 3B,C). Therefore, when MitoQ adsorbed to the inner mitochondrial membrane and remained the cationic moiety in the intermembrane space, the result was the addition of a large number of positive charges to protons. The balance of MMP was maintained by MitoQ attenuating the activity of respiratory chain complexes I, III, and IV (which are associated with proton generation) and decreasing the PPR (Figure 3C). MitoQ successfully constructed PMMP by proton displacement with exogenous positive charges based on our previous hypothesis (Figure 3D) [13]. The MMP and respiratory chain are important for mitochondrial respiration [33]. Therefore, MMP and respiratory chain complex dysfunction could lead to abnormal mitochondrial respiration.

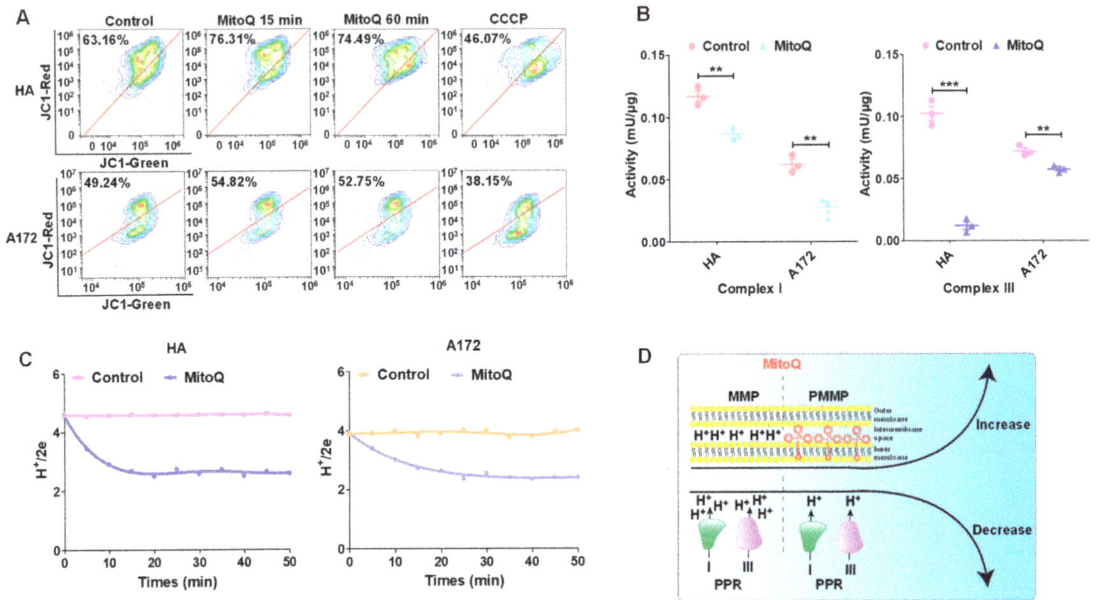

Figure 3. Construction of PMMP using MitoQ. (**A**) Fluorescence of HA and A172 cells stained using JC-1 was ascertained through flow cytometry. (**B**) The activities of respiratory chain complexes I and III associated with proton production were determined with commercial kits. (**C**) The PPR was detected through a fast-responding pH electrode system after MitoQ treatment. (**D**) The schematic diagram of mitochondrial status changes after MitoQ treatment. Each experiment was conducted at least three times. All the data were presented as mean ± SEM. Error bars represent SEM, and statistical significance between groups was analyzed using an unpaired *t*-test. ** $p < 0.01$; *** $p < 0.001$.

3.4. PMMP Disrupted Energy Metabolism in HA Cells

Metabolomics can offer insights into the cellular processes in response to stimuli or interactions, for the metabolome is considered the closest phenotype representation [34,35]. Untargeted metabolomics was performed on HA and A172 cells treated with MitoQ further to determine the effect of MitoQ on energy metabolism.

The global orthogonal partial least-squares-discriminant analysis (OPLS-DA) model revealed a clear separation between the MitoQ-treated and control groups in positive and negative ion modes (Figure S1). The permutation testing was performed on the quality of the model and revealed that the model was not over-fitted (Figure S2). The volcano maps showed that more differential metabolites appeared in MitoQ-treated HA cells than in MitoQ-treated A172 cells compared to the control group (Figure 4A,B), indicating that MitoQ had a more significant effect on HA than A172 cells.

The expressions of citrate and L-malic acid, the primary metabolites of the TCA cycle [36], were altered after HA cells were treated with MitoQ (Figure 4C and Figure S3). Moreover, their expression changes were negatively correlated (Figure 4D). The citrate expression was down-regulated, while L-malic acid was up-regulated. The results displayed that the TCA cycle of HA cells could be disturbed after treatment with MitoQ.

The KEGG enrichment pathway map results also revealed that the TCA cycle of MitoQ-treated HA cells significantly differed from the control group ($p = 0.0014$) (Figure 4E). MitoQ also had a slight effect on the TCA cycle of the A172 cells ($p = 0.0378$) (Figure 4F), only down-regulating the expression of citrate (Figure S3). Since tumor cells primarily rely on aerobic glycolysis for energy supply, a minimal change in the TCA cycle of A172 cells does not affect the energy supply of A172 cells. As shown in Figures 3 and 4, the main target of the PMMP constructed using MitoQ was mitochondrial respiration. Therefore, it could effectively disrupt the energy metabolism and subsequent energy supply of normal cells but it was ineffective against tumor cells.

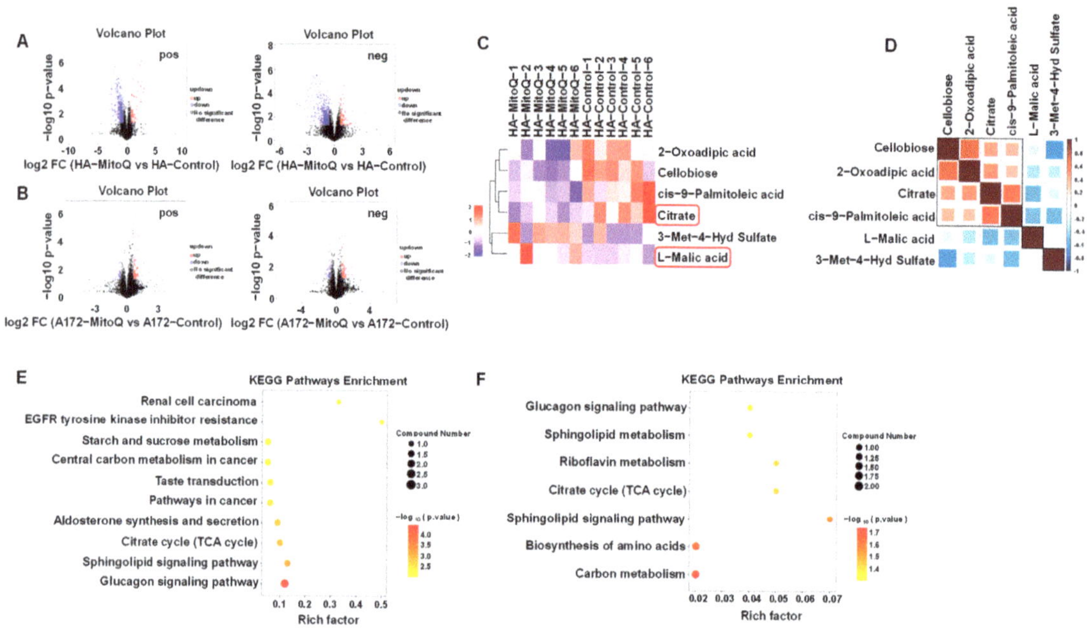

Figure 4. MitoQ disrupted the TCA cycle in HA cells. Effect of MitoQ on metabolite expression in HA (**A**) and A172 cells (**B**), $n = 6$. The Fold Change (FC) analysis and t-test were utilized in volcano plot analysis to screen the potential metabolites. The red dots indicate differential metabolites with $FC > 1.5$, and p-value < 0.05, the blue dots reveal differential metabolites with $FC < 0.67$ and p-value < 0.05, and the black dots depict no significant difference. (**C**) Influence of MitoQ on citrate and L-malic acid expression within HA cells. (**D**) The expression of citrate and L-malic acid were negatively correlated within HA cells. KEGG analysis of metabolites pathway after MitoQ treatment using HA (**E**) and A172 cells (**F**).

3.5. MitoQ Induced Autophagy in HA Cells through the AMPK/mTOR Pathway

The cellular energy sensor AMPK is activated by energy deprivation. Thus, its activity is related to cellular energy metabolism levels and autophagy [37]. Based on the Western blotting analysis results, AMPK was activated by enhancing the phosphorylation of AMPK after treatment using MitoQ in HA cells. Moreover, the phosphorylation of mTOR was significantly downregulated in HA cells (Figure 5A). However, the AMPK/mTOR pathway in A172 cells was not activated after MitoQ treatment (Figure 5A). Then, the protein levels of the autophagy markers MAP1LC3/LC3 (microtubule-associated protein 1 light chain 3) and SQSTM1/p62 [38] were tested in cells treated with MitoQ compared to control. A significant increase was observed in the LC3-II: LC3-I ratio levels, with a significant decrease in the protein levels of SQSTM1/p62 in HA cells treated with MitoQ. For A172 cells, MitoQ pre-treatment downregulated the LC3-II: LC3-I ratio level and did not affect SQSTM1/p62 (Figure 5B). Following the inhibition of MitoQ-induced autophagy flux by chloroquine

(CQ), an alkalinizing agent blocking autophagosome fusion with the lysosome [39], the expression of LC3-II was further assessed via Western blotting. As illustrated in Figure S4, the level of LC3-II in HA cells with MitoQ + CQ treatment was significantly higher than that of the MitoQ and CQ groups. The results indicated that CQ's inhibition of MitoQ-induced autophagy flux resulted in an increase in the expression level of LC3-II. The findings indicated that MitoQ was capable of stimulating autophagy flux in HA cells. In addition, the Autophagy Assay Kit staining revealed a significant increase in the number of autophagosomes in HA cells treated with MitoQ. However, the number of autophagosomes in A172 cells treated with MitoQ did not change (Figure 5C,D). The same results were also observed by TEM (Figure 5E). Furthermore, there were more autophagosome–lysosome fusions in MitoQ-treated HA cells than in control. There was no difference between the two groups of A172 cells among autophagosome–lysosome fusions (Figure 5F). The results indicated that MitoQ stimulated autophagy flux through the AMPK-mTOR pathway in HA cells.

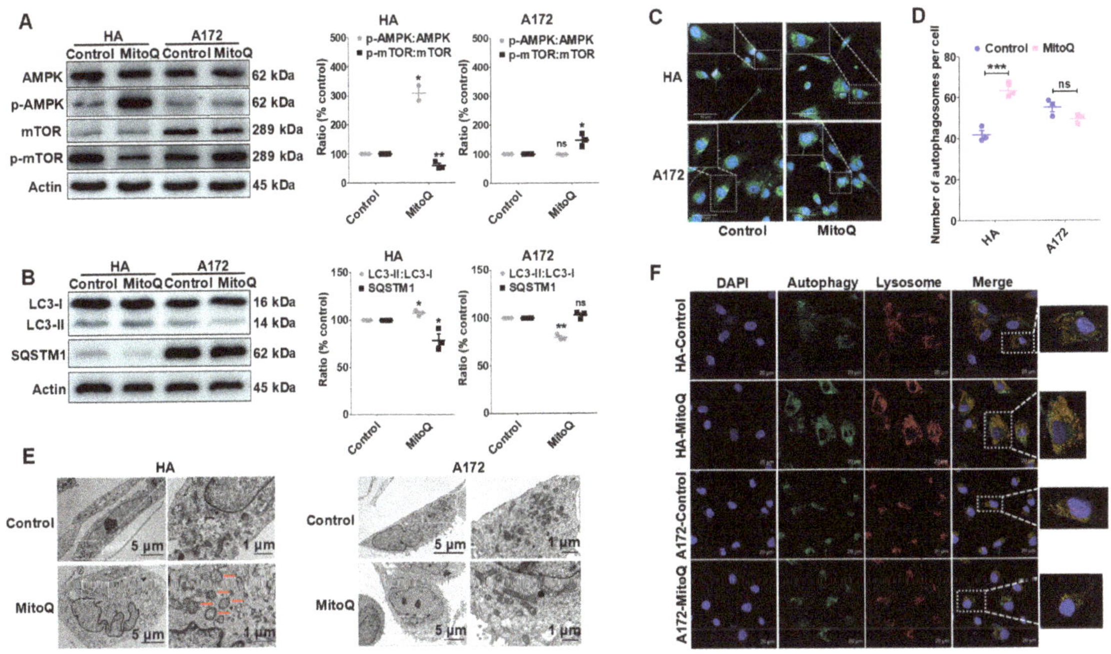

Figure 5. MitoQ induced autophagy in HA cells through the AMPK/mTOR pathway. (**A**) The phosphorylation of AMPK and mTOR in HA and A172 cells were evaluated using Western blotting after treatment with MitoQ. (**B**) The conversion of LC3-I to LC3-II and the protein levels of SQSTM1/p62 were analyzed through Western blotting. The autophagosomes induced by MitoQ were visualized (**C**) and determined (**D**) by the high content imaging system. (**E**) Transmission electron microscopy was utilized to observe the autophagic vacuoles. (**F**) The autophagosome–lysosome fusions were visualized through confocal microscopy. Representative images were provided as indicated. Each experiment was conducted at least three times. All the data are presented as mean ± SEM; error bars represent SEM. Statistical significance between the groups was analyzed using unpaired t-test. * $p < 0.05$; ** $p < 0.01$; *** $p < 0.001$; "ns" represents no statistical difference.

3.6. Protective Effect of MitoQ on Normal Cells against Radiation

The protective effect of MitoQ on normal cells by inducing autophagy during radiation was detected using CCK8 and EdU assays. MitoQ protected HA cells from X-ray exposure in a dose-dependent manner at 0–0.5 μM but did not affect A172 cells in CCK8 assays (Figure 6A). Mito effectively protected the survival of HA cells but did not affect the proliferation inhibition of A172 cells under 4Gy X-ray radiation (Figure 6B). EdU assays also demonstrated that MitoQ could protect HA cells from damage by X-rays, but not the A172 cells (Figure 6C,E,F). MitoQ protected normal cells not only under low linear energy transfer (LET) radiation but also under high LET radiation. CCK8 and EdU assays revealed that MitoQ could protect HA cells from damage by 2Gy carbon ions, which has a high LET [40], but did not affect the inhibitory effect of carbon ions on A172 cells (Figure S5). ROC-325 is a novel autophagy inhibitor that can lead to the deacidification of lysosomes, accumulation of autophagosomes, and the disruption of autophagy flux [41]. When MitoQ-induced autophagy flux was blocked within HA cells using ROC-325, the number of autophagosomes was significantly enhanced compared to that treated using MitoQ alone (Figure 6G). Correspondingly, the protective effect of MitoQ on HA cells was destroyed by ROC-325 (Figure 6D). Therefore, MitoQ-constructed PMMP-inducing autophagy played an essential role during normal cellular radioprotection.

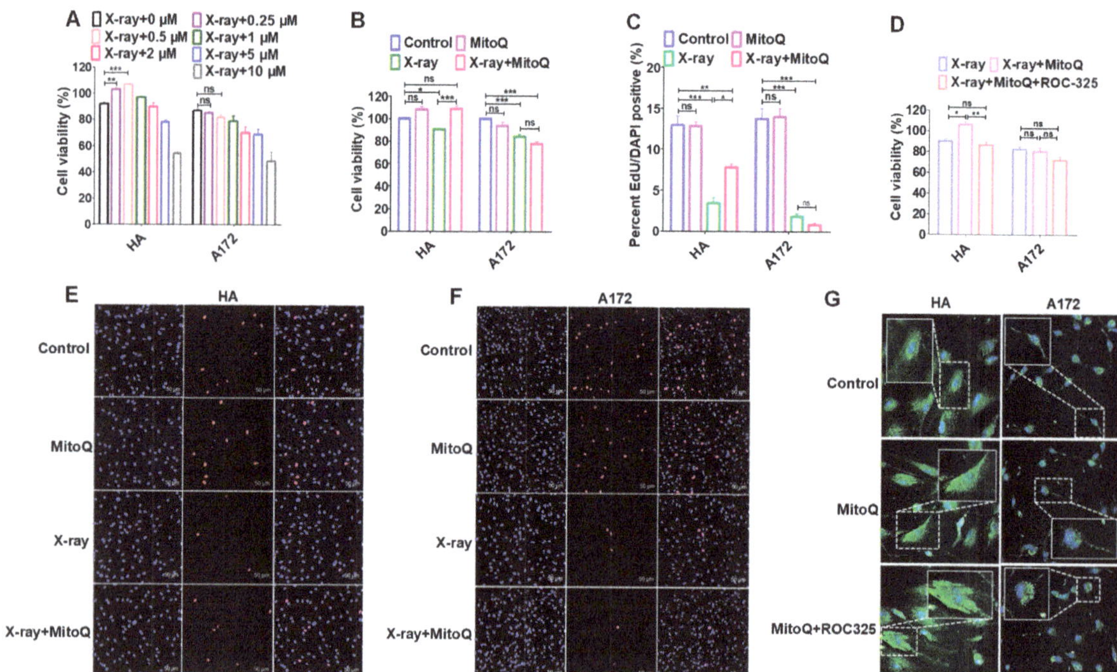

Figure 6. The protective effect of MitoQ on HA cells against X-ray radiation. (**A**) Effects of different concentrations of MitoQ on HA and A172 cells during X-ray radiation were detected through CCK8 assays. (**B**) MitoQ at a concentration of 0.5 μM promoted the proliferation of HA cells during 4 Gy X-ray radiation in CCK8 assays. (**C**) EdU assay demonstrated that MitoQ could protect HA cells from damage using X-rays. Typical photos of the EdU assay were captured with confocal microscopy (**E**,**F**). (**D**) Autophagy inhibitor ROC-325 destroyed the protective effect of MitoQ on HA cells. (**G**) ROC-325 restrained the autophagy flux. Representative images were provided as indicated. All the data are presented as mean ± SEM from at least three independent experiments, and error bars represent SEM. Statistical significance between groups was analyzed using one-way ANOVA, * $p < 0.05$; ** $p < 0.01$; *** $p < 0.001$; "ns" represents no statistical difference.

3.7. *MitoQ Protected Normal Tissue against X-rays in Mice Bearing Orthotopic Glioma*

The protective effect of MitoQ on normal cells was demonstrated in mice bearing orthotopic glioma. HLPC-MS and MALDI-TOF-MS imaging were performed to render the penetration and distribution of MitoQ in the brain. MitoQ was detected in the brain and blood of the i.p. and i.g. groups using HLPC-MS, and the concentrations of MitoQ in the brain and blood in the i.p. group were higher than that in the i.g. group (Figure 7A,B). Therefore, i.p. injection was selected for MitoQ administration. MALDI-TOF-MS visualized the spatial distribution of MitoQ on the surface of brain tissue sections (8 µm). A high level of MitoQ was witnessed in the brain tissue sections, indicating that MitoQ could penetrate the brain tissue (Figure 7C,D). A portion of the MALDI-TOF-MS spectrum within the mass ranges of 0–820 m/z obtained from brain tissue sections of the control and MitoQ-treated group also revealed that the representative MALDI-TOF-MS mass profile (583 m/z) of MitoQ was detected in brain tissue sections of MitoQ-treated group, and not in the control group (Figure 7E,F). These data depicted that MitoQ could be distributed in the brain.

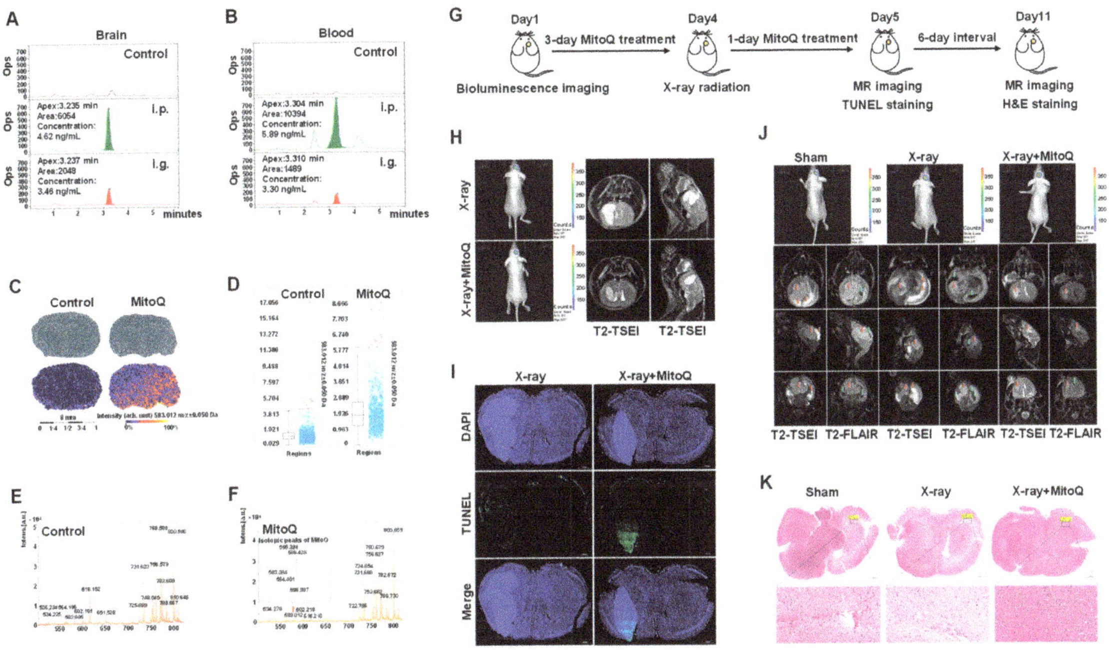

Figure 7. Protective effect of MitoQ on normal tissues against X-rays among mice bearing orthotopic glioma. HPLC-MS chromatograms of MitoQ inside the brain (**A**) and blood (**B**) of both the i.p. and ig. group, $n = 5$. (**C**) MALDI-TOF-MS was used to visualize the spatial distribution of MitoQ on the surface of brain tissue sections. The intensity box blot (**D**) and the representative MALDI-TOF-MS mass profile (583 m/z) (**E**,**F**) of MitoQ on the surface of the brain tissue sections were observed. The isotopic peaks of the MitoQ procured from MALDI-TOF-MS were also shown (**F**). (**G**) Schematic diagram illustrating the experimental design. Bioluminescence imaging was used to determine the tumor size of the nude mice bearing luciferase-positive U87MG orthotopic brain tumors on Day 1 (before treatment). The mice were intraperitoneally injected using MitoQ (10 mg/kg/day) for four days (Day 2–Day 5). The mice received 16 Gy X-ray whole-brain radiation on Day 4, two hours after MitoQ administration. MR imaging and TUNEL staining were performed on Day 5, and MR imaging and H&E staining were performed on Day 11. (**H**) The tumors were evaluated by bioluminescence before treatment, and MRI signals in the brain were monitored 24 h post-X-ray radiation. (**I**) TUNEL-stained sections were obtained at 24 h post-X-ray radiation. (**J**) The tumors were evaluated using bioluminescence before treatment, and MRI signals in the brain were detected seven days post-X-ray radiation. Red arrows indicate tumors, orange indicates edema, and green indicates hydrocephalus. (**K**) H&E staining of brain tissue on seven days post-X-ray irradiation. $n = 5$, Representative images were provided as shown.

Next, we determined whether MitoQ protected against structural damage to norma tissue due to X-rays using the nude mice bearing luciferase-positive U87MG orthotopi brain tumors. The experimental design was shown on Figure 7G. We established orthotopi brain tumors and assessed the tumors using bioluminescence before treatment (Figure 7H,J MRI signals in the brain were monitored and analyzed 24 h and seven days post-X-ra radiation. MitoQ reduced the edema in normal brain tissue after 24 h of X-ray radiatior as the MR images demonstrated that the normal brain tissue of the mice in the X-ra group was brighter than that of the mice in the X-ray + MitoQ group (Figure 7H). Th TUNEL-stained sections revealed that MitoQ promoted apoptosis inside brain tumors 24 post-X-ray radiation (Figure 7I). After seven days of X-ray radiation, MRI signals showe that MitoQ could reduce X-ray treatment-induced edema and hydrocephalus of norma tissue inside the mouse brain without affecting the X-ray tumor treatment compared to th X-ray group (Figure 7J). We also observed that normal brain tissue cells in the X-ray grou were less compacted than in the X-ray + MitoQ group. It was observed that the norma brain tissue in the X-ray group was surrounded by more edema and hydrocephalus than i the X-ray + MitoQ group (Figure 7K).

4. Discussion

RT is one of the essential treatment methods against malignant tumors, but the dam age to normal tissues is the fundamental challenge for RT [42]. In survivors, RT-induce secondary tumors, cognitive dysfunction, cardiac toxicity, gastrointestinal toxicity, radiatio dermatitis, etc., are common [4]. In 1902, the first radiation-related cancer, a radiation induced secondary primary skin cancer, was discovered [43]. Subsequently, the Gustav Roussy Institute conducted a long-term follow-up of 7711 breast cancer patients betweer 1954 and 1983 and observed that approximately 77% of the patients with secondary tumo recurrence had a history of RT [44]. With the increasing number of radiation-induce secondary primary tumors (such as leukemia, lymphoma, brain tumors, etc.), malignant tu mors are gradually recognized as a long-term effect of radiation [45]. Thus, the critical goa and principal challenge of RT is the cancellation of tumors without any concurring injurie to the surrounding tissues and organs. Modern RT methods include three-dimensiona conformal, and intensity-modulated radiotherapy (IMRT) could reduce, but not eliminate the incidence and severity of toxicities [46]. This provides a strong rationale for adding radio-protective agents. However, the only small radio-protector molecule approved b FDA, amifostine, has severe adverse effects limiting its clinical use [47]. Many antioxidant ameliorate or prevent the side effects of RT, but some even potentially promote cance development and metastasis [7–10]. Therefore, the lack of precision of radiation protectio and serious side effects have always been the main reasons restricting the developmen and application of radiation protection drugs in cancer radiotherapy.

In this study, MitoQ successfully constructed PMMP through proton displacemen with exogenous positive charges by adsorbing to the inner mitochondrial membrane and re maining the cationic moiety in the intermembrane space. The activities of respiratory chair complexes I and III were restrained, resulting in reduced PPR (Figure 3). MitoQ-constructed PMMP selectively induced autophagy within normal cells but not tumor cells (Figure 5) by interfering with the energy metabolism of normal cells (Figure 4). Autophagy can serve as a protective mechanism by executing the degradation and elimination of damaged organelles and providing energy for cellular renovation during stress conditions, including X-ray radiation [48,49]. For instance, autophagy is promoted to recycle unnecessary organelles to enhance amino acid availability in response to starvation [50]. Endothelial dysfunctior plays an essential role in liver injury. Autophagy can maintain endothelial phenotype and protect liver sinusoidal endothelial cells from oxidative stress during the early phases of liver disease [51]. In addition, autophagy is a critical protective pathway in neurons which relies on autophagy to preserve cytoplasmic homeostasis [52]. In this study, MitoQ constructed PMMP selectively induced protective autophagy to exert specific protectior for normal brain cells and tissues (Figures 6 and 7). This occurred because the difference in energy metabolism phenotypes of normal and tumor cells led to the difference in the response of the two cell types due to radiation, and this difference is precisely the entry point for specific radioprotection of normal cells.

The Warburg effect has been proposed as one of the malignant phenotypes of tumors, establishing a basis for our understanding of tumors and precision therapy [53]. For instance, an unrevealed molecular function of corannulene buckybowl glycoconjugates is to selectively annihilate tumors by targeting the cancer-specific Warburg effect [54]. Basic leucine zipper and W2 domain-containing protein 1 (BZW1) promotes tumor growth by facilitating aerobic glycolysis and serves as a therapeutic target for pancreatic cancer patients [55]. However, the value of the presence of the Warburg effect in tumor cells has never been unearthed in radioprotection. In this study, we exploited the cancer-specific aerobic glycolysis (Figure 1) to detect an entry point for precise radioprotection of normal cells. As is known, tumor treatment has entered the "Precision Era." Thus, the "Precision Era" includes "Precision Treatment" and "Precision Protection." In the precise treatment of tumors, it is fundamental to detect targeted drugs through genetic screening. Combined with our work, we suggest that the differences in energy metabolism during tumor treatment provide a precise strategy to facilitate normal tissue radiation protection. Therefore, energy metabolism screening should be paid attention to during tumor treatment.

5. Conclusions

In summary, our study demonstrated that MitoQ-constructed PMMP selectively protects normal cells and tissues in glioma RT without affecting its efficacy by inducing autophagy in normal cells by regulating the cellular energy supply. These findings provide insights into the targeted protection of MitoQ-induced PMMP on normal tissues due to the different energy phenotypes between tumor cells and normal cells. It supports a preclinical rationale to broaden the clinical evaluation of energy metabolism screening during tumor treatment. These findings also provide solid evidence to support our pioneer work on pseudo-mitochondrial membrane potential for practical application.

Supplementary Materials: The following supporting information can be downloaded at: https://www.mdpi.com/article/10.3390/antiox12020453/s1.

Author Contributions: Conceptualization, H.Z., C.S. and X.B.; Methodology, X.B. and Q.W.; Software, F.Y., Z.S., D.X. and Z.D.; Validation, H.Z., C.S., X.B., X.L., Q.W., F.Y., Z.S., D.X., J.Z. and G.H.; Formal analysis, H.Z., C.S., X.B., X.L., Q.W., J.Z., Z.D. and G.H.; Investigation, H.Z., C.S., X.B., X.L., Q.W. and G.H.; Data curation, X.B., X.L., Q.W., F.Y., Z.S., D.X. and J.Z.; Writing—original draft, H.Z., C.S., X.B. and X.L.; Writing—review and editing, F.Y., Z.S., D.X. and Z.D.; Supervision, H.Z. and C.S.; Project administration, H.Z. and C.S.; Funding acquisition, H.Z. and C.S. All authors have read and agreed to the published version of the manuscript.

Funding: This study was supported by research grants the National Key R&D project of the Chinese Ministry of Science and Technology (2018YFE0205100), the Fellow of Youth Innovation Promotion Association (Grant No. 2020414), the Key Program of the Natural Science Foundation of Gansu Province (21JR7RA097), and the National Natural Science Foundation of China (12005042).

Institutional Review Board Statement: All animal procedures strictly followed the animal experimentation regulations in China and were approved by the Animal Ethics Committee of the Institute of Modern Physics of the Chinese Academy of Science (on 8 March 2022).

Informed Consent Statement: Not applicable.

Data Availability Statement: The data generated in this study are available within the article and its supplementary data files.

Acknowledgments: We thank Boyi Yu for providing MRI technical assistance used in this study. Part of the experimental work was carried out on the biomedical platform of the Public Technology Center of the Institute of Modern Physics and Lanzhou Regional Center of Resource and Environmental Science Instrument, Chinese Academy of Sciences.

Conflicts of Interest: The authors declare no conflict of interest.

References

1. Hamblin, R.; Vardon, A.; Akpalu, J.; Tampourlou, M.; Spiliotis, I.; Sbardella, E.; Lynch, J.; Shankaran, V.; Mavilakandy, A. Gagliardi, I.; et al. Risk of second brain tumour after radiotherapy for pituitary adenoma or craniopharyngioma: A retrospective multicentre, cohort study of 3679 patients with long-term imaging surveillance. *Lancet Diabetes Endocrinol.* **2022**, *10*, 581–588 [CrossRef]
2. Hong, Z.; Liu, T.; Wan, L.; Fa, P.; Kumar, P.; Cao, Y.; Prasad, B.; Qiu, Z.; Liu, J.; Wang, H.; et al. Targeting Squalene Epoxidase Interrupts Homologous Recombination via the ER Stress Response and Promotes Radiotherapy Efficacy. *Cancer Res.* **2022**, *82* 1298–1312. [CrossRef]
3. Kim, S.H.; Lim, K.H.; Yang, S.; Joo, J.Y. Long non-coding RNAs in brain tumors: Roles and potential as therapeutic targets. *J Hematol. Oncol.* **2021**, *14*, 77–94. [CrossRef]
4. De Ruysscher, D.; Niedermann, G.; Burnet, N.G.; Siva, S.; Lee, A.W.M.; Hegi-Johnson, F. Radiotherapy toxicity. *Nat. Rev. Dis Prim.* **2019**, *5*, 13–33. [CrossRef]
5. ClinicalTrials.gov: A Database of Privately and Publicly Funded Clinical Studies Conducted around the World. Available online: https://clinicaltrials.gov/ct2/results?cond=Radiation+Protection+and+cancer%09&term=&cntry=&state=&city=&dist= (accessed on 13 December 2022).
6. Ashcraft, K.A.; Boss, M.K.; Tovmasyan, A.; Roy Choudhury, K.; Fontanella, A.N.; Young, K.H.; Palmer, G.M.; Birer, S.R.; Landon C.D.; Park, W.; et al. Novel Manganese-Porphyrin Superoxide Dismutase-Mimetic Widens the Therapeutic Margin in a Preclinical Head and Neck Cancer Model. *Int. J. Radiat. Oncol.* **2015**, *93*, 892–900. [CrossRef]
7. Forman, H.J.; Zhang, H. Targeting oxidative stress in disease: Promise and limitations of antioxidant therapy. *Nat. Rev. Drug Discov.* **2021**, *20*, 689–709. [CrossRef]
8. Fischer, N.; Seo, E.J.; Efferth, T. Prevention from radiation damage by natural products. *Phytomedicine* **2018**, *47*, 192–200. [CrossRef]
9. Bansal, A.; Simon, M.C. Glutathione metabolism in cancer progression and treatment resistance. *J. Cell Biol.* **2018**, *217*, 2291–2298 [CrossRef]
10. Wiel, C.; Le Gal, K.; Ibrahim, M.X.; Jahangir, C.A.; Kashif, M.; Yao, H.; Ziegler, D.V.; Xu, X.; Ghosh, T.; Mondal, T.; et al. BACH1 Stabilization by Antioxidants Stimulates Lung Cancer Metastasis. *Cell* **2019**, *178*, 330–345. [CrossRef]
11. Ismail, H.; Shakkour, Z.; Tabet, M.; Abdelhady, S.; Kobaisi, A.; Abedi, R.; Nasrallah, L.; Pintus, G.; Al-Dhaheri, Y.; Mondello, S.; et al. Traumatic Brain Injury: Oxidative Stress and Novel Anti-Oxidants Such as Mitoquinone and Edaravone. *Antioxidants* **2020**, *9*, 943. [CrossRef]
12. Souza-Neto, F.V.; Islas, F.; Jiménez-González, S.; Luaces, M.; Ramchandani, B.; Romero-Miranda, A.; Delgado-Valero, B.; Roldan-Molina, E.; Saiz-Pardo, M.; Cerón-Nieto, M.; et al. Mitochondrial Oxidative Stress Promotes Cardiac Remodeling in Myocardial Infarction through the Activation of Endoplasmic Reticulum Stress. *Antioxidants* **2022**, *11*, 1232. [CrossRef] [PubMed]
13. Sun, C.; Liu, X.; Di, C.; Wang, Z.; Mi, X.; Liu, Y.; Zhao, Q.; Mao, A.; Chen, W.; Gan, L.; et al. MitoQ regulates autophagy by inducing a pseudo-mitochondrial membrane potential. *Autophagy* **2017**, *13*, 730–738. [CrossRef] [PubMed]
14. Pietrocola, F.; Bravo-San Pedro, J.M. Targeting Autophagy to Counteract Obesity-Associated Oxidative Stress. *Antioxidants* **2021**, *10*, 102. [CrossRef] [PubMed]
15. Ulland, T.K.; Song, W.M.; Huang, S.C.; Ulrich, J.D.; Sergushichev, A.; Beatty, W.L.; Loboda, A.A.; Zhou, Y.; Cairns, N.J.; Kambal, A.; et al. TREM2 Maintains Microglial Metabolic Fitness in Alzheimer's Disease. *Cell* **2017**, *170*, 649–663. [CrossRef]
16. Su, K.H.; Dai, S.; Tang, Z.; Xu, M.; Dai, C. Heat Shock Factor 1 Is a Direct Antagonist of AMP-Activated Protein Kinase. *Mol. Cell* **2019**, *76*, 546–561. [CrossRef]
17. Liu, Q.; Chen, Y.; Zhou, L.; Chen, H.; Zhou, Z. From Intestinal Epithelial Homeostasis to Colorectal Cancer: Autophagy Regulation in Cellular Stress. *Antioxidants* **2022**, *11*, 1308. [CrossRef]
18. Degterev, A.; Huang, Z.; Boyce, M.; Li, Y.; Jagtap, P.; Mizushima, N.; Cuny, G.D.; Mitchison, T.J.; Moskowitz, M.A.; Yuan, J. Chemical inhibitor of nonapoptotic cell death with therapeutic potential for ischemic brain injury. *Nat. Chem. Biol.* **2005**, *1*, 112–119. [CrossRef]
19. Perlmutter, D.H. Autophagic disposal of the aggregation-prone protein that causes liver inflammation and carcinogenesis in alpha-1-antitrypsin deficiency. *Cell Death Differ.* **2009**, *16*, 39–45. [CrossRef]
20. Kang, R.; Zeh, H.J.; Lotze, M.T.; Tang, D. The Beclin 1 network regulates autophagy and apoptosis. *Cell Death Differ.* **2011**, *18*, 571–580. [CrossRef]
21. Rumyantseva, A.; Popovic, M.; Trifunovic, A. CLPP deficiency ameliorates neurodegeneration caused by impaired mitochondrial protein synthesis. *Brain* **2022**, *145*, 92–104. [CrossRef]
22. Luo, Y.; Ma, J.; Lu, W. The Significance of Mitochondrial Dysfunction in Cancer. *Int. J. Mol. Sci.* **2020**, *21*, 5598. [CrossRef] [PubMed]
23. Warburg, O. On respiratory impairment in cancer cells. *Science* **1956**, *124*, 269–270. [CrossRef] [PubMed]
24. Riester, M.; Xu, Q.; Moreira, A.; Zheng, J.; Michor, F.; Downey, R.J. The Warburg effect: Persistence of stem-cell metabolism in cancers as a failure of differentiation. *Ann. Oncol.* **2018**, *29*, 264–270. [CrossRef] [PubMed]
25. Zhou, Z.; Ibekwe, E.; Chornenkyy, Y. Metabolic Alterations in Cancer Cells and the Emerging Role of Oncometabolites as Drivers of Neoplastic Change. *Antioxidants* **2018**, *7*, 16. [CrossRef] [PubMed]

5. Schulz, S.; Becker, M.; Groseclose, M.R.; Schadt, S.; Hopf, C. Advanced MALDI mass spectrometry imaging in pharmaceutical research and drug development. *Curr. Opin. Biotech.* **2019**, *55*, 51–59. [CrossRef] [PubMed]
7. Wang, S.; Cao, S.; Arhatte, M.; Li, D.; Shi, Y.; Kurz, S.; Hu, J.; Wang, L.; Shao, J.; Atzberger, A.; et al. Adipocyte Piezo1 mediates obesogenic adipogenesis through the FGF1/FGFR1 signaling pathway in mice. *Nat. Commun.* **2020**, *11*, 2303–2316. [CrossRef] [PubMed]
8. Vander Heiden, M.G.; Cantley, L.C.; Thompson, C.B. Understanding the Warburg effect: The metabolic requirements of cell proliferation. *Science* **2009**, *324*, 1029–1033. [CrossRef]
9. Wen, J.; Zhang, L.; Wang, J.; Wang, J.; Wang, L.; Wang, R.; Wang, R.; Li, R.; Liu, H.; Wei, S.; et al. Therapeutic effects of higenamine combined with [6]-gingerol on chronic heart failure induced by doxorubicin via ameliorating mitochondrial function. *J. Cell. Mol. Med.* **2020**, *24*, 4036–4050. [CrossRef]
0. Oh, S.; Yeom, J.; Cho, H.J.; Kim, J.H.; Yoon, S.J.; Kim, H.; Sa, J.K.; Ju, S.; Lee, H.; Oh, M.J.; et al. Integrated pharmaco-proteogenomics defines two subgroups in isocitrate dehydrogenase wild-type glioblastoma with prognostic and therapeutic opportunities. *Nat. Commun.* **2020**, *11*, 3288–3304. [CrossRef]
1. Kalyanaraman, B. Teaching the basics of cancer metabolism: Developing antitumor strategies by exploiting the differences between normal and cancer cell metabolism. *Redox Biol.* **2017**, *12*, 833–842. [CrossRef]
2. Battogtokh, G.; Choi, Y.S.; Kang, D.S.; Park, S.J.; Shim, M.S.; Huh, K.M.; Cho, Y.-Y.; Lee, J.Y.; Lee, H.S.; Kang, H.C. Mitochondria-targeting drug conjugates for cytotoxic, anti-oxidizing and sensing purposes: Current strategies and future perspectives. *Acta Pharm. Sin. B* **2018**, *8*, 862–880. [CrossRef]
3. Nolfi-Donegan, D.; Braganza, A.; Shiva, S. Mitochondrial electron transport chain: Oxidative phosphorylation, oxidant production, and methods of measurement. *Redox Biol.* **2020**, *37*, 101674–101683. [CrossRef] [PubMed]
4. Bauermeister, A.; Mannochio-Russo, H.; Costa-Lotufo, L.V.; Jarmusch, A.K.; Dorrestein, P.C. Mass spectrometry-based metabolomics in microbiome investigations. *Nat. Rev. Microbiol.* **2022**, *20*, 143–160. [CrossRef] [PubMed]
5. Nemet, I.; Saha, P.P.; Gupta, N.; Zhu, W.; Romano, K.A.; Skye, S.M.; Cajka, T.; Mohan, M.L.; Li, L.; Wu, Y.; et al. A Cardiovascular Disease-Linked Gut Microbial Metabolite Acts via Adrenergic Receptors. *Cell* **2020**, *180*, 862–877. [CrossRef] [PubMed]
6. Hui, S.; Ghergurovich, J.M.; Morscher, R.J.; Jang, C.; Teng, X.; Lu, W.; Esparza, L.A.; Reya, T.; Zhan, L.; Guo, J.Y.; et al. Glucose feeds the TCA cycle via circulating lactate. *Nature* **2017**, *551*, 115–118. [CrossRef] [PubMed]
7. Hu, Y.; Chen, H.; Zhang, L.; Lin, X.; Li, X.; Zhuang, H.; Fan, H.; Meng, T.; He, Z.; Huang, H.; et al. The AMPK-MFN2 axis regulates MAM dynamics and autophagy induced by energy stresses. *Autophagy* **2021**, *17*, 1142–1156. [CrossRef]
8. Luo, R.; Fan, Y.; Yang, J.; Ye, M.; Zhang, D.F.; Guo, K.; Li, X.; Bi, R.; Xu, N.; Yang, L.-X.; et al. A novel missense variant in ACAA1 contributes to early-onset Alzheimer's disease, impairs lysosomal function, and facilitates amyloid-beta pathology and cognitive decline. *Signal Transduct. Target. Ther.* **2021**, *6*, 325–341. [CrossRef]
9. Mauthe, M.; Orhon, I.; Rocchi, C.; Zhou, X.; Luhr, M.; Hijlkema, K.J.; Coppes, R.P.; Engedal, N.; Mari, M.; Reggiori, F. Chloroquine inhibits autophagic flux by decreasing autophagosome-lysosome fusion. *Autophagy* **2018**, *14*, 1435–1455. [CrossRef]
0. Wang, F.; Xiao, Y.; Yan, J.-F.; Huang, G.-M.; Zhang, J.-H.; Di, C.-X.; Si, J.; Zhang, H.; Mao, A.-H. Carbon ion irradiation-induced DNA damage evokes cell cycle arrest and apoptosis via the pRb/E2F1/c-Myc signaling pathway in p53-deficient prostate cancer PC-3 cells. *Nucl. Sci. Tech.* **2021**, *32*, 30–42. [CrossRef]
1. Carew, J.S.; Espitia, C.M.; Zhao, W.; Han, Y.; Visconte, V.; Phillips, J.; Nawrocki, S.T. Disruption of Autophagic Degradation with ROC-325 Antagonizes Renal Cell Carcinoma Pathogenesis. *Clin. Cancer Res.* **2017**, *23*, 2869–2879. [CrossRef]
2. Sun, Y.M.; Wang, X.Y.; Zhou, X.R.; Zhang, C.; Liu, K.J.; Zhang, F.Y.; Xiang, B. Salidroside Ameliorates Radiation Damage by Reducing Mitochondrial Oxidative Stress in the Submandibular Gland. *Antioxidants* **2022**, *11*, 1414. [CrossRef] [PubMed]
3. Chauveinc, L.; Giraud PFau-Dahnier, S.; Dahnier SFau-Mounier, N.; Mounier NFau-Cosset, J.M.; Cosset, J.M. Radiotherapy-induced solid tumors: Review of the literature and risk assessment. *Cancer Radiother.* **1998**, *2*, 12–18. [CrossRef] [PubMed]
4. Rubino, C.; de Vathaire, F.; Shamsaldin, A.; Labbe, M.; Le, M.G. Radiation dose, chemotherapy, hormonal treatment and risk of second cancer after breast cancer treatment. *Brit. J. Cancer* **2003**, *89*, 840–846. [CrossRef]
5. Dorr, W.; Herrmann, T. Second tumors after oncologic treatment. *Strahlenther. Onkol.* **2008**, *184*, 67–72. [CrossRef]
6. Huh, J.W.; Tanksley, J.; Chino, J.; Willett, C.G.; Dewhirst, M.W. Long-term Consequences of Pelvic Irradiation: Toxicities, Challenges, and Therapeutic Opportunities with Pharmacologic Mitigators. *Clin. Cancer Res. Off. J. Am. Assoc. Cancer Res.* **2020**, *26*, 3079–3090. [CrossRef] [PubMed]
7. Zhang, J.; Li, K.; Zhang, Q.; Zhu, Z.; Huang, G.; Tian, H. Polycysteine as a new type of radio-protector ameliorated tissue injury through inhibiting ferroptosis in mice. *Cell Death Dis.* **2021**, *12*, 195–211. [CrossRef] [PubMed]
8. Xu, Q.; Zhang, H.; Liu, H.; Han, Y.; Qiu, W.; Li, Z. Inhibiting autophagy flux and DNA repair of tumor cells to boost radiotherapy of orthotopic glioblastoma. *Biomaterials* **2022**, *280*, 121287–121299. [CrossRef] [PubMed]
9. Wang, M.; Zeng, L.; Su, P.; Ma, L.; Zhang, M.; Zhang, Y.Z. Autophagy: A multifaceted player in the fate of sperm. *Hum. Reprod. Update* **2022**, *28*, 200–231. [CrossRef]
0. Kim, D.; Kim, J.; Yu, Y.S.; Kim, Y.R.; Baek, S.H.; Won, K.J. Systemic approaches using single cell transcriptome reveal that C/EBPgamma regulates autophagy under amino acid starved condition. *Nucleic Acids Res.* **2022**, *50*, 7298–7309. [CrossRef]
1. Ruart, M.; Chavarria, L.; Camprecios, G.; Suarez-Herrera, N.; Montironi, C.; Guixé-Muntet, S.; Bosch, J.; Friedman, S.L.; Garcia-Pagán, J.C.; Hernández-Gea, V. Impaired endothelial autophagy promotes liver fibrosis by aggravating the oxidative stress response during acute liver injury. *J. Hepatol.* **2019**, *70*, 458–469. [CrossRef]

52. Park, S.J.; Frake, R.A.; Karabiyik, C.; Son, S.M.; Siddiqi, F.H.; Bento, C.F.; Sterk, P.; Vicinanza, M.; Pavel, M.; Rubinsztein, D.C. Vinexin contributes to autophagic decline in brain ageing across species. *Cell Death Differ.* **2022**, *29*, 1055–1070. [CrossRef]
53. Jing, Z.; Liu, Q.; He, X.; Jia, Z.; Xu, Z.; Yang, B.; Liu, P. NCAPD3 enhances Warburg effect through c-myc and E2F1 and promotes the occurrence and progression of colorectal cancer. *J. Exp. Clin. Cancer Res.* **2022**, *41*, 198–216. [CrossRef] [PubMed]
54. Liu, S.; Sun, Z.; Liang, M.; Song, W.; Zhang, R.; Shi, Y.; Cui, Y.; Gao, Q. An Unrevealed Molecular Function of Corannulene Buckybowl Glycoconjugates in Selective Tumor Annihilation by Targeting the Cancer-Specific Warburg Effect. *Adv. Sci.* **2022**, *9*, 2105315–2105330. [CrossRef] [PubMed]
55. Li, Z.; Ge, Y.; Dong, J.; Wang, H.; Zhao, T.; Wang, X.; Liu, J.; Gao, S.; Shi, L.; Yang, S.; et al. BZW1 Facilitates Glycolysis and Promotes Tumor Growth in Pancreatic Ductal Adenocarcinoma Through Potentiating eIF2alpha Phosphorylation. *Gastroenterology* **2022**, *162*, 1256–1271. [CrossRef] [PubMed]

antioxidants

Systematic Review

Coenzyme Q10 and Dementia: A Systematic Review

Félix Javier Jiménez-Jiménez [1,*], Hortensia Alonso-Navarro [1], Elena García-Martín [2] and José A. G. Agúndez [2]

1 Section of Neurology, Hospital Universitario del Sureste, Arganda del Rey, Ronda del Sur 10, E-28500 Arganda del Rey, Spain
2 University Institute of Molecular Pathology Biomarkers, Universidad de Extremadura, E-10071 Cáceres, Spain
* Correspondence: fjavier.jimenez@salud.madrid.org or felix.jimenez@sen.es; Tel.: +34-636-968395; Fax: +34-91-328-0704

Abstract: It is well known that coenzyme Q_{10} (CoQ_{10}) has important antioxidant properties. Because one of the main mechanisms involved in the pathogenesis of Alzheimer's disease (AD) and other neurodegenerative diseases is oxidative stress, analysis of the concentrations of CoQ_{10} in different tissues of AD patients and with other dementia syndromes and the possible therapeutic role of CoQ_{10} in AD have been addressed in several studies. We performed a systematic review and a meta-analysis of these studies measuring tissue CoQ_{10} levels in patients with dementia and controls which showed that, compared with controls, AD patients had similar serum/plasma CoQ_{10} levels. We also revised the possible therapeutic effects of CoQ_{10} in experimental models of AD and other dementias (which showed important neuroprotective effects of coenzyme Q_{10}) and in humans with AD, other dementias, and mild cognitive impairment (with inconclusive results). The potential role of CoQ_{10} treatment in AD and in improving memory in aged rodents shown in experimental models deserves future studies in patients with AD, other causes of dementia, and mild cognitive impairment.

Keywords: coenzyme Q_{10}; tissue concentrations; therapeutics; Alzheimer's disease; dementia; vascular dementia; Lewy body dementia

Citation: Jiménez-Jiménez, F.J.; Alonso-Navarro, H.; García-Martín, E.; Agúndez, J.A.G. Coenzyme Q10 and Dementia: A Systematic Review. *Antioxidants* **2023**, *12*, 533. https://doi.org/10.3390/antiox12020533

Academic Editor: Yan-Zhong Chang

Received: 27 January 2023
Revised: 9 February 2023
Accepted: 17 February 2023
Published: 20 February 2023

1. Introduction

The 1,4-benzoquinone ubiquinone or coenzyme Q_{10} (CoQ_{10}), which is present in the majority of tissues in the human body, is an important component of the mitochondrial electron transport, participating in the generation of cellular energy through oxidative phosphorylation, and can be present in tissues in three different redox states: fully reduced (ubiquinol), fully oxidized (ubiquinone), and partially oxidized (semiquinone or ubisemiquinone). In addition to mitochondria, CoQ_{10} is present in peroxisomes, lysosomes, and the Golgi apparatus. CoQ_{10} has important antioxidant properties, with both a direct antioxidant effect of scavenging free radicals, and an indirect one of participating in the regeneration of other antioxidants such as ascorbic acid and alpha-tocopherol, offering protection to cells against oxidative stress processes [1,2].

It is well known that one of the most important pathogenetic mechanisms of Alzheimer's disease (AD) and other neurodegenerative disorders is oxidative stress [3–5]. Due to the important antioxidant functions of CoQ_{10}, several publications over the last two decades have addressed the issues of both determinations of CoQ_{10} levels in different tissues of patients diagnosed with AD or other types of dementia and on the potential therapeutic role of CoQ_{10} in these diseases (considering experimental studies in animal models of dementia in humans suffering from AD or other dementias). This systematic review and meta-analysis aims to analyze the results of studies addressing the tissular concentrations of CoQ_{10} in patients diagnosed with AD and other dementia syndromes compared with healthy controls, and the results of therapeutic trials of CoQ_{10} in AD (including experimental models of this disease) and in other causes of dementia.

2. Methods

2.1. Search Strategy and Criteria for Eligibility of Studies

We undertook a literature search using 3 well-known databases (PubMed, EMBASE, Web of Science-WOS-Main Collection) from 1966 until 31 December 2022. We crossed the term "coenzyme Q_{10}" with "Alzheimer's disease" (188, 403, and 171 items found in PubMed, EMBASE, and WOS, respectively), "dementia" (222, 79, and 86 items found in PubMed, EMBASE, and WOS, respectively), "vascular dementia" (13, 9, and 8 items found in PubMed, EMBASE, and WOS, respectively), "Lewy body dementia" (9, 8, and 11 items found in PubMed, EMBASE, and WOS, respectively) "Lewy body disease" (9, 15, and 25 items found in PubMed, EMBASE, and WOS, respectively), and "mild cognitive impairment" (68, 19, and 34 items found in PubMed, EMBASE, and WOS, respectively). The search retrieved 477 references which were examined one by one by the authors in order to select exclusively those strictly related to the proposed topic. Duplicated articles and abstracts were excluded. We did not apply any language restrictions. Figure 1 represents the flowcharts for the selection of eligible studies which analyzed tissue CoQ_{10} concentrations in patients with different types of dementia, and therapeutic trials with CoQ_{10} in experimental models of dementia or in patients with AD or other dementias according to the PRISMA guidelines [6].

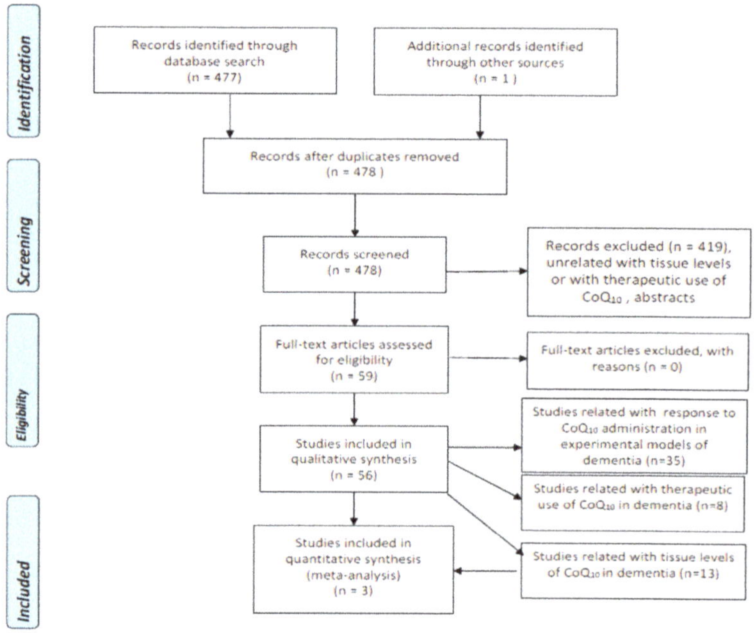

Figure 1. Flowchart for studies assessing tissue concentrations of coenzyme Q10 in dementia (PRISMA) (6, 17).

2.2. Selection of Studies and Methodology for the Meta-Analyses

We performed a meta-analysis of observational eligible studies assessing the concentrations of CoQ_{10} in tissues of patients diagnosed with AD and/or other causes of dementia and in controls. We extracted the following information: first author, year of publication, country, study design, and quantitative measures. We analyzed the risk for bias with the Newcastle–Ottawa Scale [7]. Table 1 summarizes data from selected studies analyzing tissular concentrations of CoQ_{10} in patients diagnosed with AD, Lewy body dementia (LBD), vascular dementia (VD), and dementia without specification of etiology compared with controls (with the exception of one study that compares the serum/plasma CoQ_{10} of patients with dementia with reference values).

Table 1. Coenzyme Q_{10} concentrations in several tissues from dementia patients and healthy controls (HC).

Tissue	Author, Year [Ref]	Parameter	AD N	AD Mean ± SD	HC N	HC Mean ± SD	Difference in Means (95% C.I.), *p*
		Alzheimer's Disease (AD)					
Serum/plasma	De Bustos et al., 2000 [8]	Total CoQ_{10} (nmol/L)	44	1262 ± 389	21	1209 ± 457	53.00 (−165.26 to 271.26); 0.629
	Battino et al., 2003 [9]	Total CoQ_{10} (nmol/L)	18	1216 ± 69	53	1278 ± 104	−62.00 (−114.55 to −9.45), 0.021
	Giavarotti et al., 2013 [10]	Total CoQ_{10} (nmol/L)	23	130 ± 30	42	115 ± 20	15.00 (2.57 to 27.43), 0.019
	Total series	**Total CoQ_{10} (nmol/L)**	**85**	**945.95** ± 573.50	**116**	**844.42** ± 588.71	**Random effects model *p* = 0.911**
	De Bustos et al., 2000 [8]	Total CoQ_{10} / cholesterol (µmol/mmol)	44	0.24 ± 0.05	21	0.23 ± 0.05	0.01 (−0.02 to 0.04), 0.454
CSF	Isobe et al., 2009, 2010 [11,12]	Oxidized CoQ_{10} (nmol/L)	30	5.2 ± 1.5	30	1.9 ± 1.3	3.30 (2.69 to 3.91), <0.0001
	Isobe et al., 2009, 2010 [11,12]	Reduced CoQ_{10} (nmol/L)	30	1.4 ± 0.6	30	2.7 ± 0.7	−1.30 (−1.64 to −0.96), <0.0001
	Isobe et al., 2009, 2010 [11,12]	Total CoQ_{10} (nmol/L)	30	6.6 ± 1.2	30	4.6 ± 1.0	2.00 (1.43 to 2.57), <0.0001
	Isobe et al., 2009, 2010 [11,12]	Oxidized/total CoQ_{10}	30	0.782 ± 0.188	30	0.413 ± 0.104	0.37 (0.29 to 0.45), <0.0001
		Lewy Body Dementia (LBD)					
Tissue	**Author, Year [Ref]**	**Parameter**	**LBD N**	**LBD Mean ± SD**	**HC N**	**HC Mean ± SD**	**Difference in Means (95% C.I.), *p***
Serum/plasma	Molina et al., 2002 [13]	Total CoQ_{10} (nmol/L)	18	960.6 ± 359.1	20	1205.2 ± 362.2	−244.60 (−482.30 to −6.90 to); 0.044
	Molina et al., 2002 [13]	Total CoQ_{10}/cholesterol	18	4.67 ± 1.75	20	5.05 ± 1.52	−0.38 (−1.46 to 0.70); 0.478
		Vascular Dementia (VD)					
Tissue	**Author, Year [Ref]**	**Parameter**	**VD N**	**VD Mean ± SD**	**HC N**	**HC Mean ± SD**	**Difference in Means (95% C.I.), *p***
Serum/plasma	De Bustos et al., 2000 [8]	Total CoQ_{10} (nmol/L)	17	1130 ± 452	21	1209 ± 457	−79.00 (−379.92 to 221.92), 0.598
	De Bustos et al., 2000 [8]	Total CoQ_{10}/cholesterol (µmol/mmol)	44	0.22 ± 0.06	21	0.23 ± 0.05	−0.01 (−0.04 to 0.02), 0.511
		Dementia without Specific Etiologic Diagnosis (DEM)					
Tissue	**Author, Year [Ref]**	**Parameter**	**DEM N**	**DEM Mean ± SD**	**HC N**	**HC Mean ± SD**	**Difference in Means (95% C.I.), *p***
Serum/plasma	Yamagishi et al., 2014 [14]	Total CoQ_{10} (nmol/L)	65	731 ± NA	130	762 ± NA	*p* = 0.32 (according to the authors, SD not provided)
Serum/plasma	Chang et al., 2022 [15]	Total CoQ_{10} (nmol/L)	80	410 ± 21	NA	NA	73% of patients showed low CoQ_{10} status using as reference values 500–1700 nmol/L
Serum/plasma	Yamagishi et al., 2014 [14]	Total CoQ_{10}/cholesterol (µmol /mmol)	65	0.14 ± NA	130	0.15 ± NA	*p* = 0.15 (according to the authors, SD not provided)
Serum/plasma	Chang et al., 2022 [15]	Total CoQ_{10}/cholesterol (µmol/mmol)	80	0.09 ± 0.04	NA	NA	73% of patients showed low CoQ_{10} status using as reference values 500–1700 nmol/L

We converted plasma/serum and CSF CoQ_{10} concentrations to nmol/mL when ne essary. The meta-analyses were carried out using the R software package meta [16] an following both the PRISMA [6] (Table S1) and the MOOSE guidelines [17] (Table S2). Be cause of the high heterogeneity across studies, we applied the random-effects model an used the inverse variance method for the meta-analytical procedure, the DerSimonian Laird as an estimator for Tau^2 [18], the Jackson method for the confidence interval of tau and tau [19], and the Hedges' g (bias-corrected standardized mean difference) [20]. Th statistical power to detect differences in mean values (alpha = 0.05) for the pooled sample was calculated when stated in the text. The meta-analysis was finally only applicabl to three studies on serum/plasma CoQ_{10} concentrations in patients with AD compare with controls.

3. Results

3.1. Studies Assessing Tissular CoQ_{10} Concentrations

3.1.1. Alzheimer's Disease

A total of three studies that assessed the serum/plasma levels of CoQ_{10} in patient with AD and controls failed to detect significant differences between the two stud groups (Table 1, Figure 2) [8–10]. One of these studies showed a similar serum/plasm CoQ_{10}/cholesterol ratio between AD patients and controls [8].

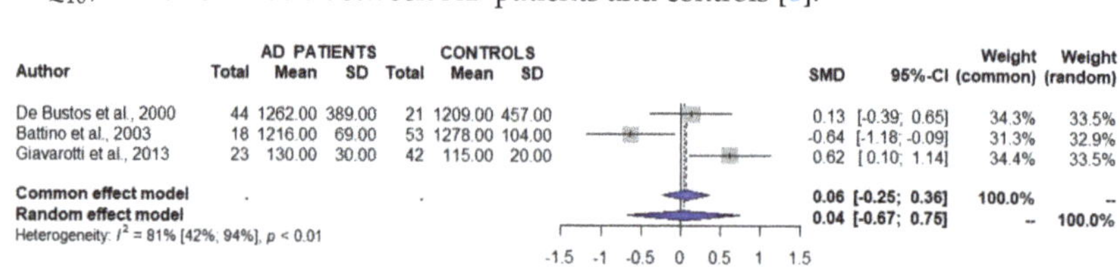

Figure 2. Studies assessing the serum/plasma levels of CoQ_{10} in patients with Alzheimer's diseas (AD) and controls show a lack of significant differences between the two groups. 95% CI 95 confidence intervals; SMD standard mean difference [8–10].

Isobe et al. [11,12] reported increased total CoQ_{10} and oxidized CoQ_{10} concentra tions in the cerebrospinal fluid from AD patients compared with controls, and a negativ correlation between oxidized/total coenzyme Q_{10} and duration of the disease.

To date, only two studies have addressed brain CoQ_{10} concentrations in patients wit AD. Edlund et al. [21] described the mean values (without SD) of CoQ_{10} in frontal, precen tral, temporal, and occipital cortex, and in nucleus caudate, hippocampus, pons, cerebellum and medulla oblongata of AD patients and controls. They reported a 30–100% increase i CoQ_{10} concentrations in most of these regions; however, the number of AD patients an controls involved in the study was not stated. Kim et al. [22] described a decreased activit of the 25 kDa subunit nicotinamide adenine dinucleotide + hydrogen (NADH):ubiquinon oxidoreductase (complex I) in the temporal and occipital cortex and of the 75 kDa subuni of this enzyme in the parietal cortex of patients with AD compared with controls, bu specific measures of CoQ_{10} were not performed.

Santa-Mara et al. [23] reported the presence of CoQ_{10} in paired helical filaments (aber rant protein aggregates containing tau protein) and in Hirano bodies (neuronal inclusion that are mainly observed in hippocampal neurons and are composed of actin either associ ated with or not associated with tau) in brain patients with AD, and state that CoQ_{10} wa able to induce the formation of aggregates when it was mixed with tau and actin.

3.1.2. Other Causes of Dementia

Serum CoQ_{10} concentrations and CoQ_{10}/cholesterol ratios from patients diagnose with LBD [13] and VD [8] did not differ significantly from those of controls according t two single studies.

Yamagishi et al. [14], in a community-based cohort study in Japan involving 6000 Japanese participants aged 40–69 years at baseline, described an inverse association between serum CoQ_{10} concentrations and the risk for disabling dementia, although serum CoQ_{10} levels and serum CoQ_{10}/cholesterol ratio did not differ significantly between 65 incident cases and 130 controls.

Finally, Chang et al. [15] reported "low CoQ_{10} status" in 73% of 80 patients diagnosed with dementia (they used reference values of their laboratory of 0.5-1-7 μM). In addition, they described a correlation between CoQ_{10} status and values of total antioxidant capacity, MiniMental State Examination, amyloid β-42 (Aβ-42), and Aβ-42/40 ratio, but not with tau protein.

3.2. Studies Assessing Therapeutic Response to CoQ_{10} Administration in Experimental Models of AD and Other Dementias

The results of studies assessing the response to the administration of COQ_{10} in different experimental models are summarized in Table 2.

Table 2. Studies on the effects of coenzyme Q_{10} in different experimental models of Alzheimer's disease.

Experimental Model	Author, Year [Ref]	Main Findings
AGED RATS	McDonald et al., 2005 [24]	Coadministration of CoQ_{10} and alpha-tocopherol (but not administration of each of these compounds alone) improved learning and memory tasks (assessed by a test that required the mice to rapidly identify and remember the correct arm of a T-maze, and to respond preemptively in order to avoid an electric shock).
AGED MICE	Wadsworth et al., 2008 [25]	Administration of CoQ_{10} decreased protein carbonyls in the brain but had no effect on lipid peroxidation, brain ATP levels, and mitochondrial membrane potential.
	Sumien et al., 2009 [26]	Intake of a low-CoQ_{10} diet did not change age-associated decrements in muscle strength, balance, coordinated running, or learning/memory, whereas intake of CoQ_{10} at a higher amount increased spontaneous activity, worsened age-related losses in acuity to auditory and shock stimuli, and impaired spatial learning/memory of old mice.
	Shetty et al., 2013 [27]	Intake of a low-CoQ_{10} diet did not change age-associated decrements in tests for spatial learning (Morris water maze), spontaneous locomotor activity, motor coordination, and startle reflex. However, intake of high-CoQ_{10} improved spatial learning and decreased protein oxidative damage in the heart, liver, skeletal muscle, and to a lesser extent, in the brain mitochondria.
	Shetty et al., 2014 [28]	Administration of α-tocopherol or α-tocopherol + CoQ_{10} diets improved coordinated running performance. The α -tocopherol + CoQ_{10} diet improved performance in a discriminated avoidance task (α-tocopherol and CoQ_{10} diets alone improved this task to a lesser degree). Both α-tocopherol and CoQ_{10} diets decreased protein damage, this effect being more marked with the α-tocopherol + CoQ_{10} combination.
HYPERCHOLESTE-ROLEMIA-INDUCED AD IN RATS	Ibrahim Fouad, 2020 [29]	Treatment with omega-3 and CoQ_{10} alone or in combination decreased markers of brain oxidative stress and inflammation and serum Aβ levels, regulated cholinergic functioning, and enhanced the functional outcome.
ALUMINIUM-INDUCED AD IN RATS	Ali et al., 2019 [30]	Treatment with CoQ_{10} in combination with vinpocetine partially reversed the changes induced by aluminium chloride (AlCl3) by decreasing malonyl-dialdehyde (MDA), increasing superoxide dismutase (SOD) and total antioxidant total capacity, decreasing IL1β, TNFα, chitinase, β-secretase, Aβ, tau protein, acetyl-cholinesterase, increasing catecholamine and brain-derived neurotrophic factor (BDNF) levels in brain tissue.
	Attia et al., 2020 [31]	Treatment with CoQ_{10} alone or in combination with biotin attenuated the changes induced by AlCl3 (impaired memory, a significant increase in Aβ, lipid peroxides, inflammatory markers—TNFα, IL6, IL1, nuclear factor κB-, caspase-3, and pSer-IRS-1, significant reduction in the antioxidants reduced glutathione and SOD-, pTyr-IRS-1, and p-Akt, reflecting Aβ-induced inflammation and defective insulin signaling, focal aggregations of inflammatory cells and neuronal degeneration).

Table 2. *Cont.*

Experimental Model	Author, Year [Ref]	Main Findings
	Ali et al., 2022 [32]	Treatment with CoQ_{10} reversed changes induced by aluminium by decreasing Aβ and acetylcholinesterase expression, increasing monoamine levels, restoring levels of total antioxidant capacity and superoxide-dismutase, and decreasing MDA, TNFα, and IL6.
FOREBRAIN LESIONED RATS	Nitta et al., 1994 [33]	Administration of CoQ_{10} to forebrain lesioned rats caused an increase in nerve growth factor (NGF) protein and mRNA and in choline acetyltransferase activity, and improved memory tasks such as behavioral deficits in habituation, water maze, and passive avoidance tasks in these animals.
INTRACEREBRO-VENTRICULAR INFUSION OF Aβ(1-42) IN RATS	Yamada et al., 1999 [34]	Coadministration of CoQ_{10} prevented some learning and memory deficits (Y-maze and water maze, but not passive avoidance tasks) in this model without affecting lipid peroxide levels in the hippocampus and cerebral cortex.
INTRAHIPPO-CAMPAL INJECTION OF Aβ(1-42) IN RATS	Singh et al., 2015 [35]	Treatment with CoQ_{10} and minocycline alone improved cognitive performance (reduced transfer latency and increased time spent in the target quadrant in the Morris Water Maze), reduced acetyl–cholinesterase activity, decreased oxidative damage (by reducing lipoperoxide and nitrite level and restoring superoxide, catalase, and reduced glutathione levels), decreased TNFα level, and restored mitochondrial respiratory enzyme complex activities and histopathological alterations induced by Aβ(1-42) in a dose-dependent and synergistic manner.
	Komaki et al., 2019 [36]	Treatment with CoQ_{10} reversed the decreased excitatory postsynaptic potential (EPSP) slope and population spike (PS) amplitude in the hippocampal dentate gyrus after induction of long-term potentiation (LTP) induced by injection of Aβ, reversed the decrease in serum MDA levels and total oxidant levels induced by injection of Aβ, and increased total antioxidant capacity levels.
INTRACEREBRO-VENTRICULAR INFUSION OF STREPTOZOTOCIN IN RATS	Ishrat et al., 2006 [37]	Coadministration of CoQ_{10} prevented learning and memory deficits (loss of cognitive performance in Morris water maze and passive avoidance tests), the increase in markers of oxidative damage (thiobarbituric acid reactive substances, reduced glutathione, protein carbonyl, activities of glutathione peroxidase and glutathione reductase), the decline of ATP in the hippocampus and cerebral cortex, the decrease in choline-acetyl-transferase activity and the increase in acetyl-cholinesterase activity induced by this neurotoxin.
	Sheykhhasan et al., 2022 [38]	Administration of CoQ10-loaded exosomes derived from adipose-derived stem cells improved memory impairment (assessed with the Morris water maze and passive avoidance task), increased BDNF expression, and increased cell density and the transcription factor *SOX2* gene expression in comparison with the administration of CoQ_{10} exosomes derived from adipose-derived stem cells alone.
TRANSGENIC MICE: AD PRESENILIN 1 MUTATION L235P	Yang et al., 2008 [39]	CoQ_{10} administration partially attenuated Abeta overproduction and intracellular Aβ deposit, partially decreased MDA increase, and up-regulated the decreased activity of SOD [24].
	Yang et al., 2010 [40]	CoQ_{10} administration reduced the burden of the amyloid plaques (assessed by immunohistochemistry and magnetic resonance imaging)
TRANSGENIC MICE: TG19959 MUTATION	Dumont et al., 2011 [41]	CoQ_{10} administration improved cognitive performance during Morris water maze testing, decreased brain levels of protein carbonyls (a marker of oxidative stress), decreased brain Aβ42 levels and Aβ protein precursor (AβPP), β-carboxyterminal fragments, and decreased plaque area and number in the hippocampus and the overlying cortex (assessed by immunostained with an Aβ42-specific antibody).
TRANSGENIC MICE: P301 TAU MUTATION (FRONTO-TEMPORAL DEMENTIA)	Elipenahli et al., 2012 [42]	CoQ_{10} administration improved survival and behavioral deficits (it increased locomotor activity and anxiety in open field testing), caused a modest reduction in phosphorylated tau, a significant increase in complex I activity and protein levels, and a reduction in lipid peroxidation in the cortex.
DOUBLE TRANSGENIC MICE: MUTATIONS *TGAPESWE* AND *PSEN1DE9*	Muthukumaran et al., 2018 [43]	Administration of ubisol-Q_{10} (a water-soluble form of coenzyme Q_{10}) improved long-term memory, preserved working spatial memory, and inhibited Aβ plaque formation in 18-month-old transgenic mice compared to an untreated transgenic group.

Table 2. *Cont.*

Experimental Model	Author, Year [Ref]	Main Findings
TRIPLE TRANSGENIC MICE: MUTATIONS *PS1M146V*, *APPSWE*, AND *TAUP301L*	Sui et al., 2014 [44]	The administration of CoQ_{10} altered changes in the differentially expressed serum proteins in the transgenic compared with wild-type mice by up-regulating 10 proteins and down-regulating another 10 proteins. Among the proteins modulated by CoQ10, clusterin and α-2-macroglobulin were validated via ELISA assay.
CELL CULTURES: HUMAN SKIN FIBROBLASTS FROM PS1 MUTATED FAMILIAL AD	Ma et al., 2014 [45]	CoQ_{10} treatment decreased reactive oxygen species generation, increased population doublings, and postponed stress-induced premature senescence. CoQ_{10} treatment increased proliferating cell nuclear antigen expression, and decreased levels of manganese-SOD (MnSOD), p21, p16Ink4A and cell cycle regulatory protein retinoblastoma (suggesting a resumption of autophagy).
	Vegh et al., 2019 [46]	Administration of ubisol-Q_{10} caused enhanced expression autophagy-related genes such as *beclin-1* (a major autophagy regulator) and *mitogen-activated protein kinase 8* (*MAPK8/JNK1*, a major activator of beclin-1) avoiding resumption of premature senescence. Withdrawal of ubisol-Q10 treatment led to the return of the senescence phenotype in AD fibroblasts.
CELL CULTURES: HUMAN SH-SY5Y NEUROBLASTOMA CELLS	Qi et al., 2005 [47]	Exposure of these cells to Aβ(1-42) caused, among other effects, enhanced lipid peroxidation and protein oxidation and significant reductions in the total contents of phospholipids, ubiquinone-10, and alpha3 and alpha7 subunit proteins of nicotinic acetylcholine receptors.
CELL CULTURES: HUMAN MC65 NEUROBLASTOMA CELLS	Wadsworth et al., 2008 [25]	Administration of CoQ_{10} showed a neuroprotective effect on the neurotoxic effects induced by the Aβ protein precursor C-terminal fragment (APP CTF).
CELL CULTURES: HUMAN UMBILICAL VEIN ENDOTHELIAL CELLS (HUVECS)	Durán-Prado et al., 2014 [48]	CoQ_{10} pretreatment delayed Aβ incorporation into the plasma membrane and mitochondria, reduced the influx of extracellular Ca^{2+} and Ca^{2+} release from mitochondria due to opening the mitochondrial transition pore after Aβ administration, decreasing O_2^- and hydrogen peroxide (H_2O_2) levels, prevented Aβ-induced necrosis and apoptosis, and restored the ability to proliferate, migrate and form tube-like structures in vitro.
CELL CULTURES: RAT BRAIN ENDOTHELIAL CELLS	Frontiñán-Rubio et al., 2021 [49]	CoQ_{10} pretreatment protected endothelial brain cells from Aβ(25–35)-induced damage, preventing nicotinamide adenine dinucleotide phosphate (NADPH) oxidase activity and reducing both reactive oxygen species generation and increase in free cytosolic Ca^{2+} induced by Aβ(25–35) (this prevented apoptosis and necrosis).
CELL CULTURES: PRIMARY CULTURED RAT CORTICAL NEURONS	Choi et al., 2012 [50]	CoQ_{10} protected neuronal cells against Aβ(25–35)-induced neurotoxicity in a concentration-dependent manner by increasing the expression levels of proteins related to neuronal cell survival (p85aPI3K, phosphorylated protein kinase B–Akt-, phosphorylated glycogen synthase kinase-3β, and heat shock transcription factor), and decreasing the levels of proteins associated with neuronal death (cytosolic cytochrome c and cleaved caspase-3). This protective effect was blocked by a phosphatidylinositol 3-kinase (PI3K) inhibitor.
	Wang et al., 2020 [51]	CoQ_{10} pretreatment significantly prevented neurons from Aβ-induced collapse of mitochondrial bioenergetics and perturbations of the protein kinase A (PKA)/cAMP response element-binding protein (CREB) signaling.
CELL CULTURES: CULTURED NEURAL STEM CELLS	Choi et al., 2013 [52]	Co-administration of CoQ_{10} restored the Aβ(25–35) oligomer-inhibited proliferation of neural stem cells by increasing the expression levels of proteins related to the PI3K pathway (p85α PI3K, phosphorylated Akt-Ser473-, phosphorylated glycogen synthase kinase-3β-Ser9-, and heat shock transcription factor). This protective effect was blocked by a phosphatidylinositol 3-kinase (PI3K) inhibitor.
CELL CULTURES: PRIMARY CULTURED HIPPOCAMPAL NEURONS FROM FETAL MICE	Yang et al., 2020 [53]	Administration of CoQ_{10} reversed all the effects induced by sevoflurane anesthesia (decrease in ATP and SOD levels, increase in apolipoprotein E (ApoE) mRNA, total ApoE protein, full-length ApoE, and ApoE fragments, increase in phosphorylated tau and neuroinflammatory factor (TNFα, IL6, and IL1β) expression levels.
CELL CULTURES: BRAIN MITOCHON-DRIA ISOLATED FROM AGED DIABETIC RATS	Moreira et al., 2005 [54]	CoQ_{10} treatment attenuated the decrease in oxidative phosphorylation efficiency and avoided the increase in H_2O_2 production induced by Aβ1-40.

Table 2. *Cont.*

Experimental Model	Author, Year [Ref]	Main Findings
CELL CULTURES: RAT PHEOCHROMO-CYTOMA (PC12) CELL LINE	Li et al., 2017 [55]	CoQ_{10} treatment suppressed the protein expression of COX-2 and the level of PGE2 in Aβ(25–35)-injured PC12 cells (this effect was correlated with the suppression of NF-κB activation by CoQ_{10}, attenuating neuroinflammation).

In general, the administration of CoQ_{10} alone or in combination with other substance (mainly other antioxidants) has been useful to improve the results of clinical tasks re lated to learning and memory and to improve or prevent oxidative stress, inflammatio and cellular death in different models of AD and frontotemporal dementia includin aged rodents [24–29], aluminium-induced AD in rats [30–32], forebrain lesioned rats [33 intracerebroventricular infusion of Aβ-42 [34] or streptozotocin [37,38] or intrahippocam pal injection of Aβ-42 [35,36] in rats, transgenic mice with different mutations inducin AD [39–41,43–46] or frontotemporal dementia [42], and cell cultures using different hu man [25,45,46,48] or rodent cells [49–55]. On the other hand, Aβ(1-42) decreased CoQ_1 concentrations in human SH-SY5Y neuroblastoma cells in culture [47].

3.3. *Studies Assessing Therapeutic Response to CoQ_{10} Administration in Patients with Dementia*

3.3.1. Alzheimer's Disease

Table 3 summarizes the results of the eight eligible studies addressing the therapeuti response to CoQ_{10} administration in patients with AD [56–63], although in one of them an important percentage of patients included were diagnosed with mixed dementia [61 Two of these studies used an open-label design [56,61] while the others were randomized clinical trials [33].

Table 3. Studies describing the effects of COQ_{10} supplementation in patients with AD. AD Alzheimer's disease; ADAS: Alzheimer's Disease Assessment Scale; ADAS-Cog: ADAS cognitive score; ADAS-Noncog: ADAS non-cognitive scores; ADCS-ADL: Alzheimer's Disease Cooperative Study Activities of Daily Living; ADL: activities of daily living; CGI-I: clinical global impression improvement; CGI-C: clinical global impression change; CMT: Central macular thickness; DAT: de mentia of Alzheimer type; DSS: Digit Symbol Substitution test; GCIPL: Ganglion cell-inner plexiform layer; MMSE: MiniMental State Examination; OCT: optic coherence tomography; RNFL: Retina nerve fiber layer; SCT: Subfoveal choroidal thickness.

Authors, Year [Ref]	Study Setting	Type of Study	Main Findings	Level of Evidence (Quality Score)
Imagawa et al., 1992 [56]	Combined therapy with CoQ10, iron, and vitamin B6 in 27 AD patients.	Open-label study	• Treatment was as effective as mitochondrial activation therapy in 27 AD patients. • Treatment induced significant clinical improvement in two genetically confirmed AD patients.	II (NA)
Weyer et al., 1997 [57]	Three hundred patients with mild to moderate degree DAT were prescribed idebenone 30 mg t.i.d. (*n* = 100), idebenone 90 t.i.d. (*n* = 100), or placebo (*n* = 100). Evaluation at baseline, 1, 3, and 6 months including a total score of the ADAS-Total, ADAS cognitive (ADAS-Cog) and noncognitive scores (ADAS-Noncog), CGI-I, MMSE, Digit Symbol Substitution test (DSS) and several scales for the assessment of daily activities (the self- and observer-rating scales NAA and NAB of the Nuremberg Age Inventory NAI and Greene's Assessment).	Multicenter, randomized, double-blind, placebo-controlled, dosage-ranging trial	• Idebenone 90 mg t.i.d. improved significantly and was superior to placebo and idebenone 60 mg t.i.d. in ADAS-Total, ADAS-Cog, ADAS-Noncog, and CGI-global improvement. • Safety results (adverse events, vital signs, ECG, and clinical laboratory parameters) were similar for the three groups.	I (>50%)

Table 3. *Cont.*

Authors, Year [Ref]	Study Setting	Type of Study	Main Findings	Level of Evidence (Quality Score)
Gutzmann and Hadler D, 1998 [58]	Four hundred and fifty patients with mild to moderate degree DAT were prescribed placebo for 12 months, followed by idebenone 90 mg for another 12 months (n = 153) or idebenone 90 mg tid for 24 months (n = 148) or 120 mg ti for 24 months (n = 149). Evaluation included a total score of the Alzheimer's Disease Assessment Scale (ADAS-Total), ADAS cognitive (ADAS-Cog) and noncognitive scores (ADAS-Noncog), CGI-Improvement), the SKT neuropsychological test battery, and the Nurses' Observation Scale for Geriatric Patients (NOSGER-Total and IADL subscale).	Prospective, randomized, double-blind multicentre study in three parallel groups	• During the placebo-controlled period, idebenone showed statistically significant dose-dependent improvement in all the efficacy variables. • A further improvement of most efficacy variables was determined in the second year in comparison to the results at the 12-month visit, with a clear dose–effect relationship (placebo < idebenone 90 mg < idebenone 120 mg). • Safety results (adverse events, vital signs, ECG, and clinical laboratory parameters) were similar for the three groups.	I (>50%)
Gutzmann et al., 2002 [59]	Two hundred and three patients with mild to moderate degree DAT were prescribed idebenone 360 mg/day (n = 104) or tacrine up to 160 mg/day (n = 99) for 60 weeks. Evaluation included the Efficacy Index Score (EIS, a combination of improvement in cognitive function, activities of daily living, and global function), the ADAS-Cog score, the NOSGER-IADL score, and the CGI-I.	Prospective, randomized, double-blind, parallel-group multicenter study	• A total of 28.8% of the patients on idebenone and 9.1% of the patients on tacrine finalized the follow-up. • A total of 50% of the patients on idebenone and 39.4% of the patients on tacrine showed an improvement in at least one of the other (secondary) outcome variables. • Patients on idebenone showed a higher benefit from treatment than patients on tacrine.	I (>50%)
Thal et al., 2003 [60]	Five hundred and thirty-six patients diagnosed with probable AD aged over 50 with MMSE scores between 12 and 25 were prescribed idebenone 120, 240, or 360 mg, or placebo (n = 136, 138, 133, and 126, respectively) during 1 year. Evaluation included ADAS-Cog, CGIC (primary outcome measures), and measurements of ADL, Behavioral Pathology in Alzheimer's Disease Rating Scale, and MMSE (secondary outcomes).	Multicenter, double-blind, placebo-controlled, randomized trial	• The study was completed by 95, 94, 92, and 96 of the patients assigned to idebenone 120, 240, or 360 mg, or placebo, respectively. • Primary outcome measures did not differ significantly between the four groups. • In an exploratory two-group analysis comparing all three treated groups combined with a placebo, drug-treated patients performed better on the ADAS-Cog, although CGIC scores did not differ significantly.	I (>50%)
Voronkova and Meleshkov, 2009 [61]	Thirty-five patients were diagnosed with AD (n = 9), mixed dementia (n = 21), or memory impairment not reaching dementia (n = 5). Treatment with CoQ_{10} 120 mg/day for 6 months. Assessment with the Luriya method (memory and especially auditory-speech memory), Clinical Dementia Rating scale (CDRS), CGIC, and MMSE.	Open-label study	• Improvement in the MMSE score in patients with mild and moderate dementia. • Improvement in daily activities in 27% of patients, including improvement in short-term and long-term memory and attention, speech functions, the performance of kinesthetic, spatial, and dynamic praxis tests, visuospatial gnosis, thought, and writing. • Improvement on the CGI scale in 37% of patients.	II (NA)

Table 3. *Cont.*

Authors, Year [Ref]	Study Setting	Type of Study	Main Findings	Level of Evidence (Quality Score)
Galasko et al., 2012 [62]	Seventy-eight patients with mild to moderate AD (66 of them provided serial CSF specimens adequate for biochemical analyses). Random assignment to treatment for 16 weeks with 800 IU/d of vitamin E (α-tocopherol) plus 500 mg/d of vitamin C plus 900 mg/d of α-lipoic acid (E/C/ALA); 400 mg of coenzyme Q_{10} 3 times/day; or placebo (26 to each group; 24, 20, and 12, respectively, provided CSF). Evaluation at baseline and 16 weeks of MMSE and ADCS-ADL scale, and CSF biomarkers related to AD.	Monocenter, randomized, placebo-controlled, double-blind clinical trial	• Accelerated decline in MMSE scores occurred in the E/C/ALA group. • Changes in CSF Aβ42, tau, and P-tau(181) levels did not differ between the three groups. • Cerebrospinal fluid F2-isoprostane levels decreased on average by 19% from baseline to week 16 in the E/C/ALA group but were unchanged in the other groups. • Drugs used were well tolerated.	I (>50%)
Karakahya and Özcan, 2020 [63]	Sixty-two patients diagnosed with AD (31 randomized to the treatment group and 31 to the observational group), and 31 healthy controls. The treatment group received topical application of CoQ10 on the retina and choroids. Assessment of CMT, RNFL thickness, GCIPL thickness, and SCT with OCT at baseline and after 6 months.	Monocenter, randomized clinical trial	• Increased RNFL thickness in all quadrants in the treatment group, but only significant in the temporal sector (inversely correlated with AD duration). • Increased GCIPL thickness in the treatment on average and superonasal sector (inversely correlated with AD severity). • Increased ganglion cell-inner plexiform layer in the treatment group.	I (>50%)

Imagawa et al. [56], after a preliminary report indicating that therapy with CoQ10, iron, and vitamin B6 was effective as mitochondrial activation therapy in 27 AD patients, reported a significant clinical improvement with this therapy in two genetically confirmed AD patients. Three of the randomized clinical trials showed improvement in neuropsychological assessments in patients treated with CoQ_{10} compared, respectively, with placebo [57,58] or with tacrine [59], while the other two did not show any improvement in comparison with placebo [60,62], although in one of them, the patients treated with CoQ_{10} showed a better outcome than those treated with a combination of α-tocopherol, vitamin C, and α-lipoic acid [62].

Karakahya and Özcan [63], in a study using optic coherence tomography (OCT) reported an improvement in retinal ganglion cell loss related to AD with short-term topical administration of CoQ10. Finally, an open-label study showed some degree of improvement in the MMSE score and other neuropsychological tests in patients with AD or mixed dementia [61].

3.3.2. Vascular Dementia (VD)

Kawakami et al. [64] measured CSF levels of homovanillic acid (HVA), 5-hydroxyindole acetic acid (5-HIAA), 3-methoxy-4-hydroxyphenylethylenglycol (MHPG), and noradrenalin (NA) in six patients with cerebrovascular dementia. CoQ_{10} administration during 1–2 months returned to normal CSF levels of HVA, 5-HIAA, and MHPG, which had been previously decreased compared to control values.

Qi et al. [65], in a randomized clinical trial involving 88 patients diagnosed with VD (44 of them assigned to treatment with butylphthalide plus idebenone as the observational group, and 44 to idebenone as the control group), showed a higher degree of improvement in MMSE, clinical dementia rating scale (CDRS), and ability of daily life (ADL), and a higher decrease in serum IL6, C reactive protein, TNFα, IL1β, CD31+, CDl44+, and endothelin-1 levels in the observational group compared with the control group.

3.3.3. Mild Cognitive Impairment and Normal Aging

García-Carpintero et al. [66], in a 1-year randomized, double-blind, placebo-controlled observational analytical study involving 69 patients diagnosed with mild cognitive impairment (MCI) assigned to CoQ_{10} 200 mg/day (n = 33) or placebo (n = 36) showed that although CoQ_{10} treatment improved cerebral vasoreactivity (assessed by transcranial Doppler sonography) and inflammatory markers, it did not display any significant improvement in the results of an extensive neuropsychological assessment.

Finally, Stough et al. [67] designed a 90-day randomized, double-blind, placebo-controlled, parallel group clinical trial involving 104 healthy subjects aged 60 years and over randomized to either CoQ_{10} 200 mg/day or placebo (52 per group), aiming to evaluate the effects of CoQ_{10} in the amelioration of cognitive decline that it should be undergoing. Interestingly, a recent study described a significant association of plasma CoQ_{10} concentrations with cognitive functioning and executive function in elderly people [68].

4. Discussion and Conclusions

The possible role of CoQ_{10} in the pathogenesis of AD and other causes of dementia, if any, is far from established with the current evidence. The studies addressing the serum/plasma levels of CoQ_{10}, which are scarce and based on a relatively small sample size, were similar for AD patients and controls [8–10]. The increased values of total and oxidized CoQ_{10} concentrations in the cerebrospinal fluid from patients [11,12] and in certain brain areas from patients with AD [13], described in single studies, have not had further replication studies and await confirmation. Studies on human AD brain are restricted to a single report of a 30–100% increase in CoQ_{10} concentrations in most of the regions studied (which included the frontal, precentral, temporal and occipital cortex, nucleus caudate, hippocampus, pons, cerebellum, and medulla oblongata) in an unspecified number of AD patients compared with controls [18]. The possibility of induction of aggregates of tau protein and actine by CoQ_{10} and the finding of the presence of this coenzyme in paired helical filaments and Hirano bodies in the hippocampus [23] lends support to the hypothesis of the possible role of CoQ_{10} in AD. Studies reporting on CoQ_{10} concentrations in other causes of dementia are restricted to the measurements in serum/plasma from patients with Lewy body dementia (LBD) [13], vascular dementia [8], and dementia without specification of etiologic diagnosis [14,15].

Because of their antioxidant actions, it was proposed that CoQ_{10} administration could be a potential protective therapy in AD [69,70]. Moreover, an important number of studies have shown a significant neuroprotective and/or clinical effect of the administration of CoQ_{10} in different experimental models of AD, as was previously commented in more detail in the Section 3 [24–46,49–55] (Table 2). Interestingly, most of the studies performed using cell cultures including human neuroblastoma SH-SY5Y [47] and human MC 65 neuroblastoma cells [25], human umbilical vein endothelial cells [48], rat endothelial [49], cortical [50,51] and brain stem cells [52], hippocampal neurons from fetal mice [53], brain mitochondria isolated from aged diabetic rats [54], and rat pheochromocytoma (PC12) cells [55] have shown a protective effect of CoQ_{10} on the neurotoxic effects of different types of Aβ. In addition, it has been shown that oral administration of CoQ_{10} results in an important increase in serum/plasma CoQ_{10} concentrations in humans [71–73] and in rats [73].

The potential beneficial effects of CoQ_{10} administration, its good absorption, and the lack of important adverse effects led to some initial short-term randomized clinical trials that showed improvement in several neuropsychological tests in patients with AD treated with CoQ_{10} in comparison with those assigned to placebo [57,58] or the anticholinesterase drug tacrine [59]. However, a further short-term randomized clinical trial failed to determine any benefit except a mild improvement in the ADAS-Cog scores [60].

In conclusion, according to the data from the results presented in this review, there are still important knowledge gaps regarding both the suitability of CoQ_{10} as a biomarker of AD and other causes of dementia (studies on this issue in brain, cerebrospinal fluid, and other tissues are scarce) and the possible usefulness of treatment with CoQ_{10} in patient with AD (controversial results of randomized controlled trials with a maximum of 1 yea of follow-up) despite the promising neuroprotective effects of CoQ_{10} detected in differen models of AD. The design of further studies with a longer-term follow-up period is neede

Supplementary Materials: The following supporting information can be downloaded at: https://www. mdpi.com/article/10.3390/antiox12020533/s1, Table S1: PRISMA Checklist; Table S2: MOOSE Checklis

Author Contributions: F.J.J.-J.: Conceptualization, Methodology, Investigation, Validation, Forma analysis, Investigation, Writing—original draft, Writing—review and editing, Project administratio H.A.-N.: Conceptualization, Methodology, Investigation, Validation, Formal analysis, Investigatio Writing—original draft, Writing—review and editing, Project administration. E.G.-M.: Conceptua ization, Methodology, Investigation, Validation, Formal analysis, Investigation, Writing—origina draft, Writing—review and editing, Project administration, Obtaining funding. J.A.G.A.: Conceptua ization, Methodology, Investigation, Validation, Formal analysis, Investigation, Writing—origina draft, Writing—review and editing, Project administration, Obtaining funding. All authors have rea and agreed to the published version of the manuscript.

Funding: The work at the authors' laboratory is supported in part by Grants PI18/00540 an PI21/01683 from Fondo de Investigación Sanitaria, Instituto de Salud Carlos III, Madrid, Spain, an IB20134 and GR21073 from Junta de Extremadura, Mérida, Spain. Partially funded with FEDER fund

Acknowledgments: We thank James McCue for assistance with the English language. We als appreciate the efforts of the personnel of the Library of Hospital Universitario del Sureste, Argand del Rey, who retrieved an important number of papers for us.

Conflicts of Interest: The authors declare that the research was conducted in the absence of an commercial or financial relationships that could be construed as a potential conflict of interest.

References

1. Crane, F.L. Biochemical functions of coenzyme Q10. *J. Am. Coll. Nutr.* **2001**, *20*, 591–598. [CrossRef]
2. Mantle, D.; Heaton, R.A.; Hargreaves, I.P. Coenzyme Q10, Ageing and the Nervous System: An Overview. *Antioxidants* **202** *11*, 2. [CrossRef] [PubMed]
3. Jiménez-Jiménez, F.J.; Alonso-Navarro, H.; Ayuso-Peralta, L.; Jabbour-Wadih, T. Estrés oxidativo y enfermedad de Alzheime *Rev. Neurol.* **2006**, *42*, 419–427. [CrossRef] [PubMed]
4. Kumaran, K.R.; Yunusa, S.; Perimal, E.; Wahab, H.; Müller, C.P.; Hassan, Z. Insights into the Pathophysiology of Alzheimer' Disease and Potential Therapeutic Targets: A Current Perspective. *J. Alzheimers Dis.* **2023**, *91*, 507–530. [CrossRef] [PubMed]
5. Jurcau, A. Insights into the Pathogenesis of Neurodegenerative Diseases, Focus on Mitochondrial Dysfunction and Oxidativ Stress. *Int. J. Mol. Sci.* **2021**, *22*, 11847. [CrossRef]
6. Moher, D.; Liberati, A.; Tetzlaff, J.; Altman, D.G.; Group, P. Preferred reporting items for systematic reviews and meta-analyse the PRISMA statement. *PLoS Med.* **2009**, *6*, e1000097. [CrossRef]
7. Wells, G.A.; Shea, B.; O'Connell, D.; Peterson, J.; Welch, V.; Losos, M.; Tugwell, P. The Newcastle-Ottawa Scale (NOS) fo Assessing the Quality If Nonrandomized Studies in Meta Analyses. Available online: http://www.ohri.ca/programs/clinical epidemiology/oxford.asp (accessed on 3 January 2023).
8. de Bustos, F.; Molina, J.A.; Jiménez-Jiménez, F.J.; García-Redondo, A.; Gómez-Escalonilla, C.; Porta-Etessam, J.; Berbel, A Zurdo, M.; Barcenilla, B.; Parrilla, G.; et al. Serum levels of coenzyme Q10 in patients with Alzheimer's disease. *J. Neural. Transm* **2000**, *107*, 233–239. [CrossRef]
9. Battino, M.; Bompadre, S.; Leone, L.; Devecchi, E.; Degiuli, A.; D'Agostino, F.; Cambiè, G.; D'Agostino, M.; Faggi, L Colturani, G.; et al. Coenzyme Q, Vitamin E and Apo-E alleles in Alzheimer Disease. *Biofactors* **2003**, *18*, 277–281. [CrossRef]
10. Giavarotti, L.; Simon, K.A.; Azzalis, L.A.; Fonseca, F.L.; Lima, A.F.; Freitas, M.C.; Brunialti, M.K.; Salomão, R.; Moscardi, A.A. Montaño, M.B.; et al. Mild systemic oxidative stress in the subclinical stage of Alzheimer's disease. *Oxid. Med. Cell. Longev.* **2013** *2013*, 609019. [CrossRef]
11. Isobe, C.; Abe, T.; Terayama, Y. Increase in the oxidized/total coenzyme Q-10 ratio in the cerebrospinal fluid of Alzheimer' disease patients. *Dement. Geriatr. Cogn. Disord.* **2009**, *28*, 449–454. [CrossRef]
12. Isobe, C.; Abe, T.; Terayama, Y. Levels of reduced and oxidized coenzyme Q-10 and 8-hydroxy-2'-deoxyguanosine in the CSF o patients with Alzheimer's disease demonstrate that mitochondrial oxidative damage and/or oxidative DNA damage contribute to the neurodegenerative process. *J. Neurol.* **2010**, *257*, 399–404. [CrossRef] [PubMed]
13. Molina, J.A.; de Bustos, F.; Ortiz, S.; Del Ser, T.; Seijo, M.; Benito-Léon, J.; Oliva, J.M.; Pérez, S.; Manzanares, J. Serum levels o coenzyme Q in patients with Lewy body disease. *J. Neural. Transm.* **2002**, *109*, 1195–1201. [CrossRef]

14. Yamagishi, K.; Ikeda, A.; Moriyama, Y.; Chei, C.L.; Noda, H.; Umesawa, M.; Cui, R.; Nagao, M.; Kitamura, A.; Yamamoto, Y.; et al. Serum coenzyme Q10 and risk of disabling dementia, the Circulatory Risk in Communities Study (CIRCS). *Atherosclerosis* **2014**, *237*, 400–403. [CrossRef]
15. Chang, P.S.; Chou, H.H.; Lai, T.J.; Yen, C.H.; Pan, J.C.; Lin, P.T. Investigation of coenzyme Q10 status.; serum amyloid-β.; and tau protein in patients with dementia. *Front. Aging Neurosci.* **2022**, *14*, 910289. [CrossRef]
16. Balduzzi, S.; Rücker, G.; Schwarzer, G. How to perform a meta-analysis with R: A practical tutorial. *BMJ Ment. Health* **2019**, *22*, 153–160. [CrossRef]
17. Stroup, D.F.; Berlin, J.A.; Morton, S.C.; Olkin, I.; Williamson, G.D.; Rennie, D.; Moher, D.; Becker, B.J.; Sipe, T.A.; Thacker, S.B. Meta-analysis of observational studies in epidemiology: A proposal for reporting. Meta-analysis Of Observational Studies in Epidemiology (MOOSE) group. *JAMA* **2009**, *283*, 2008–2012. [CrossRef]
18. Der Simonian, R.; Laird, N. Meta-analysis in clinical trials. *Control. Clin. Trials* **1986**, *7*, 177–188. [CrossRef] [PubMed]
19. Jackson, D. Confidence intervals for the between-study variance in random effects meta-analysis using generalised Cochran heterogeneity statistics. *Res. Synth. Methods* **2013**, *4*, 220–229. [CrossRef] [PubMed]
20. Hedges, L.V. Meta-Analysis. *J. Educ. Stat.* **1992**, *17*, 279–296. [CrossRef]
21. Edlund, C.; Söderberg, M.; Kristensson, K.; Dallner, G. Ubiquinone, dolichol, and cholesterol metabolism in aging and Alzheimer's disease. *Biochem. Cell. Biol.* **1992**, *70*, 422–428. [CrossRef]
22. Kim, S.H.; Vlkolinsky, R.; Cairns, N.; Fountoulakis, M.; Lubec, G. The reduction of NADH ubiquinone oxidoreductase 24- and 75-kDa subunits in brains of patients with Down syndrome and Alzheimer's disease. *Life Sci.* **2001**, *68*, 2741–2750. [CrossRef]
23. Santa-Mara, I.; Santpere, G.; MacDonald, M.J.; Gomez de Barreda, E.; Hernandez, F.; Moreno, F.J.; Ferrer, I.; Avila, J. Coenzyme q induces tau aggregation.; tau filaments.; and Hirano bodies. *J. Neuropathol. Exp. Neurol.* **2008**, *67*, 428–434. [PubMed]
24. McDonald, S.R.; Sohal, R.S.; Forster, M.J. Concurrent administration of coenzyme Q10 and alpha-tocopherol improves learning in aged mice. *Free Radic. Biol. Med.* **2005**, *38*, 729–736. [CrossRef]
25. Wadsworth, T.L.; Bishop, J.A.; Pappu, A.S.; Woltjer, R.L.; Quinn, J.F. Evaluation of coenzyme Q as an antioxidant strategy for Alzheimer's disease. *J. Alzheimers Dis.* **2008**, *14*, 225–234. [PubMed]
26. Sumien, N.; Heinrich, K.R.; Shetty, R.A.; Sohal, R.S.; Forster, M.J. Prolonged intake of coenzyme Q10 impairs cognitive functions in mice. *J. Nutr.* **2009**, *139*, 1926–1932. [CrossRef]
27. Shetty, R.A.; Forster, M.J.; Sumien, N. Coenzyme Q(10) supplementation reverses age-related impairments in spatial learning and lowers protein oxidation. *Age* **2013**, *35*, 1821–1834. [CrossRef] [PubMed]
28. Shetty, R.A.; Ikonne, U.S.; Forster, M.J.; Sumien, N. Coenzyme Q10 and α-tocopherol reversed age-associated functional impairments in mice. *Exp. Gerontol.* **2014**, *58*, 208–218. [CrossRef]
29. Ibrahim Fouad, G. Combination of Omega 3 and Coenzyme Q10 Exerts Neuroprotective Potential against Hypercholesterolemia-Induced Alzheimer's-Like Disease in Rats. *Neurochem. Res.* **2020**, *45*, 1142–1155. [CrossRef]
30. Ali, A.A.; Abo El-Ella, D.M.; El-Emam, S.Z.; Shahat, A.S.; El-Sayed, R.M. Physical & mental activities enhance the neuroprotective effect of vinpocetine & coenzyme Q10 combination against Alzheimer & bone remodeling in rats. *Life Sci.* **2019**, *229*, 21–35.
31. Attia, H.; Albuhayri, S.; Alaraidh, S.; Alotaibi, A.; Yacoub, H.; Mohamad, R.; Al-Amin, M. Biotin, coenzyme Q10, and their combination ameliorate aluminium chloride-induced Alzheimer's disease via attenuating neuroinflammation and improving brain insulin signaling. *J. Biochem. Mol. Toxicol.* **2020**, *34*, e22519. [CrossRef]
32. Ali, A.A.; Khalil, M.G.; Abd El-Latif, D.M.; Okda, T.; Abdelaziz, A.I.; Abu-Elfotuh, K.; Kamal, M.M.; Wahid, A. The influence of vinpocetine alone or in combination with Epigallocatechin-3-gallate, Coenzyme COQ10, Vitamin E and Selenium as a potential neuroprotective combination against aluminium-induced Alzheimer's disease in Wistar Albino Rats. *Arch. Gerontol. Geriatr.* **2022**, *98*, 104557. [CrossRef] [PubMed]
33. Nitta, A.; Murakami, Y.; Furukawa, Y.; Kawatsura, W.; Hayashi, K.; Yamada, K.; Hasegawa, T.; Nabeshima, T. Oral administration of idebenone induces nerve growth factor in the brain and improves learning and memory in basal forebrain-lesioned rats. *Naunyn Schmiedebergs Arch. Pharmacol.* **1994**, *349*, 401–407. [CrossRef] [PubMed]
34. Yamada, K.; Tanaka, T.; Han, D.; Senzaki, K.; Kameyama, T.; Nabeshima, T. Protective effects of idebenone and alpha-tocopherol on beta-amyloid-(1-42)-induced learning and memory deficits in rats: Implication of oxidative stress in beta-amyloid-induced neurotoxicity in vivo. *Eur. J. Neurosci.* **1999**, *11*, 83–90. [CrossRef] [PubMed]
35. Singh, A.; Kumar, A. Microglial Inhibitory Mechanism of Coenzyme Q10 Against Aβ (1-42) Induced Cognitive Dysfunctions: Possible Behavioral, Biochemical, Cellular, and Histopathological Alterations. *Front. Pharmacol.* **2015**, *6*, 268. [CrossRef]
36. Komaki, H.; Faraji, N.; Komaki, A.; Shahidi, S.; Etaee, F.; Raoufi, S.; Mirzaei, F. Investigation of protective effects of coenzyme Q10 on impaired synaptic plasticity in a male rat model of Alzheimer's disease. *Brain Res. Bull.* **2019**, *147*, 14–21. [CrossRef]
37. Ishrat, T.; Khan, M.B.; Hoda, M.N.; Yousuf, S.; Ahmad, M.; Ansari, M.A.; Ahmad, A.S.; Islam, F. Coenzyme Q10 modulates cognitive impairment against intracerebroventricular injection of streptozotocin in rats. *Behav. Brain Res.* **2006**, *171*, 9–16. [CrossRef] [PubMed]
38. Sheykhhasan, M.; Amini, R.; Soleimani Asl, S.; Saidijam, M.; Hashemi, S.M.; Najafi, R. Neuroprotective effects of coenzyme Q10-loaded exosomes obtained from adipose-derived stem cells in a rat model of Alzheimer's disease. *Biomed. Pharmacother.* **2022**, *152*, 113224. [CrossRef] [PubMed]
39. Yang, X.; Yang, Y.; Li, G.; Wang, J.; Yang, E.S. Coenzyme Q10 attenuates beta-amyloid pathology in the aged transgenic mice with Alzheimer presenilin 1 mutation. *J. Mol. Neurosci.* **2008**, *34*, 165–171. [CrossRef]

40. Yang, X.; Dai, G.; Li, G.; Yang, E.S. Coenzyme Q10 reduces beta-amyloid plaque in an APP/PS1 transgenic mouse model of Alzheimer's disease. *J. Mol. Neurosci.* **2010**, *41*, 110–113. [CrossRef]
41. Dumont, M.; Kipiani, K.; Yu, F.; Wille, E.; Katz, M.; Calingasan, N.Y.; Gorras, G.K.; Lin, M.T.; Beal, M.F. Coenzyme Q10 decreases amyloid pathology and improves behavior in a transgenic mouse model of Alzheimer's disease. *J. Alzheimers Dis.* **2011**, *27* 211–223. [CrossRef]
42. Elipenahli, C.; Snack, C.; Jainuddin, S.; Gerges, M.; Yang, L.; Starkov, A.; Beal, M.F.; Dumont, M. Behavioral improvement after chronic administration of coenzyme Q10 in P301S transgenic mice. *J. Alzheimers Dis.* **2012**, *28*, 173–182. [CrossRef]
43. Muthukumaran, K.; Kanwar, A.; Vegh, C.; Marginean, A.; Elliott, A.; Guilbeault, N.; Badour, A.; Sikorska, M.; Cohen, J. Pandey, S. Ubisol-Q10 (a Nanomicellar Water-Soluble Formulation of CoQ10) Treatment Inhibits Alzheimer-Type Behavioral and Pathological Symptoms in a Double Transgenic Mouse (TgAPEswe.; PSEN1dE9) Model of Alzheimer's Disease. *J. Alzheimers Dis* **2018**, *61*, 221–236. [CrossRef]
44. Sui, X.; Ren, X.; Huang, P.; Li, S.; Ma, Q.; Ying, M.; Ni, J.; Liu, J.; Yang, X. Proteomic analysis of serum proteins in triple transgenic Alzheimer's disease mice: Implications for identifying biomarkers for use to screen potential candidate therapeutic drugs for early Alzheimer's disease. *J Alzheimers Dis.* **2014**, *40*, 575–586. [CrossRef] [PubMed]
45. Ma, D.; Stokes, K.; Mahngar, K.; Domazet-Damjanov, D.; Sikorska, M.; Pandey, S. Inhibition of stress induced premature senescence in presenilin-1 mutated cells with water soluble Coenzyme Q10. *Mitochondrion* **2014**, *17*, 106–115. [CrossRef] [PubMed]
46. Vegh, C.; Pupulin, S.; Wear, D.; Culmone, L.; Huggard, R.; Ma, D.; Pandey, S. Resumption of Autophagy by Ubisol-Q_{10} in Presenilin-1 Mutated Fibroblasts and Transgenic AD Mice, Implications for Inhibition of Senescence and Neuroprotection. *Oxid Med. Cell. Longev.* **2019**, *2019*, 7404815. [CrossRef]
47. Qi, X.L.; Xiu, J.; Shan, K.R.; Xiao, Y.; Gu, R.; Liu, R.Y.; Guan, Z.Z. Oxidative stress induced by beta-amyloid peptide(1-42) is involved in the altered composition of cellular membrane lipids and the decreased expression of nicotinic receptors in human SH-SY5Y neuroblastoma cells. *Neurochem. Int.* **2005**, *46*, 613–621. [CrossRef] [PubMed]
48. Durán-Prado, M.; Frontiñán, J.; Santiago-Mora, R.; Peinado, J.R.; Parrado-Fernández, C.; Gómez-Almagro, M.V.; Moreno, M. López-Domínguez, J.A.; Villalba, J.M.; Alcaín, F.J. Coenzyme Q10 protects human endothelial cells from β-amyloid uptake and oxidative stress-induced injury. *PLoS ONE* **2014**, *9*, e109223. [CrossRef] [PubMed]
49. Frontiñán-Rubio, J.; Rabanal-Ruiz, Y.; Durán-Prado, M.; Alcain, F.J. The Protective Effect of Ubiquinone against the Amyloid Peptide in Endothelial Cells Is Isoprenoid Chain Length-Dependent. *Antioxidants* **2021**, *10*, 1806. [CrossRef] [PubMed]
50. Choi, H.; Park, H.H.; Koh, S.H.; Choi, N.Y.; Yu, H.J.; Park, J.; Lee, Y.J.; Lee, K.Y. Coenzyme Q10 protects against amyloid beta-induced neuronal cell death by inhibiting oxidative stress and activating the P13K pathway. *Neurotoxicology* **2012**, *33*, 85–90 [CrossRef]
51. Wang, H.; Li, L.; Jia, K.; Wang, Q.; Sui, S.; Lin, Y.; He, Y. Idebenone protects mitochondrial function against amyloid beta toxicity in primary cultured cortical neurons. *Neuroreport* **2020**, *31*, 1104–1110. [CrossRef]
52. Choi, H.; Park, H.H.; Lee, K.Y.; Choi, N.Y.; Yu, H.J.; Lee, Y.J.; Park, J.; Huh, Y.M.; Lee, S.H.; Koh, S.H. Coenzyme Q10 restores amyloid beta-inhibited proliferation of neural stem cells by activating the PI3K pathway. *Stem Cells Dev.* **2013**, *22*, 2112–2120 [CrossRef] [PubMed]
53. Yang, M.; Lian, N.; Yu, Y.; Wang, Y.; Xie, K.; Yu, Y. Coenzyme Q10 alleviates sevoflurane-induced neuroinflammation by regulating the levels of apolipoprotein E and phosphorylated tau protein in mouse hippocampal neurons. *Mol. Med. Rep.* **2020**, *22*, 445–453 [CrossRef] [PubMed]
54. Moreira, P.I.; Santos, M.S.; Sena, C.; Nunes, E.; Seiça, R.; Oliveira, C.R. CoQ10 therapy attenuates amyloid beta-peptide toxicity in brain mitochondria isolated from aged diabetic rats. *Exp. Neurol.* **2005**, *196*, 112–119. [CrossRef] [PubMed]
55. Li, L.; Xu, D.; Lin, J.; Zhang, D.; Wang, G.; Sui, L.; Ding, H.; Du, J. Coenzyme Q10 attenuated β-amyloid$_{25-35}$-induced inflammatory responses in PC12 cells through regulation of the NF-κB signaling pathway. *Brain Res Bull.* **2017**, *131*, 192–198. [CrossRef]
56. Imagawa, M.; Naruse, S.; Tsuji, S.; Fujioka, A.; Yamaguchi, H. Coenzyme Q10.; iron.; and vitamin B6 in genetically-confirmed Alzheimer's disease. *Lancet* **1992**, *340*, 671. [CrossRef] [PubMed]
57. Weyer, G.; Babej-Dölle, R.M.; Hadler, D.; Hofmann, S.; Herrmann, W.M. A controlled study of 2 doses of idebenone in the treatment of Alzheimer's disease. *Neuropsychobiology* **1997**, *36*, 73–82. [CrossRef]
58. Gutzmann, H.; Hadler, D. Sustained efficacy and safety of idebenone in the treatment of Alzheimer's disease: Update on a 2-year double-blind multicentre study. *J. Neural Transm. Suppl.* **1998**, *54*, 301–310.
59. Gutzmann, H.; Kühl, K.P.; Hadler, D.; Rapp, M.A. Safety and efficacy of idebenone versus tacrine in patients with Alzheimer's disease: Results of a randomized, double-blind, parallel-group multicenter study. *Pharmacopsychiatry* **2002**, *35*, 12–18. [CrossRef]
60. Thal, L.J.; Grundman, M.; Berg, J.; Ernstrom, K.; Margolin, R.; Pfeiffer, E.; Weiner, M.F.; Zamrini, E.; Thomas, R.G. Idebenone treatment fails to slow cognitive decline in Alzheimer's disease. *Neurology* **2003**, *61*, 1498–1502. [CrossRef]
61. Voronkova, K.V.; Meleshkov, M.N. Use of Noben (idebenone) in the treatment of dementia and memory impairments without dementia. *Neurosci. Behav. Physiol.* **2009**, *39*, 501–506. [CrossRef]
62. Galasko, D.R.; Peskind, E.; Clark, C.M.; Quinn, J.F.; Ringman, J.M.; Jicha, G.A.; Cotman, C.; Cottrell, B.; Montine, T.J.; Thomas, R.G.; et al. Alzheimer's Disease Cooperative Study. Antioxidants for Alzheimer disease: A randomized clinical trial with cerebrospinal fluid biomarker measures. *Arch. Neurol.* **2012**, *69*, 836–841. [CrossRef]
63. Karakahya, R.H.; Özcan, T.Ş. Salvage of the retinal ganglion cells in transition phase in Alzheimer's disease with topical coenzyme Q10, is it possible? *Graefes Arch. Clin. Exp. Ophthalmol.* **2020**, *258*, 411–418. [CrossRef]

4. Kawakami, M.; Itoh, T. Effects of idebenone on monoamine metabolites in cerebrospinal fluid of patients with cerebrovascular dementia. *Arch. Gerontol. Geriatr.* **1989**, *8*, 343–353. [CrossRef]
5. Qi, F.X.; Hu, Y.; Kang, L.J.; Li, P.; Gao, T.C.; Zhang, X. Effects of Butyphthalide Combined with Idebenone on Inflammatory Cytokines and Vascular Endothelial Functions of Patients with Vascular Dementia. *J. Coll. Physicians Surg. Pak.* **2020**, *30*, 23–27. [CrossRef]
6. García-Carpintero, S.; Domínguez-Bértalo, J.; Pedrero-Prieto, C.; Frontiñán-Rubio, J.; Amo-Salas, M.; Durán-Prado, M.; García-Pérez, E.; Vaamonde, J.; Alcain, F.J. Ubiquinol Supplementation Improves Gender-Dependent Cerebral Vasoreactivity and Ameliorates Chronic Inflammation and Endothelial Dysfunction in Patients with Mild Cognitive Impairment. *Antioxidants* **2021**, *10*, 143. [CrossRef]
7. Stough, C.; Nankivell, M.; Camfield, D.A.; Perry, N.L.; Pipingas, A.; Macpherson, H.; Wesnes, K.; Ou, R.; Hare, D.; de Haan, J.; et al. CoQ_{10} and Cognition a Review and Study Protocol for a 90-Day Randomized Controlled Trial Investigating the Cognitive Effects of Ubiquinol in the Healthy Elderly. *Front. Aging Neurosci.* **2019**, *11*, 103. [CrossRef] [PubMed]
8. Fernández-Portero, C.; Amián, J.G.; Bella, R.; López-Lluch, G.; Alarcón, D. Coenzyme Q10 Levels Associated with Cognitive Functioning and Executive Function in Older Adults. *J. Gerontol. A Biol. Sci. Med. Sci.* **2023**, *78*, 1–8. [CrossRef] [PubMed]
9. Grundman, M.; Grundman, M.; Delaney, P. Antioxidant strategies for Alzheimer's disease. *Proc. Nutr. Soc.* **2002**, *61*, 191–202. [CrossRef] [PubMed]
0. Beal, M.F. Mitochondrial dysfunction and oxidative damage in Alzheimer's and Parkinson's diseases and coenzyme Q10 as a potential treatment. *J. Bioenerg. Biomembr.* **2004**, *36*, 381–386. [CrossRef]
1. Lönnrot, K.; Metsä-Ketelä, T.; Molnár, G.; Ahonen, J.P.; Latvala, M.; Peltola, J.; Pietilä, T.; Alho, H. The effect of ascorbate and ubiquinone supplementation on plasma and CSF total antioxidant capacity. *Free Radic. Biol. Med.* **1996**, *21*, 211–217. [CrossRef]
2. Shults, C.W.; Flint Beal, M.; Song, D.; Fontaine, D. Pilot trial of high dosages of coenzyme Q10 in patients with Parkinson's disease. *Exp. Neurol.* **2004**, *188*, 491–494. [CrossRef] [PubMed]
3. Nukui, K.; Yamagishi, T.; Miyawaki, H.; Kettawan, A.; Okamoto, T.; Belardinelli, R.; Tiano, L.; Littarru, G.P.; Sato, K. Blood CoQ10 levels and safety profile after single-dose or chronic administration of PureSorb-Q40, animal and human studies. *Biofactors* **2008**, *32*, 209–219. [CrossRef] [PubMed]

MDPI
St. Alban-Anlage 66
4052 Basel
Switzerland
www.mdpi.com

Antioxidants Editorial Office
E-mail: antioxidants@mdpi.com
www.mdpi.com/journal/antioxidants

www.ingramcontent.com/pod-product-compliance
Lightning Source LLC
Chambersburg PA
CBHW061457210926
51850CB00008B/1740